Principles of Process Engineering

ASAE *The Society for engineering in agricultural, food, and biological systems*

2950 Niles Road
St Joseph MI 49085-9659 USA

Principles of Process Engineering

Fourth Edition

S. M. Henderson
Professor of Agricultural Engineering, Emeritus
University of California
Davis, California

R. L. Perry
Professor of Agricultural Engineering, Emeritus
University of California
Riverside, California

J. H. Young
Professor of Biological and Agricultural Engineering
North Carolina State University
Raleigh, North Carolina

The American Society of Agriculture Engineers is not responsible for the statements and opinions advanced in its meetings or printed in its publications. They represent the views of the individuals to whom they are credited and are not binding on the society as a whole.

Copyright © 1997 by the American Society of Agricultural Engineers
All Rights Reserved

REVISED EDITION
LCCN 97-74549 ISBN 0-929355-85-7
ASAE Textbook 801M0297

This book may not be reproduced in whole or in part by any means (with the exception of short quotes for the purpose of review) without the permission of the publisher.

For Information, contact:

The Society for engineering in agricultural, food, and biological systems
2950 Niles Road
St. Joseph, MI 49085-9659, USA
voice: 616.429.0300 fax: 616.429.3852
World Wide Web: http://asae.org/ e-mail: hq@asae.org

Manufactured in the United States of America

Table of Contents

Preface xiii

Chapter

1	Introduction		1
	Engineering Science		1
	1.1	Basic and Supplemental Units of Measurement	1
	1.2	Derived Quantities and Their Units	2
	1.3	Alternate Units (The English System)	3
	1.4	Unit Conversion—The Unit Factor Method	5
	1.5	Analytical or Rational Relationships	5
	Engineering Art		6
	1.6	The Idealized Solution	6
	1.7	Variation or Factor Uncertainty	7
	1.8	Empirical Relationships	7
	1.9	Error Analysis	8
	1.10	Economics	9
2	Fluid Mechanics		11
	Basic Considerations		11
	2.1	Classification of Fluids	11
	2.2	Conservation of Mass	11
	2.3	Conservation of Energy	13
	2.4	Power Requirement	15
	Characteristics of Fluid Flow		16
	2.5	Laminar (Streamlined) and Turbulent Flow	16
	2.6	Distribution of Velocities	16
	2.7	Reynolds Number	16
	2.8	Viscosity	18
	2.9	Non-Newtonian Fluids	19
	Friction Losses		22
	2.10	Darcy's Formula	22
	2.11	Energy Losses Due to Fittings	24
	2.12	Energy Loss Due to Sudden Contractions	24
	2.13	Energy Loss Due to a Sudden Enlargement	25
	2.14	Energy Loss Due to Flow Through Agricultural Products	26
	2.15	Energy Loss Due to Flow Through Floors	29
	2.16	Energy Loss in Heat Exchangers	31
	2.17	System Characteristic Curves	32
	Design Considerations		37
	2.18	Branching System Design	37
	2.19	Compressibility Error	39

		2.20 Optimum Rates of Flow	39
		Flow of Granular Materials	41
		2.21 Rate of Flow	41
		2.22 Angle of Repose	41
		2.23 Coefficient of Friction	42
		References	43
3		Fluid-Flow Measurements	47
		Pressure Measurement	47
		3.1 Units of Pressure	47
		3.2 Types of Pressure Measurements	48
		3.3 Manometers	50
		3.4 Bourdon Tube	54
		3.5 Diaphragm Gages	56
		3.6 Bellows Gages	57
		3.7 Electrical Pressure Transducers	57
		Differential Pressure Flow Meters	57
		3.8 Pitot Tube	57
		3.9 Venturi Meter	61
		3.10 Nozzle Meter	65
		3.11 Orifice Meter	66
		3.12 Dall Flow Tube	69
		3.13 Laminar Flow Meter	69
		Volumetric Flow Meters	72
		3.14 Nutating Disk Meters	72
		3.15 Reciprocating Piston Meters	73
		3.16 Oscillating or Rotary Piston Meters	73
		3.17 Bellows Meters	73
		Force-Displacement Meters	74
		3.18 Rotameters	75
		3.19 Turbine or Propeller Meters and Rotating Vane or Cup-Type Anemometers	78
		3.20 Swinging Vane Anemometers	81
		Thermal Meters	81
		3.21 Thomas Meter	81
		3.22 Hot-Wire Anemometer	82
		3.23 Thermocouple Anemometer	84
		Special Metering Methods	84
		References	85
4		Pumps	89
		4.1 Evaluating Performance	89
		Displacement Pumps	90
		4.2 Reciprocating or Piston Pumps	90
		4.3 Rotary Pumps	93

	Dynamic Pumps	95
4.4	Centrifugal Pumps	95
4.5	Jet Pumps	101
4.6	Air Lift Pumps	103
	Performance	105
4.7	Performance Curves	105
4.8	Specific Speed	109
4.9	Net Positive Suction Head	111
4.10	Pump Performance on a System	113
4.11	Viscosity	114
	References	116

5 Fans 119
   5.1   Aerodynamic Classification 119
   Axial-Flow Fans 120
   5.2   Propeller Fans 120
   5.3   Tube-Axial Fans 120
   5.4   Vane-Axial Fans 122
   Centrifugal Fans 123
   5.5   Forward-Curved-Tip Fans 123
   5.6   Straight or Radial-Tip Fans 124
   5.7   Backward-Curved-Tip Fans 124
   Cross-Flow Fans 124
   5.8   Cross-Flow or Tangential Fans 124
   Mixed-Flow Fans 124
   5.9   In-Line or Tubular Centrifugal Fans 124
   Performance 125
   5.10 Fan Testing 125
   5.11 Sound Power Level Ratings 128
   5.12 Specific Speed 129
   5.13 Axial-Flow Fan Performance 130
   5.14 Backward-Curved-Tip Centrifugal Fan Performance 132
   5.15 Radial-Tip Centrifugal Fan Performance 133
   5.16 Forward-Curved-Tip Centrifugal Fan Performance 134
   5.17 Fan Laws 134
   5.18 Compressibility 136
   Fan Operation on a System 137
   5.19 System Characteristic Curves 137
   5.20 Fans in Series 140
   5.21 Fans in Parallel 141
   5.22 Systems in Series and Parallel 143
   5.23 Effect of Varying Fan Operating Speed 144
   References 146

6	Heat Transfer—Conduction		149
	6.1	Conduction	149
	6.2	Convection	150
	6.3	Radiation	150
	6.4	Fourier-Poisson Equation	152
	Steady-State Conduction		153
	6.5	Steady-State Conduction in a Plane Wall with No Heat Generation	153
	6.6	Steady-State Heat Conduction in a Plane Wall with Uniform Heat Generation	155
	6.7	Steady-State Heat Conduction in a Composite Plane Wall with Convection at the Surfaces and No Heat Generation	156
	6.8	Steady-State Heat Conduction in a Cylindrical Shell	161
	6.9	Steady-State Heat Conduction in a Composite Cylinder with Convection at the Surfaces	163
	6.10	Biot Number for Maximum Heat Transfer Through a Composite Cylindrical Shell	165
	6.11	Steady-State Heat Conduction in a Spherical Shell	166
	6.12	Steady-State Heat Conduction in a Composite Sphere with Convection at the Surfaces	168
	6.13	Biot Number for Maximum Heat Transfer Through a Composite Spherical Shell	170
	Transient Heat Flow		170
	6.14	Transient Heat Conduction with Convection at Surface	170
	Numerical Solutions		177
	6.15	Finite Difference Solutions for Steady-State Heat Transfer in Two Dimensions with No Heat Generation	177
	References		184
7	Heat Transfer—Convection		187
	7.1	Dimensional Analysis	187
	7.2	Application of Dimensional Analysis to Free Convection	189
	7.3	Application of Dimensional Analysis to Forced Convection	191
	7.4	Main Advantages of Dimensional Analysis	192
	7.5	Main Disadvantages of Dimensional Analysis	192
	7.6	Experimental Equations for Free Convection	193
	7.7	Experimental Equations for Forced Convection	197
	Heat Exchangers		198
	7.8	Types of Heat Exchangers	198
	7.9	Log Mean Temperature Difference	199
	7.10	Heat Exchanger Effectiveness Ratio	204
		7.10.1 Parallel-Flow Heat Exchanger Effectiveness Ratio	204

	7.10.2	Counter-Flow Heat Exchanger Effectiveness Ratio		208
	7.10.3	Crossflow Heat Exchanger Effectiveness Ratio		212
References				213

8 Heat Transfer—Radiation ... 215
- 8.1 The Electromagnetic Wave Spectrum ... 215
- 8.2 Emissivity ... 216
- 8.3 Absorption, Reflection, and Transmission of Radiation ... 220
- 8.4 Kirchoff's Law ... 221
- 8.5 Radiation Intensity and Total Emissive Power ... 222
- 8.6 Heat Exchange by Radiation Between Black Surfaces ... 223
- 8.7 Heat Flow by Radiation Between Gray Surfaces ... 225
- 8.8 Heat Flow Between Surfaces Separated by a Transparent Layer ... 227
- 8.9 Solar Radiation and the Greenhouse Effect ... 230
- References ... 234

9 Psychrometrics ... 237
- 9.1 Ideal Gas Law ... 237
- 9.2 Dalton's Law ... 238
- 9.3 Definitions of Psychrometric Terms ... 239
  - 9.3.1 Dry-Bulb Temperature ... 239
  - 9.3.2 Saturation Pressure ... 239
  - 9.3.3 Humidity Ratio ... 241
  - 9.3.4 Relative Humidity ... 242
  - 9.3.5 Specific Volume ... 242
  - 9.3.6 Dew-Point Temperature ... 243
  - 9.3.7 Enthalpy ... 243
  - 9.3.8 Adiabatic Saturation Temperature ... 246
  - 9.3.9 Wet-Bulb Temperature ... 248
  - 9.3.10 Density ... 251
- 9.4 The Psychrometric Chart ... 252
- 9.5 Uses of the Psychrometric Chart ... 257
  - 9.5.1 Determination of State Factors ... 257
  - 9.5.2 Cooling and Heating ... 257
  - 9.5.3 Mixtures ... 258
  - 9.5.4 Cooling and Dehumidifying ... 260
  - 9.5.5 Direct Heating by Combustion of a Fuel ... 263
  - 9.5.6 Drying Processes ... 265
  - 9.5.7 Addition of Water at any State ... 266
  - 9.5.8 Dehumidification Using an Absorbent ... 268
- References ... 270

10	Drying		273
	10.1 Moisture Content		273
	10.2 Interaction of Moisture and Materials		274
	10.3 Equilibrium Moisture Content		275
		10.3.1 Langmuir Equation	275
		10.3.2 BET Equation	276
		10.3.3 Smith Equation	276
		10.3.4 Henderson Equation	277
		10.3.5 Young Equations	279
	10.4 Moisture Content Measurement		280
		10.4.1 Oven Methods	280
		10.4.2 Desiccant Drying	282
		10.4.3 Distillation Methods	282
		10.4.4 Chemical Methods	282
		10.4.5 Gas Chromatography	282
		10.4.6 Electrical Methods	282
		10.4.7 Spectrophotometric Methods	283
		10.4.8 Nuclear Methods	283
		10.4.9 Equilibrium Relative Humidity	283
		10.4.10 Suction Methods	283
		10.4.11 Summary of Advantages and Disadvantages of Various Moisture Measuring Methods	284
	10.5 Thin-Layer Drying		286
		10.5.1 Constant Drying-Rate Period	286
		10.5.2 Falling Rate Period	288
	10.6 Bulk Drying Simulation		293
		10.6.1 Significance of Adiabatic Saturation Process	293
		10.6.2 Energy Balance	294
		10.6.3 Hukill's Analysis	296
		10.6.4 Numerical Deep-Bed Calculations	305
	10.7 Drying Procedures		310
		10.7.1 Batch or Bin Driers	310
		10.7.2 Continuous Gravity-Flow Driers	311
		10.7.3 Rotary Driers	312
		10.7.4 Tray Driers	314
		10.7.5 Spray Driers	315
		10.7.6 Concentrators	317
	References		320
11	Refrigeration		325
	11.1 Natural Refrigeration		325
	11.2 Mechanical Refrigeration		325
	11.3 Rating		328
	11.4 General Considerations		329

Components		334
11.5	Compressors	334
11.6	Condensers	335
11.7	Evaporators	336
11.8	Expansion Valves	337
System Design and Balance		339
11.9	The Evaporator	339
11.10	Defrosting	341
11.11	The Compressor	341
11.12	Condensing Unit	342
Controls		342
11.13	Motor Circuit Thermostats	343
11.14	Low-Side-Pressure Switch	343
11.15	Magnetic Valves	343
11.16	Evaporative-Pressure-Maintaining Valves	344
Heat Pump		345
Sources and Sinks		346
References		347

Index 351

Preface

This edition presents major revisions in the book. Expanded emphasis is placed on the basic transport phenomena of fluid flow, heat transfer, and mass transfer while the current edition does not include previous chapters on size reduction, cleaning and sorting, materials handling, instrumentation and controls, cost analysis, and process analysis and plant design. Some of the topics omitted (instrumentation and controls, cost analysis, and process analysis and plant design) are adequately covered by other reference materials used by biological and agricultural engineers while some topics (size reduction, cleaning and sorting, and materials handling) have simply not yet undergone the extensive review and revision of those topics which are currently included.

The chapter on heat transfer in the previous edition is expanded to three chapters in this edition separately giving the basics of conduction, convection, and radiation. This expanded emphasis on heat transfer is due to its importance not only in the farm level processing of agricultural commodities but also in food processing, environmental control, and optimization and control of other biological processes.

While an agricultural flavor is retained in this revision, an increased emphasis is placed on more general processing applications and primary emphasis is given toward those basic relationships which may be applied to a multitude of problem areas. This revision in emphasis has led to the new title *Principles of Process Engineering*.

Other major developments in this edition are: (1) a conversion to SI (System International) units throughout and (2) the use of computer software packages for problem solution. It was acknowledged in the preface of the 1976 edition that engineering should "go metric" as soon as convenient. While change never seems to be "convenient", those who make the change to the SI system find that the reduced difficulties in making unit conversions more than offset the inconvenience of learning a new system. Unfortunately, the United States of America is a society which still makes predominant use of the English system of units and thus conversion between the two systems is still necessary. This book, however, emphasizes the solution of problems in a consistent set of units rather than the nonproductive drudgery of conversion between alternative units for the same basic quantities.

Computer hardware and software is a necessary tool of engineers today. Since both hardware and software are changing rapidly, it is best to emphasize the theory behind any computational methods rather than any particular machine or software package. Thus, while the software package TK Solver* and various spreadsheets are discussed in the context of problem solutions within this edition of the book, attempts are made to explain what the software has to do to reach a solution. This should allow for the substitution of other existing software packages or packages which will be developed in the future if they have the capabilities which are needed for reaching a solution. It is simply not practical in today's world to limit the problems we attack to those which can be neatly solved with pencil and paper (or even slide rule or pocket calculator).

---

*TK Solver is a product of Universal Technical Systems, Inc.

I wish to express my sincere appreciation to Professor S. M. Henderson for allowing me to retain much of the basic outline and many of the figures of previous editions of this book. Agricultural engineers have been well served for many years by this textbook in the processing area. A wide variety of subjects have been brought together in one manuscript which provides the basics for many engineering problems. This required much study and careful planning by the original authors and they are to be highly commended.

I also wish to thank those who have helped me with these revisions. They include students at North Carolina State University over a number of years who have studied from my notes and helped to find errors and identify areas needing further clarification. It also includes a number of instructors at other universities who have reviewed the material and used some of it in their classes. Their feedback has been very helpful and much appreciated.

Finally, it is our goal at this time to distribute this edition in not only a hardcopy form, but to make it electronically available as well. It is hoped that this will make further modifications easier and that the manuscript can adapt to the changing needs of students. Thus, comments from both instructors and students using this book are highly encouraged and solicited.

April 1997 *James H. Young*

1 Introduction

Engineering has been defined as "the art and science of utilizing the forces and materials of nature for the benefit of man and the direction of man's activities toward this end." This definition implies the division of engineering into two activities: (1) science and (2) art.

Engineering Science

The *science* of engineering is that phase of the field which is exact and rational. For any set of conditions the end point will always be the same. The conditions can be related mathematically and are based on laws that can be rationalized on the pure sciences. Any constants or variables that must be determined experimentally can be defined and do not vary greatly after being established.

1.1 Basic and Supplemental Units of Measurement

In order to rationally describe the physical, thermal, electrical, and chemical phenomena to be encountered by the process engineer, it is necessary to define a consistent set of units. In problem solving it is just as important, if not more so, to use proper units as it is to make proper numerical calculations. The system of units to be used in this book is the International System of Units (SI). This system of units consists of seven base units, two supplemental units, and a series of derived units consistent with the basic and supplemental units.

The seven base units of the SI system are (for definitions refer to International Organization for Standardization ISO 1000, SI Units and Recommendations for the Use of Their Multiples and of Certain Other Units):

meter (m)	unit of length
second (s)	unit of time
kilogram (kg)	unit of mass
kelvin (K)	unit of thermodynamic temperature
ampere (A)	unit of electric current
candela (cd)	luminous intensity
mole (mol)	the amount of a substance

The two supplementary units are:

radian (rad)   plane angle
steradian (sr) solid angle

1.2 Derived Quantities and Their Units

All other quantities used in engineering can be related to those basic and supplemental units of the previous section. Some of those quantities to be encountered in this book will be briefly described, their units given, and their relationship to the base units given in the following paragraphs. It is important that the student thoroughly understand these definitions and units since they quite often point the way toward a problem solution.

Acceleration is the rate of change of *velocity* with respect to *time*. It is expressed in units of *length* per *time* squared (L/T^2) or in SI units of *meters* per *second* squared (m/s^2).

Area is a measure of the size of a surface. It always has the units of *length* squared (L^2) or in the SI system the units of meters squared (m^2).

Density is the *mass* of a substance per unit *volume*. Thus the units are *mass* per *length* cubed (M/L^3) or (kg/m^3).

Energy, *heat*, and *work* are equivalent quantities which may be related to the product of a *force* and the distance or *length* through which it acts. The defined SI unit of *energy* is the joule which is equal to a *force* of one newton multiplied by a *length* of one meter (1 J = 1 Nm). If the *energy* unit is reduced to the basic units it becomes *mass* times *length* squared per *time* squared (ML^2/T^2) or kilograms times meters squared per second squared (kg m^2/s^2).

Enthalpy is a defined property sometimes called the heat content which represents a sum of the internal *energy* and the product of the *pressure* and the *volume* of a substance. The units of *enthalpy* are then the same as those for *energy*. Enthalpy may also be expressed on a per unit mass basis in joules per kilogram (J/kg).

Entropy or more correctly the change in *entropy* is defined as the integral of the change in *heat* or *energy* divided by the *temperature*. It then has units of joules per degree kelvin (J/K). Reduction of the units of *entropy* to the basic units gives *mass* times *length* squared per *time* squared per *temperature* ($ML^2/T^2\,\Theta$) or kilograms times meters squared per second squared per degree kelvin (kg m^2/s^2 K). *Entropy* is also often expressed on a per unit mass basis in the units of joules per kilogram per degree kelvin (J/kg K).

Force is a *mass* times an *acceleration*. The defined SI unit of *force* is the newton which is a kilogram times a meter per second squared (1 N = 1 kg m/s^2) or units of *mass* times *length* per *time* squared (ML/T^2).

Head is a quantity which refers to the *energy* per unit weight or *force* of a fluid. Its units are joules per newton (J/N) or simply meters (m).

Power is *energy* per unit *time*. The defined SI unit of *power* is the watt which is one joule per second (1 W = 1 J/s). If the *power* unit is reduced to the basic units it becomes *mass* times *length* squared per *time* cubed (ML^2/T^3) or kilograms times meters squared per second cubed (kg m^2/s^3).

Introduction

Pressure and *stress* are equivalent quantities which represent a *force* per unit *area*. The defined SI unit of *pressure* is the pascal which is one newton per square meter (1 Pa = 1 N/m^2). If the *pressure* unit is reduced to the basic units it becomes *mass* per unit *length* per unit *time* squared (M/LT^2) or kilograms per meter per second squared (kg/m s^2).

Specific gravity is the ratio of the *density* of a material to the *density* of water. Thus it is a dimensionless quantity.

Specific heat is a measure of the *energy* required to raise the thermodynamic *temperature* of a *mass*. It has units in the SI system of joules per kilogram per kelvin degree (J/kg K). Reduction of the *specific heat* units to basic units results in units of *length* squared per *time* squared per *temperature* (L^2/T^2 Θ) or meters squared per second squared per degree kelvin (m^2/s^2 K).

Specific weight is the weight or *force* of a material per unit *volume*. It is the product of *density* and the *acceleration* due to gravity. It has units in the SI system of newtons per meter cubed (N/m^3). Reduction of the *specific weight* units to basic units results in units of *mass* per *length* squared per *time* squared (M/L^2T^2) or kilograms per meter squared per second squared (kg/m^{2}s^2).

Thermal conductivity is a measure of the *power* transferred per unit area per unit *temperature* gradient within a material. It has units in the SI system of watts per meter per degree (W/m K). Reduction of the *thermal conductivity* units to basic units results in units of *mass* times *length* per *time* cubed per *temperature* (ML/T^3 Θ) or kilograms times meters per second cubed per degree kelvin (kg m/s^3 K).

Velocity is a change in distance or *length* with respect to *time*. It has units of *length* per unit *time* (L/T) or meters per second (m/s).

Viscosity (dynamic) is a measure of the *force* per unit *area* due to a *velocity* gradient within a fluid. In the SI system the *dynamic viscosity* has units of pascals times seconds (Pa s). If the *dynamic viscosity* unit is reduced to basic units it becomes *mass* per *length* per *time* (M/LT) or kilograms per meter per second (kg/m s).

Viscosity (kinematic) is the ratio of the *dynamic viscosity* of a fluid to its *density*. It is expressed in terms of basic units as *length* squared per *time* (L^2/T) or meters squared per second (m^2/s).

Volume is a measure of the space occupied by an object. It has units of *length* cubed (L^3) or in the SI system units of meters cubed (m^3).

1.3 Alternate Units (The English System)

A brief review of the English System of units is included here for those students who will be forced to work with both systems. Both the base and the derived quantities are the same in the two systems and their definitions are identical. The only difference is in the system of units chosen for measurements. The English System is much more cumbersome to work with because of the many conversion factors which must be remembered. Table 1-1 gives comparisons between the two systems for the quantities discussed in the previous two sections.

A complicating factor in the English system of units is the use of the unit of pounds to represent both mass and force. This creates numerous problems for students in making

Table 1-1. Comparison between the SI and English system of units

Quantities	SI Units	English Units
Length	meter (m)	foot (ft)
Mass	kilogram (kg)	pound-mass (lbm)
Time	second (s)	second (s)
Temperature	kelvin (K)	fahrenheit (F)
Electric Current	ampere (A)	ampere (A)
Luminous Intensity	candela (cd)	candela (cd)
Amount of Substance	mole (mol)	mole (mol)
Plane angle	radian (rad)	radian (rad)
Solid angle	steradian (sr)	steradian (sr)
Acceleration	m/s^2	ft/s^2
Area	m^2	ft^2
Density	kg/m^3	lb$_m$/ft^3
Energy	J	ft lb$_f$ or B
Enthalpy	J	B
Entropy	J/K	B/F
Force	N	pound-force (lb$_f$)
Head	J/N or m	ft lb$_f$/lb$_f$ or ft
Power	W	ft lb$_f$/min or hp
Pressure	Pa	lb$_f$/ft^2
Specific Heat	J/kg K	B/lb$_m$ F
Specific Weight	N/m^3	lb$_f$/ft^3
Thermal Conductivity	W/m K	B/hr ft F
Velocity	m/s	ft/s
Viscosity (dynamic)	Pa s	lb$_m$/ft s
Viscosity (kinematic)	m^2/s	ft^2/s
Volume	m^3	ft^3

sure units are consistent. It must be understood that a pound-mass is not the same as a pound-force and thus they cannot cancel each other. A pound-mass is defined as the quantity of mass which weighs one pound-force when subjected to the normal gravitational acceleration. Thus, one pound-force equals one pound-mass times 32.2 ft per second squared (1 lb$_f$ = 1 lb$_m$ * 32.2 ft/s^2) or 1 lb$_f$ = 32.2 lb$_m$ ft/s^2.

Other conversion factors which are encountered in the English system include the following:

1 British thermal unit (B) = 778.3 ft lb$_f$
1 B = 3.931 × 10^{-4} horsepower-hours (hp h)
1 B = 2.930 × 10^{-4} kilowatt-hours (kWh)
1 bushel (bu) = 1.244 ft^3
1 ft^3 = 7.481 gal
1 hp = 42.40 B/min
1 hp = 33,000 ft lb$_f$/min
1 hp = 747.7 W

1 hp hr = 2,544 B
1 ft = 12 in
1 yd = 3 ft

1.4 Unit Conversion—The Unit Factor Method

In converting between different sets of units, the original value is multiplied by a factor (or factors) which has a value of unity when both its numerical value and units are considered. This is called a unit factor. The unit factor is chosen so that the original units cancel and the desired or final units are left. The following examples illustrate the procedure.

Example 1.1. Convert 2.5 min to units of seconds.

Solution

$$2.5 \text{ min} = (2.5 \text{ min})(60 \text{ s}/1 \text{ min}) = 150 \text{ s}$$

Example 1.2. Convert 25 047 g to kilograms.

Solution

$$25\,047 \text{ g} = (25\,047 \text{ g})(1 \text{ kg}/1\,000 \text{ g}) = 25.047 \text{ kg}$$

Example 1.3. Convert 25 kPa to kilograms per meter per second squared.

Solution

$$25 \text{ kPa} = (25 \text{ kPa})(1\,000 \text{ Pa}/1 \text{ kPa}) = 25\,000 \text{ Pa}$$
$$= (25\,000 \text{ Pa})(1 \text{ Nm}^{-2}/1 \text{ Pa}) = 25\,000 \text{ N/m}^2$$
$$= (25\,000 \text{ N/m}^2)(1 \text{ kg m s}^{-2}/1 \text{ N}) = 25\,000 \text{ kg/m s}^2$$

Example 1.4. Convert 40 500 kg times meters squared per second cubed to kilowatts.

Solution

$$40\,500 \text{ kg m}^2/s^3 = (40\,500 \text{ kgm}^2/s^3)(1 \text{N}/1 \text{ kg m s}^{-2})$$
$$= (40\,500 \text{ N m/s})(1 \text{ J}/1 \text{ N m}) = 40\,500 \text{ J/s}$$
$$= (40\,500 \text{ J/s})(1 \text{ W}/1 \text{ J s}^{-1}) = 40\,500 \text{ W}$$
$$= (40\,500 \text{ W})(1 \text{ kW}/1\,000 \text{ W}) = 40.5 \text{ kW}$$

Appendix A contains a number of conversion factors which may be useful in converting between sets of units. If the SI system of units is used exclusively, the unit conversions are greatly simplified.

1.5 Analytical or Rational Relationships

Analytical relationships between variables in engineering are based on basic laws of science and mathematics. They may be rationally deduced from basic laws without the need for further experimentation. Analytic relationships are valid in any consistent set of units. Thus, Newton's second law is expressed as:

$$F = ma \tag{1.1}$$

where

$F = $ force
$m = $ mass
$a = $ acceleration

regardless of the specific units in which force, mass, and acceleration are expressed. If an equation has numerical constants, they are either dimensionless or their dimensions must be given as a part of the equation.

Engineering Art

The *art* of engineering refers to the ability to judge, estimate, and manipulate the uncertainties of engineering to a satisfactory solution to a problem. It refers to a procedure that has been found by a series of trial and error events, carried out in as logical a sequence as possible, to produce a desired result without knowledge of the basic principles involved. It refers to the use of empiricals in an efficient manner. The Chinese made iron and steel, and the Egyptians glass; although the Chinese knew nothing of metallurgy, and the Egyptians nothing of the science of glass making. The Indians fertilized their corn with dead fish, but they knew nothing of plant and soil science.

The field of agricultural processing contains more engineering uncertainties than the more traditional engineering fields. Successful treatment of a problem frequently requires that the engineer estimate, extrapolate, or secure information empirically to solve a problem. Occasionally, decisions must be based on intuitive judgment. This procedure is hazardous but sometimes necessary.

It is this ability, the ability to evaluate the uncertainties, that differentiates an engineer from a pure scientist, and the engineer's success will depend in great measure on the skill with which he handles these uncertainties. It should be emphasized, however, that empiricism or the art of engineering is only to be used when attempts at rational analysis are unsuccessful in reaching a definitive solution.

1.6 The Idealized Solution

An engineering problem or project, in design, development, or research, can best be evaluated by establishing all known facts and procedures that are, or appear to be, related to it. If a problem is first idealized on the basis of known rules, factors, and laws, its solution will serve as a standard or measure of fit or performance of the final engineering decision.

For example, the amount of energy required to reduce the moisture of 1 000 kg of grain from 24% to 14% wet basis moisture content may be estimated by comparison with a better understood situation in which free water is to be vaporized. In making the reduction in moisture content, 117 kg of water must be removed. In the absence of specific data on the amount of energy required to remove this amount of water from the grain, the quantity required to vaporize the same amount of free water may be used as a first estimate. Existing data on the evaporation of free water indicates that approximately 272 MJ of energy would be required. The engineer should recognize, however, that the water contained within the hygroscopic material (grain) would be bound by forces in excess of those binding free water and that the true energy requirement would be somewhat greater than the above estimate. The estimate represents a lower bound on the quantity of energy which the engineer should prepare to provide.

1.7 Variation or Factor Uncertainty

Many engineering calculations are rational in concept but empirical in application because important factors must be determined experimentally. The performance of a wood member under load can be calculated on the basis of certain rational formulas that yield tension in the outer fiber, maximum horizontal shear of the member, and the amount of flexure. However, these calculations require certain "constants" that define the limits of performance, ultimate strength, elastic limit, and modulus of elasticity. These constants are averages of a great number of individual observations which may vary considerably. Consequently, since any single member may be much weaker than the average, a factor of safety of two to six is applied to the rational calculation to insure satisfactory performance.

The engineers' factor of safety is needed for two reasons: (1) insufficient or incomplete basic information and/or (2) inability to forecast future conditions related to the operation. The variations in products, weather, markets, demand, etc., which affect many of the engineering aspects of a problem, are difficult and sometimes impossible to evaluate.

A knowledge of statistical procedures will aid in providing a satisfactory answer to many problems involving variable or uncertain factors. It is especially helpful for those engaged in research who are attempting to establish basic relationships.

Statistics, especially analytical statistics, may be defined as the mathematical science of variation. The procedures that it embraces may be used to (1) show a mass of data in an easily understandable graphical form, (2) resolve the data into a mathematical formula, or (3) determine its reliability. The statistical evaluation of reliability is very important since it aids in determining the qualitative value of data, the probability of certain events, and the number and characteristics of samples that must be taken to yield significant results.

1.8 Empirical Relationships

Certain engineering relationships can be expressed graphically or mathematically even though the basis for the relationship is not known or is not apparent. This type of relationship is based entirely on experimental data and is completely empirical. Examples are the power-particle size relationship for grinding grain and the change of viscosity with temperature.

Empirical relationships have often been developed that are dimensionally inconsistent. This reduces their usefulness to the system of units in which they were originally developed. It is recommended that even in empirical relationships unit consistency should be maintained and any numerical constants should be shown with their units.

Statistics may be used to develop a mathematical relationship which fits the empirical data as suggested in the previous section. Regression techniques may be used to obtain the values of *constants* in equations which result in the best fit of the data and may determine confidence limits or estimates of the variability of such constants.

1.9 Error Analysis

The effect of an observational error, or number of errors, on a computed end point may be estimated through the following procedure using differential calculus when a known functional relationship exists. The total differential of a function composed of two or more variables is:

$$df(x, y) = (\partial f/\partial x)\,dx + (\partial f/\partial y)\,dy + \cdots \tag{1.2}$$

If the infinitesimal increments are replaced by finite increments an approximate relationship results,

$$\Delta f = (\partial f/\partial x)\Delta x + (\partial f/\partial y)\Delta y + \cdots \tag{1.3}$$

and the Δ's may be considered as errors.

If the errors are systematic, that is, carry plus or minus signs, equation (1.3) will provide the error in the function. If the errors are random (have an equal chance of being plus or minus) a statistical procedure is used and the equation becomes:

$$\Delta f = \sqrt{((\partial f/\partial x)\Delta x)^2 + ((\partial f/\partial y)\Delta y)^2 + \cdots} \tag{1.4}$$

Example 1.5. What is the error in the volume of a cylinder 40 cm high and 25 cm in radius if both measurements are in error by 0.2 cm?

Solution

$$V = \pi r^2 h = 78\,539.8 \text{ cm}^3$$
$$(\partial f/\partial r)\Delta r = 2\pi r h \Delta r = 1\,256.6 \text{ cm}^3$$
$$(\partial f/\partial h)\Delta h = \pi r^2 \Delta h = 392.7 \text{ cm}^3$$

If both errors are systematic and carry plus signs, the error in the volume is:

$$\Delta V = 1\,256.6 \text{ cm}^3 + 392.7 \text{ cm}^3 = 1\,649.3 \text{ cm}^3$$

or 2.1% of the true volume. If the errors are random,

$$\Delta V = \sqrt{(1\,256.6 \text{ cm}^3)^2 + (392.7 \text{ cm}^3)^2}$$
$$= 1\,316.5 \text{ cm}^3$$

or 1.7% of the true volume. Note that the systematic error in the radius has more than three times the effect on the calculated volume as does the error in the height.

This procedure is useful in determining the relative effect of errors and, as a result, observations that require a high degree of accuracy can be isolated.

1.10 Economics

The economic phase of an engineering problem must never be overlooked. Many engineering processes are designed especially to reduce production costs, usually by speeding up the process, eliminating or making manual labor more efficient, or reducing overhead costs. A new or improved engineering procedure must always be judged by its economic value. The effect may be indirect in that a particular machine or operation may contribute to better application of another unit.

In processing work there is often a small difference between the cost of the raw products and the selling price. The processing operation must be performed well within this economic bracket if a fair return on the investment is to be assured. Economic improvement of an operation is usually produced in one of two ways: (1) by reducing the cost of production per unit or (2) by raising the net return per unit. Increased net return could result from reducing waste, using by-products more effectively, or raising the quality of the product. The activities of the process engineer are usually directed toward one of the above-mentioned objectives.

2 Fluid Mechanics

Basic Considerations

A complete study of fluid mechanics would cover two areas: (1) fluids at rest or hydrostatics and (2) fluids in motion or hydrodynamics. We will assume that the first, treating fluids at rest, was covered in required physics, chemistry, or basic engineering courses. The second part, which deals with the various factors affecting the relationship between the rate of flow and the various pressures tending to cause or inhibit flow, will be treated in detail here.

2.1 Classification of Fluids

Fluids may be classified as either compressible (gases) or incompressible (liquids). Liquids are actually compressible to a small degree, but are normally treated as incompressible since errors introduced by that assumption are usually insignificant.

Although gases are compressible, the agricultural process engineer will generally work with air at such low pressures that compressibility may be neglected. Thus, compressibility, the change of density with pressure, will not be considered here.

2.2 Conservation of Mass

Consider the fluid system shown in figure 2.1. If the rate of flow is constant at any point and there is no accumulation or depletion of fluid within the system, the principle of conservation of mass within the system requires:

$$\dot{m}_1 = \dot{m}_2 = \dot{m}_3 = \cdots \tag{2.1}$$

where $\dot{m}_i$ = mass flow rate of fluid at point i (kg/s).

The mass flow rate at a point in the system is given by:

$$\dot{m}_i = A_i V_i \rho_i \tag{2.2}$$

where

A_i = cross-sectional area of flow system at point i (m^2)
V_i = average fluid velocity at point i (m/s)
ρ_i = fluid density at point i (kg/m^3)

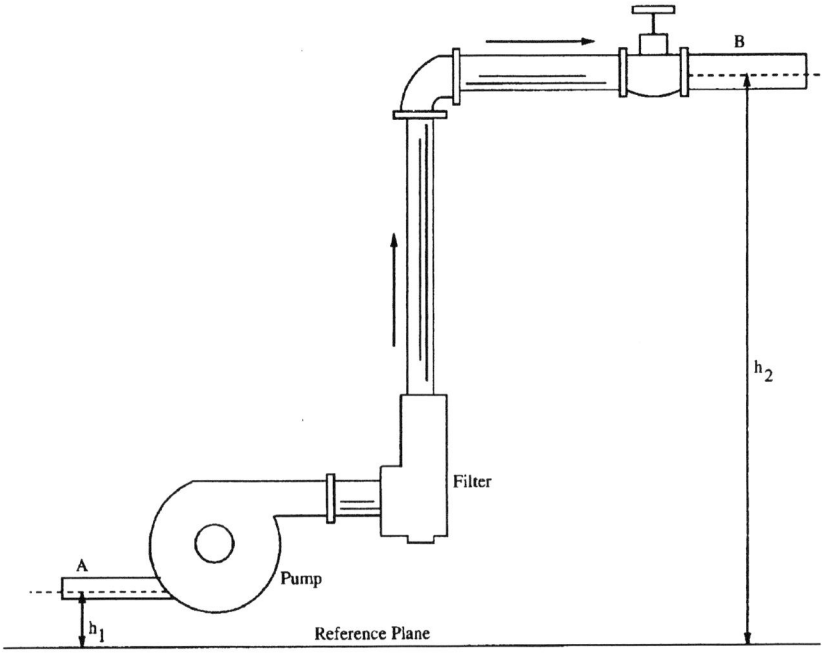

Figure 2.1. A hydraulic system.

Substitution of equation 2.2 into 2.1 yields:

$$A_1 V_1 \rho_1 = A_2 V_2 \rho_2 = A_3 V_3 \rho_3 = \cdots = \dot{m} \qquad (2.3)$$

Equation 2.3 may be further simplified for the case of incompressible fluids. For incompressible fluids the density remains constant and the equation becomes:

$$A_1 V_1 = A_2 V_2 = A_3 V_3 = \cdots = Q \qquad (2.4)$$

Example 2.1. Water is flowing in a 15-cm inside diameter pipe at a velocity of 0.3 m/s. The pipe enlarges to an inside diameter of 30 cm. What is the velocity in the larger section, the volumetric flow rate, and the mass flow rate?

Solution

$$A_1 V_1 = A_2 V_2$$
$$V_2 = V_1(A_1/A_2) = V_1(D_1/D_2)^2$$
$$V_2 = (0.3 \text{ m/s})(15 \text{ cm}/30 \text{ cm})^2 = (0.3 \text{ m/s})(1/2)^2$$
$$V_2 = (0.3 \text{ m/s})/4 = 0.075 \text{ m/s}$$
$$Q = A_1 V_1 = \left(\pi D_1^2/4\right) V_1$$
$$Q = \pi (15 \text{ cm})^2 (0.3 \text{ m/s})(0.01 \text{ m/cm})^2/4 = 0.0053 \text{ m}^3/\text{s}$$

Fluid Mechanics

$$\dot{m} = Q\rho$$
$$\dot{m} = (0.0053 \text{ m}^3/\text{s})(1\,000 \text{ kg/m}^3) = 5.30 \text{ kg/s}$$

2.3 Conservation of Energy

Since energy is neither created nor destroyed within the fluid system, the total energy of the fluid at one point in the system must equal the total energy at any other point plus any transfers of energy either into or out of the system. Referring again to figure 2.1, if we consider a small element of fluid (dm) passing through the system, the following relationship holds:

$$\begin{aligned}&(\text{Potential Energy at 1}) + (\text{Kinetic Energy at 1}) + (\text{Energy Added}) \\ &- (\text{Energy Loss Due to Friction}) \\ &= (\text{Potential Energy at 2}) + (\text{Kinetic Energy at 2})\end{aligned} \quad (2.5)$$

The potential energy at any point in the system may be divided as follows:

$$(\text{Potential Energy}) = (\text{Elevation Energy}) + (\text{Pressure Energy}) \quad (2.6)$$

The elevation energy at any point is proportional to the elevation above some reference plane as follows:

$$\text{Elevation energy} = dmgh \quad (2.7)$$

where

dm = mass of the fluid element (kg)
g = acceleration due to gravity (m/s^2)
h = elevation above reference plane (m)

The pressure energy is equal to the work that must be done on the fluid to move the element of fluid against the pressure. If we refer to the element dm in figure 2.1, the element of mass dm moves through a distance dl in passing point 1 in the system. The force acting on the element is:

$$\text{Force} = pA \quad (2.8)$$

where

p = fluid pressure (Pa)
A = cross-sectional area of flow stream (m^2)

The distance over which the force must act is:

$$dl = dm/(\rho A) = (dmg)/(\gamma A) \quad (2.9)$$

where

dl = distance over which the force must act (m)
ρ = fluid density (kg/m^3)
γ = specific weight of fluid (N/m^3)

The pressure energy is equal to the product of the force and the distance over which it acts:

$$\text{Pressure Energy} = (\text{Force})\, dl = pA(dmg)/(\gamma A) = (dmgp)/\gamma \tag{2.10}$$

The kinetic energy of the element dm is given by:

$$\text{K.E.} = \tfrac{1}{2}\, dm V^2 \tag{2.11}$$

Substitution of equations 2.7, 2.10, and 2.11 into equation 2.5 gives:

$$dmgh_1 + (dmgp_1)/\gamma + \tfrac{1}{2} dm V_1^2 + dmg W - dmg F$$
$$= dmgh_2 + (dmgp_2)/\gamma + \tfrac{1}{2} dm V_2^2 \tag{2.12}$$

where

W = energy added by pumps, fans, etc. per unit weight of fluid (J/N)
F = energy lost due to friction per unit weight of fluid (J/N)

Dividing through by dmg yields:

$$h_1 + p_1/\gamma + V_1^2/(2g) + W - F = h_2 + p_2/\gamma + V_2^2/(2g) \tag{2.13}$$

Equation 2.13 is known as the Bernoulli theorem for incompressible flow. The terms of the left side of the equation are frequently referred to as the elevation head, pressure head, velocity head, work head, and friction head, respectively.

Example 2.2. A frictionless, incompressible fluid with a density of 1 000 kg/m³ flows in a pipe line. At a point in the line where the pipe diameter is 18 cm, the fluid velocity is 4 m/s and its pressure is 350 kPa. Determine the pressure at a point 15 m downstream where the diameter is 8 cm if the pipe is (a) horizontal or (b) vertical with flow downward.

Solution

(a) $h_1 + p_1/\gamma + V_1^2/(2g) + W - F = h_2 + p_2/\gamma + V_2^2/(2g)$

where h_1 and h_2 cancel since they are equal and W and F are zero. Then

$$p_2/\gamma = p_1/\gamma + V_1^2/(2g) - V_2^2/(2g)$$
$$p_2 = p_1 + (\rho/2)(V_1^2 - V_2^2)$$
$$A_1 V_1 = A_2 V_2$$
$$V_2 = V_1 (D_1/D_2)^2$$
$$p_2 = p_1 + (\rho/2)(V_1^2 - V_1^2 (D_1/D_2)^4)$$
$$p_2 = p_1 + (\rho V_1^2/2)(1 - (D_1/D_2)^4)$$
$$p_2 = 350 \text{ kPa} + ((1\,000 \text{ kg/m}^3)\,(4 \text{ m/s})^2/2)\,(1 - (18 \text{ cm}/8 \text{ cm})^4)$$
$$p_2 = 350 \text{ kPa} + (8\,000 \text{ kg/m s}^2)\,(1 - 25.629)$$
$$p_2 = 350 \text{ kPa} - 197\,031 \text{ N/m}^2$$
$$p_2 = 350 \text{ kPa} - 197 \text{ kPa} = 153 \text{ kPa}$$

Fluid Mechanics

(b) $h_1 = 15$ m and $h_2 = 0$

Then

$$p_2/\gamma = h_1 + p_1/\gamma + V_1^2/(2g) - V_2^2/(2g)$$
$$p_2 = h_1\gamma + p_1 + (\rho/2)(V_1^2 - V_2^2)$$
$$p_2 = (15 \text{ m})(1\,000 \text{ kg/m}^3)(9.81 \text{ m/s}^2) + 153 \text{ kPa}$$
$$p_2 = 147\,150 \text{ N/m}^2 + 153 \text{ kPa} = 147 \text{ kPa} + 153 \text{ kPa} = 300 \text{ kPa}$$

2.4 Power Requirement

In section 2.3, W was defined as the energy supplied per unit of weight passing through the system. Thus, to determine power supplied to the system we must multiply by the weight flow rate. Then,

$$P_s = W\dot{m}g \tag{2.14}$$

The mass flow rate may be expressed in terms of the volume flow rate by:

$$\dot{m} = Q\rho \tag{2.15}$$

Then, substituting into equation 2.14 gives:

$$P_s = WQ\rho g = WQ\gamma \tag{2.16}$$

The power which is required by the pump, fan, etc. is greater than that supplied to the fluid system due to the inefficiency of the device. This may be expressed as:

$$P_r = P_s/e = WQ\gamma/e \tag{2.17}$$

Example 2.3. A storage tank is located 30 m above a body of water. Assume no energy loss in the connecting pipe. If the pipe has an inside diameter of 5 cm, what power would have to be supplied to the fluid in order to pump the water into the tank at a rate of 0.03 m³/s? If the pump has an efficiency of 70%, what will be the power input required by the pump?

Solution

Let the reference elevation be at the surface of the body of water. Then $h_1 = 0$, $V_1 = 0$, and $p_1 = p_2 = p_{atm}$. The Bernoulli equation then becomes:

$$W = h_2 + V_2^2/(2g)$$
$$A_2 V_2 = Q$$
$$V_2 = Q/A_2 = 4Q/(\pi D_2^2)$$
$$W = h_2 + 16Q^2/(2g\pi^2 D_2^4)$$
$$W = h_2 + 8Q^2/(g\pi^2 D_2^4)$$
$$W = 30 \text{ m} + 8(0.03 \text{ m}^3/\text{s})^2/((9.81 \text{ m/s}^2)\pi^2(.05 \text{ m})^4)$$
$$W = 30 \text{ m} + 11.9 \text{ m} = 41.9 \text{ m} = 41.9 \text{ J/N}$$
$$P_s = WQ\rho g$$

$$P_s = (41.9 \text{ J/N}) (.03 \text{ m}^3/\text{s}) (1\,000 \text{ kg/m}^3) (9.81 \text{ m/s}^2)$$
$$P_s = 12\,330 \text{ W} = 12.3 \text{ kW}$$
$$P_r = P_s/e = 12.3 \text{ kW}/0.7 = 17.6 \text{ kW}$$

Characteristics of Fluid Flow

The manner in which a fluid flows through a system is dependent upon the characteristics of the fluid, the size, shape, and condition of the inside surface of the pipe or tube, and the fluid velocity. The frictional resistance F in the Bernoulli equation and the performance of devices for measuring flow rates which are discussed in Chapter 3 are intimately related to the characteristics of flow.

2.5 Laminar (Streamlined) and Turbulent Flow

In laminar flow the fluid moves in parallel elements, the direction of motion of each element being parallel to that of any other element. The velocity of any element is constant but not necessarily the same as that of an adjacent element. In turbulent flow the fluid moves in elemental swirls or eddies, both velocity and direction of each element changing with time. A violent mixing results, whereas there is no significant mixing in the case of laminar flow.

2.6 Distribution of Velocities

A velocity traverse of a fluid (liquid or gas) flowing in a pipe will show that the velocity is highest at the center and decreases toward the surface of the container, the velocity at the surface being zero. This characteristic, which holds for both laminar and turbulent flow, is illustrated in figure 2.2.

The velocity profile for laminar flow in a long circular conduit is parabolic in shape; and the average velocity is one half the maximum, which is at the center. For turbulent flow, the profile flattens and the relationship between the maximum and average velocity changes, its exact value being a function of a number of conditions under which flow results.

2.7 Reynolds Number

Reynolds, an English investigator who was the first to demonstrate the existence of laminar and turbulent flow, also developed a mathematical relationship defining the conditions under which flow changes from laminar to turbulent. Reynolds introduced a thin stream of colored liquid into the bell inlet of a pipe as shown in figure 2.3. He found that the colored thread persisted under low velocities, but as the velocity was increased there was a definite point at which the thread broke and the coloring filled the tube due to eddies or turbulent flow. The velocity at which transition results is called the critical velocity. Reynolds found four factors that affect the critical velocity. These factors and their mathematical relationship follow:

$$Re = DV\rho/\mu \qquad (2.18)$$

Fluid Mechanics

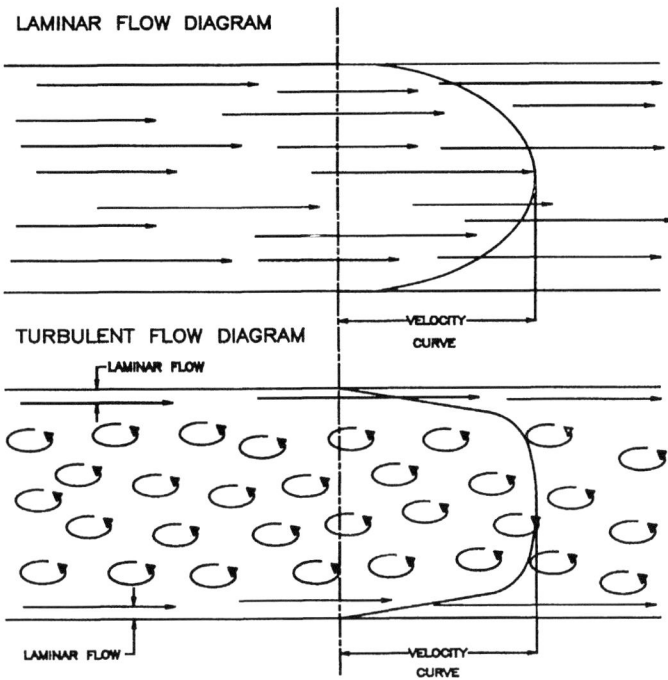

Figure 2.2. Laminar and turbulent flow.

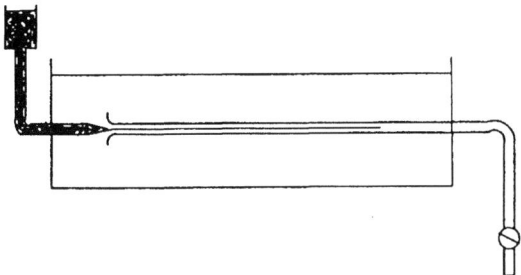

Figure 2.3. Reynolds device for studying the transition from streamlined to turbulent flow.

where

Re = Reynolds number (dimensionless)
D = diameter of pipe (m)
V = average velocity of fluid (m/s)
ρ = fluid density (kg/m^3)
μ = dynamic viscosity of fluid (Pa s)

It can be shown that the Reynolds number (eq. 2.18) is proportional to the ratio of the inertia forces of an element of fluid to the viscous force acting on the fluid. It has been found experimentally that if Re is less than 2130, flow will be laminar and, if over 4000, turbulent. For values between 2130 and 4000, the characteristics of flow will depend upon the details of the structure and any definite prediction is impossible. The above conditions hold for straight circular pipe with isothermal flow.

The above discussion considered only circular pipes. The equation for Reynolds number (eq. 2.18) can be used satisfactorily for rectangular and other shaped conduits by introducing the hydraulic diameter. The definition of the hydraulic diameter varies in the literature but will be defined in this text as follows:

$$D_h = 4(\text{Area of cross section})/(\text{Wetted perimeter}) \tag{2.19}$$

For a conduit filled with a gas or completely filled with a liquid, the complete perimeter is used. If the conduit, a flume for example, is only partially filled, only the "wetted" portion of the perimeter, that contacting the liquid, is used. With the above definition of hydraulic diameter, D_h for a circular pipe is:

$$D_h = 4(\pi D^2/4)/(\pi D) = D \tag{2.20}$$

The hydraulic diameter may be used in Reynolds number as follows:

$$Re = D_h V \rho / \mu \tag{2.21}$$

The student should be aware that the hydraulic diameter defined above is *four times* the hydraulic radius traditionally used in many texts (especially for open channel flow).

Equation 2.21 can be used with fair results for turbulent flow but should not be used under laminar conditions except for nearly square or nearly circular ducts.

2.8 Viscosity

Fluid viscosity μ in equation 2.18 refers to the internal resistance of fluids to shear. The coefficient may be considered as the coefficient of friction of fluid on fluid. The latter consideration is not strictly true since one fluid layer does not actually move over another, but the analogy will serve to give the reader a physical concept of the meaning of viscosity.

Consider two layers of fluid y meters apart, the inner space being filled with fluid, as shown in figure 2.4. Because of the resistance to motion offered by the fluid, a force P is required to maintain a constant velocity V of the top layer relative to the lower layer. Experimental results have shown that for most fluids the required force, P, is directly proportional to the velocity gradient within the fluid, dV/dy, and the surface area, A, of the plate. Thus for figure 2.4:

$$P = \mu A (dV/dy) \tag{2.22}$$

Solving for the units of μ in equation 2.22 results in units of $N \cdot s/m^2$ or Pa s.

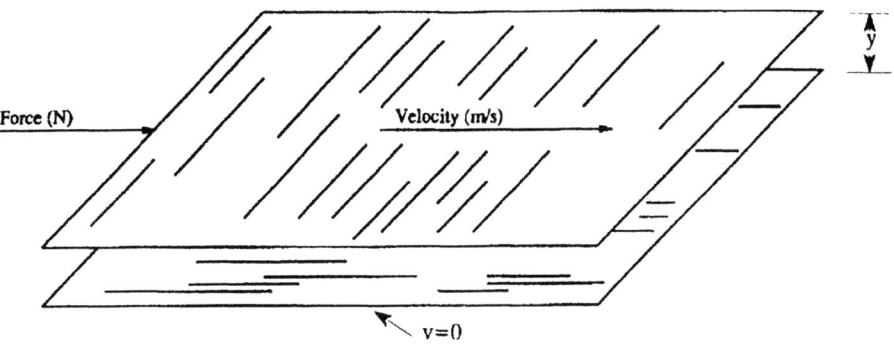

Figure 2.4. Viscosity elements visualized.

A fluid which behaves according to equation 2.22 is called a Newtonian fluid. The equation may be rewritten as:

$$P/A = \mu(dV/dy) \tag{2.23}$$

The left side of equation 2.23 represents the shear force and dV/dy represents the fluid shear. Gases under normal pressures and most liquids are Newtonian fluids. However, materials which consist of long chain molecules, polymers, colloidal suspensions, or non-miscible materials in liquids are usually non-Newtonian. The viscosity of a fluid may also be expressed as a kinematic viscosity given by:

$$\nu = \mu/\rho \tag{2.24}$$

Table 2-1 gives both dynamic and kinematic viscosity values for a number of materials.

2.9 Non-Newtonian Fluids

If the linear relationship noted in equation 2.23 between the shear force and the fluid shear does not hold, the fluid is known as non-Newtonian and the shear rate (for most fluids) may be represented by the following relationship:

$$P/A = a(dV/dy)^b + c \tag{2.25}$$

where a, b, and c are constants for the fluid.

Non-Newtonian fluids are composed of long chain molecules, polymers, colloidal suspensions, and non-miscible materials in liquids, for example, slurries, food purees, paints, mayonnaise, and butter. Some types of non-Newtonian fluids are illustrated in figure 2.5 and are discussed in the following paragraphs.

Bingham fluids are true plastics (figure 2.5) which seldom, if ever, exist in a real state. The true shear rate might be represented best by the dotted curve. Some materials that may approach the Bingham concept are damp clay, concentrated slurries, modeling clay, window putty, and cheese.

Pseudoplastics are assumed to be represented by equation 2.25 with c being zero. The viscosity decreases as the shear rate increases, a feature that is very important when the fluid is pumped or involved in heat transfer. The Reynolds number, required

Table 2-1. Viscosity of various materials

	Temp. (C)	Mass Density (Approx. kg/m^3)	Dynamic Viscosity (Pa s)	Kinematic Viscosity (m^2/s)
Air	0.0	1.295	1.71×10^{-5}	1.33×10^{-5}
	21.0	1.202	1.82×10^{-5}	1.51×10^{-5}
	100.0	0.947	2.19×10^{-5}	2.30×10^{-5}
CaCl$_2$ brine, 24% sol.	−23.0	1238.0	1.25×10^{-2}	1.01×10^{-5}
	−18.0	1234.0	8.80×10^{-3}	7.13×10^{-6}
	2.0	1227.0	3.69×10^{-3}	3.01×10^{-6}
Cotton-seed oil	16.0	920.0	9.10×10^{-2}	9.89×10^{-5}
Cream, 20% fat, past.	3.0	1010.0	6.19×10^{-3}	6.13×10^{-6}
30% fat, past.	3.0	1000.0	1.39×10^{-2}	1.39×10^{-5}
Freon-12	−15.0	1440.0	3.30×10^{-4}	2.29×10^{-7}
Liquid Ammonia	−15.0	660.0	2.50×10^{-4}	3.79×10^{-7}
	27.0	660.0	2.10×10^{-4}	3.50×10^{-7}
Lub. oil, SAE 10	16.0	900.0	1.00×10^{-1}	1.11×10^{-4}
	66.0	870.0	1.00×10^{-2}	1.15×10^{-5}
SAE 30	16.0	900.0	4.00×10^{-1}	4.44×10^{-4}
	66.0	870.0	2.69×10^{-2}	3.09×10^{-5}
Milk, skim	25.0	1040.0	1.37×10^{-3}	1.32×10^{-6}
whole	0.0	1035.0	4.29×10^{-3}	4.14×10^{-6}
	20.2	1030.0	2.13×10^{-3}	2.07×10^{-6}
Molasses, heavy dark	21.0	1430.0	6.59	4.61×10^{-3}
	38.0	1380.0	1.88	1.36×10^{-3}
	49.0	1310.0	9.20×10^{-1}	7.02×10^{-4}
	66.0	1160.0	3.74×10^{-1}	3.22×10^{-4}
NaCl brine, 22% sol.	−18.0	1190.0	6.10×10^{-3}	5.13×10^{-6}
	2.0	1170.0	2.69×10^{-3}	2.30×10^{-6}
Soybean oil	30.0	920.0	4.06×10^{-2}	4.41×10^{-5}
Sucrose, 20% sol.	0.0	1086.0	3.81×10^{-3}	3.51×10^{-6}
	21.0	1082.0	1.92×10^{-3}	1.77×10^{-6}
	80.0	1055.0	5.92×10^{-4}	5.61×10^{-7}
60% sol.	21.0	1289.0	6.01×10^{-2}	4.66×10^{-5}
	80.0	1252.0	5.42×10^{-3}	4.33×10^{-6}
Water	0.0	1000.0	1.80×10^{-3}	1.80×10^{-6}
	21.0	998.0	9.84×10^{-4}	9.86×10^{-7}
	49.0	987.0	5.58×10^{-4}	5.65×10^{-7}

for computations in both cases, is a function of a variable viscosity in addition to the variation in fluid velocity.

Examples of pseudoplastic materials are paints, mayonnaise, heavy slurries, human blood, some honeys, melts and solutions of high molecular weight substances, tomato paste, and purees. Many materials in this list are also often classed as Bingham fluids

Fluid Mechanics

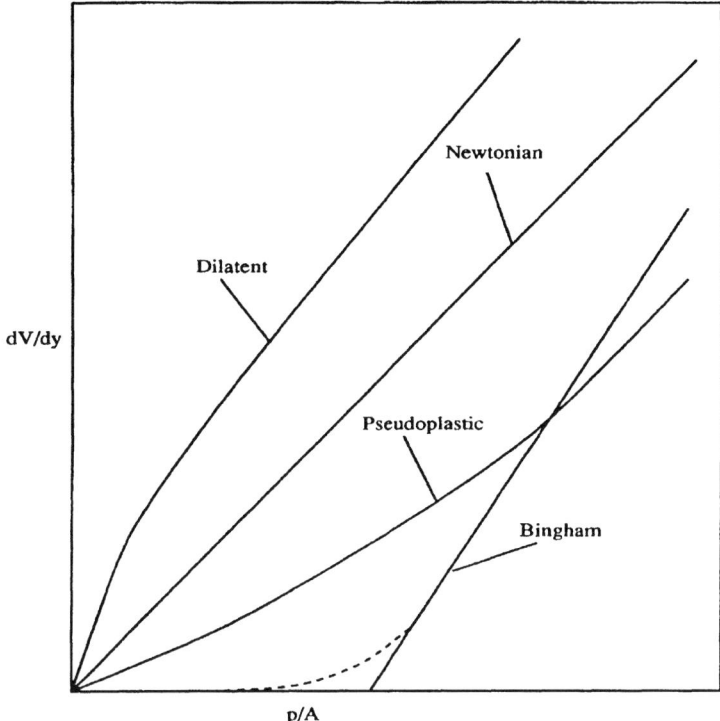

Figure 2.5. Relative relationships of Newtonian and non-Newtonian fluids.

which, in effect, emphasizes the observation that Bingham fluids seldom exist in a real state.

Some constants for equation 2.25 by Harper (1960) are:

Material	Temp. (C)	a	b
Peach puree	39.4	58.5	0.35
Peach puree	43.3	44.0	0.38
Apricot puree	47.8	56.0	0.37
Apricot puree	76.1	11.2	0.35

Dilatent fluids have a viscosity that increases as the shear rate increases. Examples are moist sand, quick-sand, heavy starch suspensions, many honeys, buckwheat, sumac, orange, tarweed, Hawaiian Honeydew, and others.

Thixiotropic fluids become more fluid and viscosity decreases with time as they are stirred. Although extensive experimental data are not available, the time function seems to be logarithmic. When motion stops, the fluid will return to its original viscosity although a hysteresis may be formed on the return. Bread dough, heather honey, some soil suspensions, and certain paints are examples.

Rheopectic fluids exhibit an opposite effect to thixiotropic fluids because they become less fluid with time. Otherwise, the characteristics are comparable.

The science that considers stress-strain relations where flow proceeds irreversibly with time is called *rheology*. Creep in metals and concrete and the strength properties of fruits and vegetables are examples. Strictly recognized, thixiotropic and rheopectic fluids are not rheological materials because there is no permanent deformation. Unfortunately, many writers and groups use the term *rheology* to include a variety of stress-strain processes that may or may not be time dependent or irreversible. The engineer should be aware that the term is frequently used to cover processes outside those in the definition.

Friction Losses

The friction term, F, in the Bernoulli equation (eq. 2.13) represents the energy lost per unit weight by the fluid due to effects such as internal fluid resistance, turbulence, and resistance of the flow retainer. Evaluation of this factor involves the Reynolds number, the dimensions of the conduit, and certain empirically determined information.

2.10 Darcy's Formula

The most widely used equation for determining the friction loss in a pipe was developed by Darcy:

$$F = f(L/D)(V^2/(2g)) \tag{2.26}$$

where

F = energy lost in a pipe due to friction per unit weight of fluid (J/N or m)
L = length of pipe (m)
D = diameter of pipe (m)
V = average fluid velocity (m/s)
g = gravitational acceleration (m/s^2)
f = friction factor (dimensionless)

The friction factor f is a function of the Reynolds number and the relative roughness of the pipe. The relative roughness is the ratio of the roughness factor, ε, to the pipe diameter, D. Table 2-2 gives roughness factors for a number of pipe materials.

Table 2-2. Roughness factor for various types of pipe

Pipe Material	Roughness Factor, ε (mm)
Riveted steel	0.9–9.0
Concrete	0.3–3.0
Wood stave	0.2–0.9
Cast iron	0.26
Galvanized iron	0.15
Asphalted cast iron	0.12
Commercial steel or wrought iron	0.046
Drawn tubing	0.0015

Fluid Mechanics

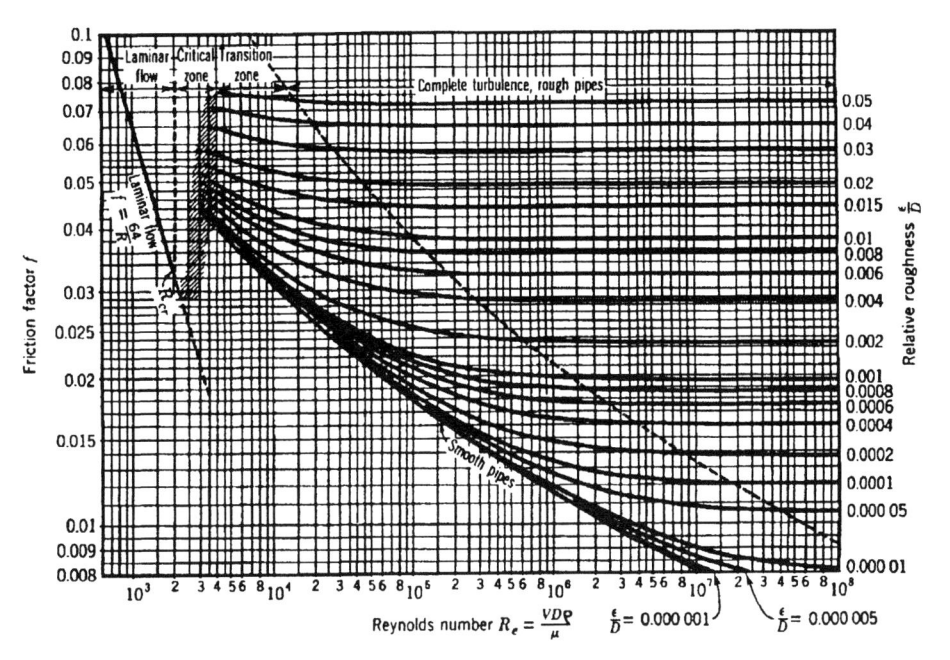

Figure 2.6. Friction factor, f, vs. Reynolds number for pipe of various relative roughness.

The Reynolds number and the ε/D ratio are calculated and the friction factor, f, determined from one of the following equations:

$$f = 64/Re \tag{2.27}$$

when Re is less than 2130 and by the Colebrook equation:

$$f = 0.25/(\log_{10}(\varepsilon/(3.7D) + 2.51/(Re f^{0.5})))^2 \tag{2.28}$$

when Re is greater than 2130.

Note that f appears on both sides of equation 2.28 and must be evaluated by a trial and error procedure or manually from plots of the relationship. Figure 2.6 is a graphical representation of equations 2.27 and 2.28 and is called a Moody diagram. Unfortunately, there is within the engineering literature an alternate representation of those parameters in equations 2.26 to 2.28. In some engineering literature, primarily that of the chemical engineering profession, equation 2.26 is modified by placing a coefficient of 4 (four) in front of the friction factor, f. The friction factor is then defined such that its value is exactly 1/4 of the values given in equations 2.27 and 2.28. Use of this alternate set of equations results in identical evaluations of the F term in the Bernoulli equation. However, the student must be careful to obtain the friction factor value from the appropriate equations or graphs within the literature. Within this text, the procedure represented by equations 2.26 to 2.28 will be used exclusively for both fluid flow within pipes and other Darcy-like relationships to be discussed later (i.e., resistance to flow through beds of granular material and through heat exchangers).

Example 2.4. Soybean oil at 30 C is pumped through a 5-cm-inside-diameter stainless steel pipe at a flow rate of 0.03 m^3/s. Determine the friction loss, F, for the fluid in a 15 m length of pipe.

Solution

$$Q = VA$$
$$V = Q/A = 4Q/(\pi D^2)$$
$$V = 4(0.03 \text{ m}^3/\text{s})/(\pi (.05 \text{ m})^2) = 15.28 \text{ m/s}$$
$$Re = \rho DV/\mu$$
$$Re = (920 \text{ kg/m}^3)(.05 \text{ m})(15.28 \text{ m/s})/(4.06 \times 10^{-2} \text{ Pa s})$$
$$Re = 17\,312$$
$$\varepsilon/D = (.000046 \text{ m})/(.05 \text{ m}) = 0.00092$$

From figure 2.6, $f = 0.0286$.
Then,

$$F = f(L/D)(V^2/(2g))$$
$$F = 0.0286((15 \text{ m})/(0.05 \text{ m}))((15.28 \text{ m/s})^2/(2(9.81 \text{ m/s}^2)))$$
$$F = 102.1 \text{ m} = 102.1 \text{ J/N}$$

2.11 Energy Losses Due to Fittings

Pipe and conduit fittings or transitions in dimensions cause energy losses which contribute to the total value of F in the Bernoulli equation. The characteristics of this loss have not been sufficiently treated in most cases and must be evaluated using empirical data. The friction loss term is usually expressed as:

$$F = K(V^2/2g) \qquad (2.29)$$

where

F = energy lost in fittings due to friction per unit weight of fluid (J/N)
K = friction loss factor (dimensionless)
V = average fluid velocity (m/s)
g = acceleration due to gravity (m/s^2)

Friction loss factors for some types of resistances are given in table 2-3.

2.12 Energy Loss Due to Sudden Contractions

The friction loss due to sudden contractions of the fluid may also be evaluated using equation 2.29 where the value of K changes with the degree of contraction. Table 2-4 gives K values for different ratios of the contracted and initial areas.

Fluid Mechanics

Table 2.3. Friction loss factors, K

Nature of Resistance		K
Valves, fully open	Gate	0.15
	Globe	7.5
	Angle	4.0
Elbows		A: 0.50
		B: 0.25
		C: 0.90
		D: $1.25(\alpha/90)^2$
Tees		XA: 1.50
		XB: 0.50
Discharge nozzles		A: 0.01–0.03
		B: 0.01–0.04*
		C: *
Entrances		A: 0.50
		B: 0.05
		C: 1.00

* Varies, use manufacturers values.

Table 2.4. Friction loss factors for sudden contractions*

A_2/A_1	K
0	0.5
0.1	0.46
0.3	0.36
0.5	0.24
0.7	0.12
0.9	0.02
1.0	0.00

*From Vennard, "*Elementary Fluid Mechanics*".

2.13 Energy Loss Due to a Sudden Enlargement

The friction loss due to a sudden enlargement such as shown in figure 2.7 is given by:

$$F = (V_1 - V_2)^2/2g \qquad (2.30)$$

where

F = friction loss due to sudden enlargement per unit weight of fluid (J/N)
V_1 = average velocity of fluid prior to enlargement (m/s)

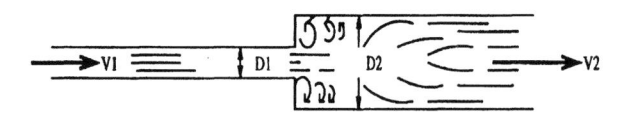

Figure 2.7. Sudden enlargement.

V_2 = average velocity of fluid after enlargement (m/s)
g = acceleration due to gravity (m/s^2)

Note that if the expanded area is very large, the velocity after the expansion becomes zero and the friction loss becomes the velocity head prior to the expansion:

$$F = V_1^2/2g \qquad (2.31)$$

2.14 Energy Loss Due to Flow Through Agricultural Products

The resistance of fixed beds of granular materials to air flow is a function of the size and shape, surface configuration, and size distribution of the granular particles and to the method of pouring the material into the container which affects the amount of void space in the bed.

Zenz and Othmer (1960) and Leva (1959) have reviewed a large number of carefully conducted experimental studies. Considerable variation exists between the observations of the investigators and there is some variation in the general recommendations by the respective investigators. One procedure outlined by Bird, Stewart and Lightfoot (1960) is based on the Ergun (1952) equation for estimating the friction factor in the Darcy equation:

$$F = \Delta p/\gamma = f(L/D_p)(V^2/2g) \qquad (2.32)$$

where

F = friction loss due to resistance of bed of material per unit weight of fluid (J/N)
Δp = change in fluid pressure in passing through bed (Pa)
γ = specific weight of fluid (N/m^3)
f = friction factor (dimensionless)
L = depth of bed of material (m)
D_p = diameter of particles in bed (m)
V = volume flow rate of fluid per unit cross-sectional area of bed, sometimes called superficial velocity (m^3/s m^2 or m/s)
g = acceleration due to gravity (m/s^2)

When the particles are not spheres, the particle diameter used in equation 2.32 and in the Reynolds number is computed as:

$$D_p = 6v_p/s_p \qquad (2.33)$$

Fluid Mechanics

ERGUN EQUATION

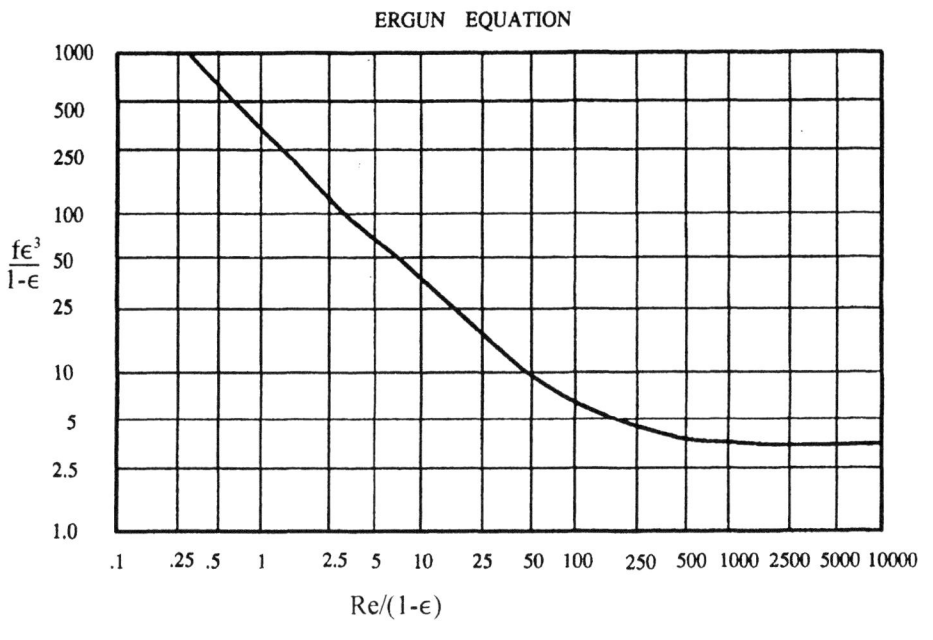

[Figure: Plot with y-axis $\frac{f\epsilon^3}{1-\epsilon}$ ranging from 1.0 to 1000, x-axis $Re/(1-\epsilon)$ ranging from .1 to 10000, showing a decreasing curve that levels off.]

Figure 2.8. Plot of the Ergun equation for friction factor for fluid flowing through a bed of particles.

where

v_p = particle volume (m³)
s_p = particle surface area (m²)

The Ergun equation for estimating the friction factor f as a function of the Reynolds number and the voidage, ε_p, of the bed is as follows:

$$f = ((1 - \varepsilon_p)/\varepsilon_p^3)(300(1 - \varepsilon_p)/Re + 3.5) \quad (2.34)$$

where ε_p = bed voidage (dimensionless).

A plot of $f\varepsilon_p^3/(1 - \varepsilon_p)$ versus $Re/(1 - \varepsilon_p)$ results in the single curve shown in figure 2.8. Laminar flow is expected when the $Re/(1-\varepsilon_p)$ term is less than approximately 10.

Agricultural particles are usually of an irregular shape and frequently have a surface configuration which makes use of the Ergun equation difficult. However, Bakker-Arkema et al. (1969) found that the relationship was successful in fitting resistance data of cherry pits after adjusting the equation by multiplying by a constant of approximately 1.22. It was recommended that resistance data for agricultural products be fitted to the Ergun equation and appropriate effective diameters, voidage fractions, and correction factors be tabulated. Apparently this has not as yet been done.

Resistance data for a number of agricultural products have been observed by various investigators and much of it is included as ASAE Data D272 in *ASAE Standards, 1996* of the American Society of Agricultural Engineers. Figures 2.9 and 2.10 give some

28   Principles of Process Engineering

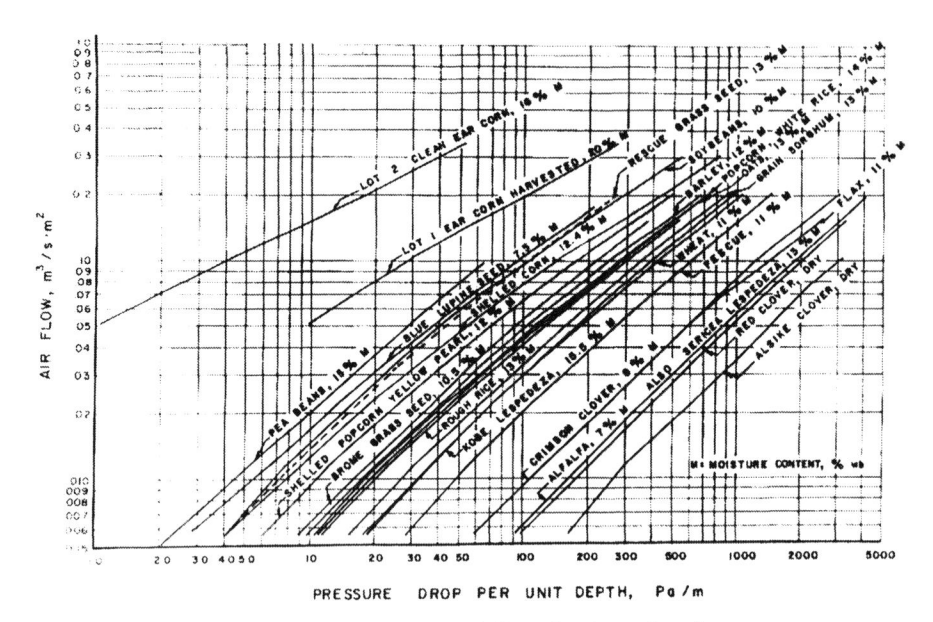

Figure 2.9. Resistance to airflow of grains and seeds.

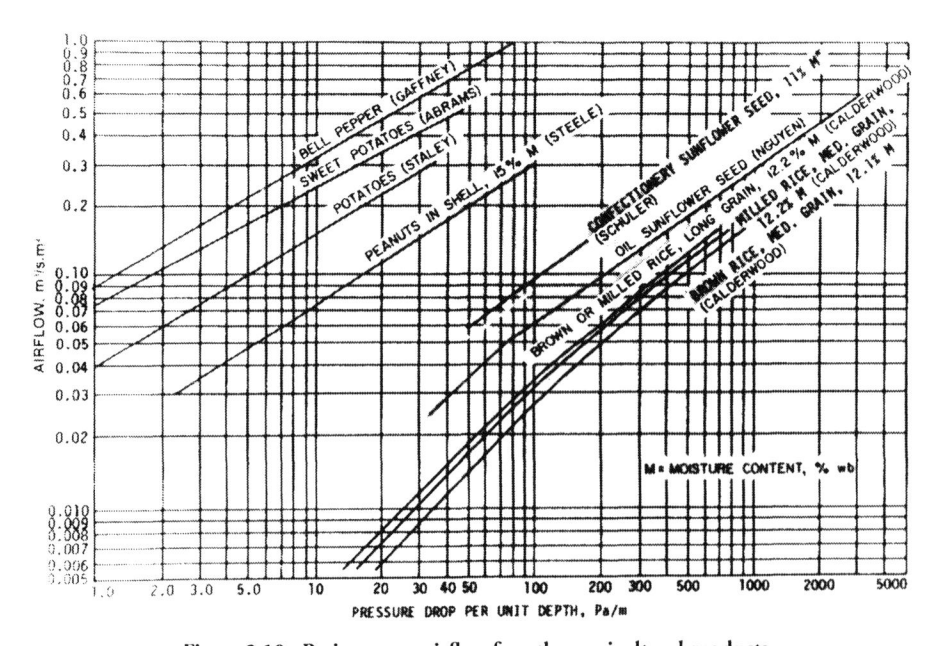

Figure 2.10. Resistance to airflow for other agricultural products.

Fluid Mechanics

experimental plots of superficial velocity versus the pressure drop per unit depth for a number of agricultural materials.

Hukil and Ives (1955) suggested the following relationship for describing the pressure loss data for agricultural materials:

$$\Delta p / L = a V^2 / \ln(1 + bV) \qquad (2.35)$$

where

Δp = pressure loss through bed (Pa)
L = depth of bed of material (m)
V = superficial velocity (m/s or m^3/s m^2)
a = constant for material (Pa s^2/m^3)
b = constant for material (m^2 s/m^3)

Table 2-5 gives the values of a and b which best fit the data shown in figures 2.9 and 2.10. The experimental data on pressure drop through beds of material is for clean, loosely packed beds. Therefore, it is common practice to apply a correction factor to the estimates obtained from equation 2.35 in order to more nearly predict the pressure drop in a real situation where foreign material is in the bed and the material may be more densely packed due to handling operations. Though the correction factor may vary widely for given situations, a value of 1.5 is often used for engineering design. Then:

$$\Delta p = a C_f L V^2 / \ln(1 + bV) \qquad (2.36)$$

where C_f = correction factor (dimensionless).

Once the pressure drop is computed from equation 2.36, the friction term F of the Bernoulli equation may be determined as:

$$F = \Delta p / \gamma \qquad (2.37)$$

While the empirical relationship represented by the Hukil and Ives (1955) equation is quite useful in predicting the pressure drop through agricultural products for the conditions under which experimental data have been obtained, it does not allow for predictions under varying voidage fractions, particle diameters, or fluids. Further work on development of an Ergun-type relationship to satisfactorily describe resistance in beds of agricultural products appears needed.

2.15 Energy Loss Due to Flow Through Floors

The perforated floor or wall which retains an agricultural product being dried or aerated offers resistance to air flow in addition to the resistance of the material. Henderson (1943) found the following experimental relationship for perforated floors:

$$F = (1.071 \text{ Pa s}^2/\text{m}^2)(V/O_f)^2/(\rho g) \qquad (2.38)$$

where

F = friction loss due to floor (J/N)
V = superficial velocity (m/s)

Table 2-5. Values for constants in airflow resistance equation

Material	Value of a (Pa s^2/m^3)	Value of b (m^2 s/m^3)	Range of V (m^3/m^2 s)	Ref.
Alfalfa	6.40×10^4	3.99	.0056–.152	Shedd
Barley	2.14×10^4	13.2	.0056–.203	Shedd
Brome grass	1.35×10^4	8.88	.0056–.152	Shedd
Clover, alsike	6.11×10^4	2.24	.0056–.101	Shedd
crimson	5.32×10^4	5.12	.0056–.203	Shedd
red	6.24×10^4	3.55	.0056–.152	Shedd
Corn, ear	1.04×10^4	325	.051–.353	Shedd
shelled	2.07×10^4	30.4	.0056–.304	Shedd
sh. low flow	9.77×10^3	8.55	.00025–.0203	Sheldon
Fescue	3.15×10^4	6.70	.0056–.203	Shedd
Flax	8.63×10^4	8.29	.0056–.152	Shedd
Lespedeza, Kobe	1.95×10^4	6.30	.0056–.203	Shedd
Sericea	6.40×10^4	3.99	.0056–.152	Shedd
Lupine	1.07×10^4	21.1	.0056–.152	Shedd
Oats	2.41×10^4	13.9	.0056–.203	Shedd
Peanuts	3.80×10^3	111	.030–.304	Steele
Peppers	5.44×10^2	868	.030–1.00	Gaffney
Popcorn, white	2.19×10^4	11.8	.0056–.203	Shedd
yellow	1.78×10^4	17.6	.0056–.203	Shedd
Potatoes	2.18×10^3	824	.030–.300	Staley
Rescue	8.11×10^3	11.7	.0056–.203	Shedd
Rice, rough	2.57×10^4	13.2	.0056–.152	Shedd
long brown	2.05×10^4	7.74	.0055–.164	Calderwood
long milled	2.18×10^4	8.34	.0055–.164	Calderwood
med. brown	3.49×10^4	10.9	.0055–.164	Calderwood
med. milled	2.90×10^4	10.6	.0055–.164	Calderwood
Sorghum	2.12×10^4	8.06	.0056–.203	Shedd
Soybeans	1.02×10^4	16	.0056–.304	Shedd
Sunflower, conf	1.10×10^4	18.1	.0055–.178	Shuler
oil	2.49×10^4	23.7	.025–.570	Nguyen
Sweet Potato	3.40×10^3	6.1×10^8	.050–.499	Abrams
Wheat	2.70×10^4	8.77	.0056–.203	Shedd
Wheat, low flow	8.41×10^3	2.72	.00025–.0203	Sheldon

O_f = percent floor opening expressed as a decimal
ρ = density of fluid (kg/m^3)
g = acceleration due to gravity (m/s^2)

When a bed of material is placed on a perforated floor, the effective amount of floor opening is decreased. Theoretically, we would expect a decrease in the effective area by a fraction equal to the voidage fraction of the material. Henderson (1943) confirmed this expectation for shelled corn having a voidage fraction of 40%. Assuming this condition to hold for all material, the expression for energy loss through a perforated floor supporting a bed of material becomes:

$$F = (1.071 \text{ Pa s}^2/\text{m}^2)(V/(O_f \varepsilon_p))^2/(\rho g) \qquad (2.39)$$

where ε_p = voidage fraction of the material.

2.16 Energy Loss in Heat Exchangers

The resistance to air flow or pressure drop through heat exchangers is important in installations such as refrigeration plants, air conditioning units, and driers. In general, it is expedient to use the pressure drop data supplied by the manufacturer of the heat exchanger in question. However, a general method for estimating this effect may be found useful for certain conditions.

If the exchanger is made of a series of parallel tubes, the friction energy loss can be calculated from the following Darcy-type equation based on the results of studies by a number of investigators:

$$F = \Delta p/\gamma = fE(V^2/2g) \qquad (2.40)$$

where

F = friction loss due to heat exchanger (J/N)
Δ_p = pressure drop through heat exchanger (Pa)
γ = specific weight of fluid (N/m^3)
f = friction factor (dimensionless)
E = number of rows of tubes normal to fluid stream,
V = maximum velocity through the minimum cross-section (m/s)
g = acceleration due to gravity (m/s^2)

The friction factor f is a function of the Reynolds number as follows:

$$f = 3Re^{-0.2} \qquad (2.41)$$

where

$Re = CV\rho/\mu$
C = clearance between tubes in a row, A through D of figure 2.11 (m)
ρ = fluid density (kg/m^3)
μ = dynamic viscosity of fluid (Pa s)

Equation 2.40 is probably reliable to within 25% for pitch distances A of 1.25 to 1.50 tube diameters D, which is the normal commercial spacing. Flow is probably turbulent if the Reynolds number is greater than 40.

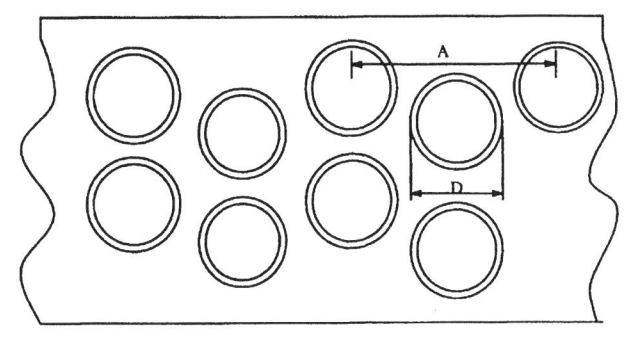

Figure 2.11. Cross section of a heat exchanger.

Baffled or finned exchangers require an involved method of calculation and will not be discussed here.

2.17 System Characteristic Curves

If the resistance to flow through a given system is plotted versus the volumetric flow rate through the system, the plot is called a system characteristic curve. The resistance to flow through the system may be characterized by one of several indices. It is conventional practice to use total head, W, as the index of system resistance for systems involving liquid flow and to use either total pressure, p_t, or static pressure, p_s, in systems involving flow of gases.

Example 2.5. Water is to be pumped from a reservoir at atmospheric pressure to a second reservoir at atmospheric pressure with an elevation 5 m above the first. The water must flow through 20 m of 3-cm-inside-diameter galvanized iron pipe containing two Type A elbows. The water enters the pipe through a type A entrance. Plot a system characteristic curve for the flow system for flow rates between 0 and 0.005 m³/s.

Solution
First note that nothing is said about a pump in the statement of the problem. The system characteristic curve is not dependent upon the pump, but only on those components of the system which offer resistance to the flow. A pump, however, is the component which must provide the necessary fluid energy to overcome the friction forces in the system. The system characteristic curve tells us how much energy must be supplied by a pump as a function of flow rate through the system.

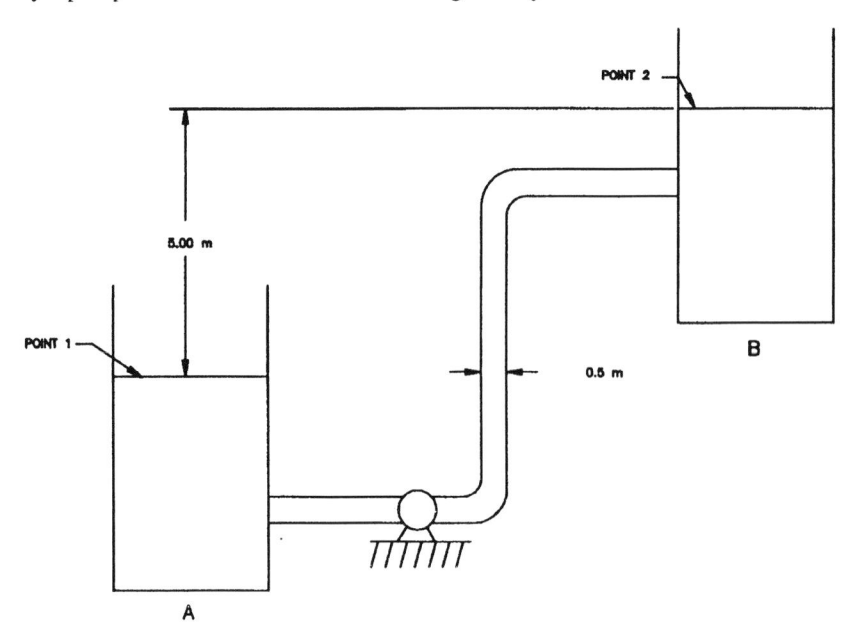

Figure 2.12. Sketch of example 2.5 with liquid flow from reservoir A to reservoir B.

Figure 2.12 sketches one system arrangement with the properties stated in the problem. The Bernoulli equation 2.13 may be written between points 1 and 2 as follows:

$$h_1 + p_1/\gamma + v_1^2/(2g) + W - F = h_2 + p_2/\gamma + v_2^2/(2g)$$

The pressure at both points 1 and 2 is atmospheric (or $p_1 = p_2$) and the velocity at both points is zero. Therefore, the equation reduces to:

$$W = F + h_2 - h_1 = F + 5\,\text{m}$$

The friction term is the sum of friction terms due to the entrance, the pipe, the elbows, and the exit (sudden enlargement). Thus,

$$F = F_{entrance} + F_{pipe} + F_{elbows} + F_{exit}$$

and the friction terms are given by:

$$F_{entrance} = K_{entrance}(v^2/(2g))$$
$$F_{pipe} = f(l/d)(v^2/(2g))$$
$$F_{elbows} = 2K_{elbows}(v^2/(2g))$$
$$F_{exit} = (v - v_2)^2/(2g) = v^2/(2g)$$

where

v = average velocity of fluid in the pipe (m/s)

and

$$v = Q/A \quad \text{(equation 2.4)}$$

Then,

$$W = 5\,\text{m} + (K_{entrance} + f(l/d) + 2K_{elbows} + 1)(v^2/(2g))$$

or

$$W = 5\,\text{m} + (K_{entrance} + f(l/d) + 2K_{elbows} + 1)(Q^2/(2gA^2))$$

A plot of W versus Q for this situation is given in figure 2.13 and is the *system characteristic curve*. Note that the intercept is at 5 m (the difference in elevation of the reservoirs) and that W increases with roughly the square of the flow rate at flows above zero. (Note: Deviation from this is due to the fact that the friction factor, f, is a decreasing function of flow rate.)

Example 2.6. Air is to be forced through a grain drying bin similar to that of figure 2.14. The air flows through 5 m of 0.5-m-diameter galvanized iron conduit, exhausts into a plenum beneath the grain, passes through a perforated metal floor having a percent opening of 10%, and is finally forced through a 1m depth of wheat having a void fraction of 0.4. The area of the bin floor is 20 m². Plot a system characteristic curve (Note: Plot both total and static pressure curves.) for this system for flow rates between 0 and 4 m³/s.

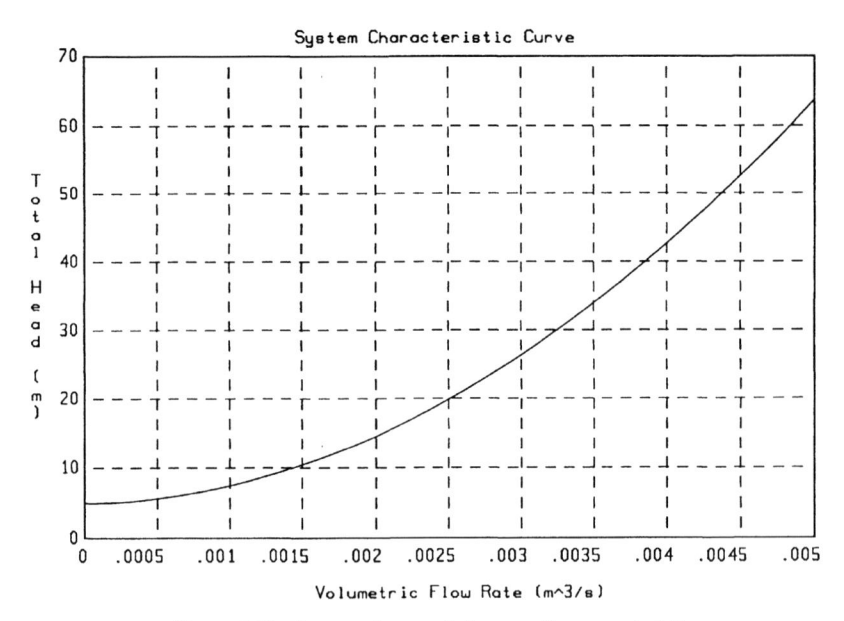

Figure 2.13. System characteristic curve for example 2.5.

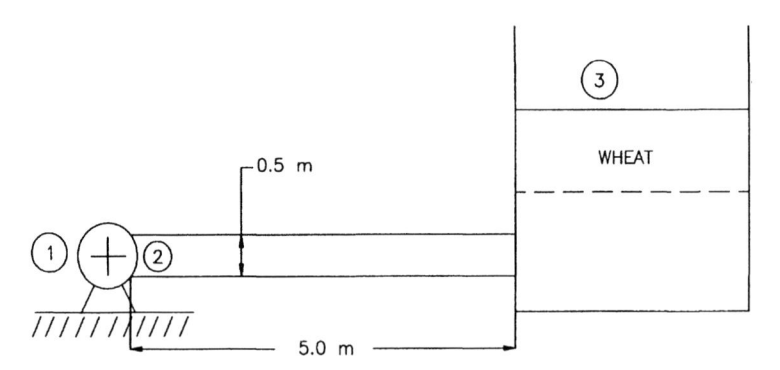

Figure 2.14. Sketch of grain drying bin for example 2.6.

Solution
As in example 2.5, nothing is said about a fan. A fan is the component which must provide the energy to the air to overcome the friction losses in passing through the system. The system characteristic curve tells us how much energy must be supplied by the fan as a function of volumetric flow rate. It is conventional for air flow systems to use either the static pressure or the total pressure which must be developed by the fan as an indicator of the energy per unit weight of air which must be supplied.

An important feature of an air flow system such as this, in which *both the entrance and exit are exposed to atmospheric pressures*, is that the difference in elevation of the entrance and exit is not important. This may be shown by writing Bernoulli's

Fluid Mechanics

equation for still air at points 1 and 3 in figure 2.14 to obtain:

$$h_1 + p_1/\gamma + v_1^2/(2g) + W - F = h_3 + p_3/\gamma + v_3^2/(2g)$$

or

$$h_1 + p_{1\text{atm}}/\gamma = h_3 + p_{3\text{atm}}/\gamma$$

where

$p_{1\text{atm}}$ = atmospheric pressure at point 1 (Pa)
$p_{3\text{atm}}$ = atmospheric pressure at point 3 (Pa)

and

$$v_1 = v_3 = W = F = 0$$

Thus, the difference in elevation is exactly offset by the small difference in atmospheric pressure head due to the column of air between the two elevations.

We can now write Bernoulli's equation between points 2 and 3 in the air flow system. Point 2 is taken just downstream from the fan (i.e., the exit of the fan or beginning of the system). Then,

$$h_2 + p_2/\gamma + v_2^2/(2g) + W - F = h_3 + p_3/\gamma + v_3^2/(2g)$$

where

$W = 0$ since no energy is added between points 2 and 3,
$v_3 = 0$
$h_2 = h_1$
$p_2 = p_{1\text{atm}} + p_s$
$p_3 = p_{3\text{atm}}$
p_s = static pressure = pressure added by fan (Pa)

Then,

$$h_1 + p_{1\text{atm}}/\gamma + p_s/\gamma + v_2^2/(2g) - F = h_3 + p_{3\text{atm}}/\gamma$$

and

$$p_s/\gamma + v_2^2/(2g) = F$$

The friction term is the sum of friction losses in the pipe, the sudden expansion at the plenum, the floor, and the bed of wheat.

$$F = F_{\text{pipe}} + F_{\text{expansion}} + F_{\text{floor}} + F_{\text{wheat}}$$

and these are given by,

$$F_{\text{pipe}} = f(l/d)(v^2/(2g))$$
$$F_{\text{expansion}} = (v - v_{\text{plenum}})^2/(2g)$$
$$F_{\text{floor}} = (1.071 \text{ Pa s}^2/\text{m}^2)(V/(O_f \varepsilon_p))^2/(\rho g)$$
$$F_{\text{wheat}} = \Delta p/\gamma = a\, C_f L\, V^2/(\gamma \ln(1 + bV))$$

where

v = velocity in pipe, m/s = $Q/A_{pipe} = v_2$
v_{plenum} = 0 = velocity in the plenum
V = superficial velocity through floor and bed of wheat (m/s) = Q/A_{floor}
A_{pipe} = $\pi d^2/4$ = cross-sectional area of pipe (m^2)
A_{floor} = area of bin floor (m^2)
Q = volumetric flow rate through the system (m^3/s)

Thus, for a given flow rate, all the velocities and friction terms can be evaluated and the static pressure computed. By solving for a series of flow rates ranging between 0 and 4.0 m^3/s, a series of corresponding static pressures may be obtained and plotted. Figure 2.15 gives a static pressure curve for this example problem.

The total pressure is defined by the following relationship:

$$p_t/\gamma = p_s/\gamma + v^2/(2g)$$

where v is the fluid velocity at the point where the static and total pressures are evaluated.

Thus, at point 2 of figure 2.14, the total pressure is:

$$p_t = p_s + \gamma v_2^2/(2g) = F\gamma$$

This relationship may be used to solve for total pressure at each volumetric flow rate and may then be plotted as is shown for this problem in figure 2.15.

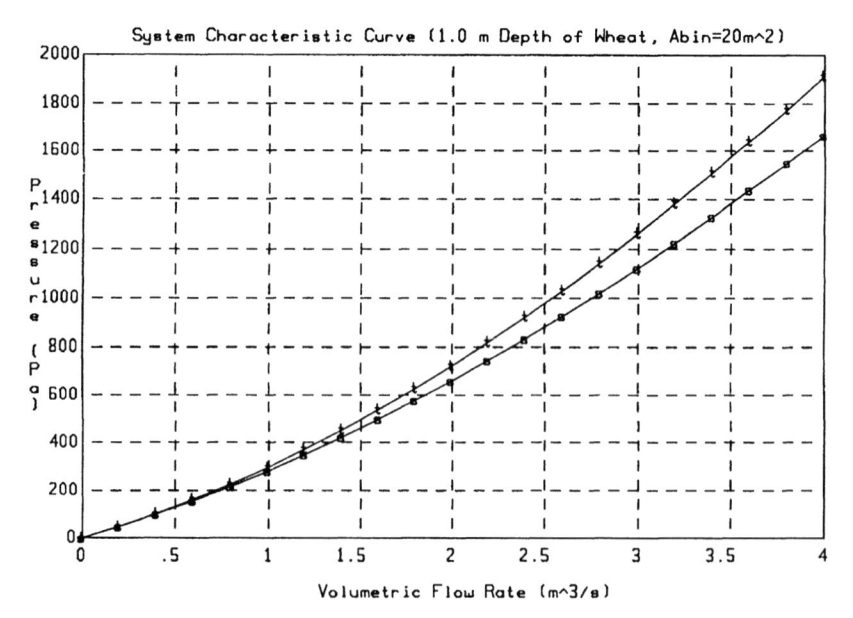

Figure 2.15. Total pressure (*t*) and static pressure (*s*) curves for example 2.6

Fluid Mechanics

Design Considerations

2.18 Branching System Design

Frequently a system of conduits must be designed so that the flowing fluid is divided in some proportion among a number of branching lines. An air-conditioning or ventilating system serving a number of locations is an example.

Where a dividing system of conduits is to be designed the *equal-pressure-drop* method is the most usable. This method is illustrated by the following example.

Example 2.7. A seed cleaning house is to install a hood over each machine to exhaust dust arising from the cleaning operation. A schematic plan of the system is shown in figure 2.16. Assume that the hoods have a friction loss equivalent to 20 diameters of the connecting pipe and that the optimum velocity in the pipes for dust removal is 5 m/s.

Solution
The objective in this problem is to determine pipe diameters which give desirable air velocities and offer the same friction loss through each flow path. This must be done by simultaneously solving (usually iteratively) a set of equations describing the friction losses in the system. The equations to be solved are given in figure 2.17 in the form of a rules sheet and function sheets for a TK SOLVER program which is used here to iteratively determine a solution. Figure 2.18 is the variables sheet for the program and gives the definition and units for all variables. There are 59 variables and only 39 independent rules in the listings of figures 2.17 and 2.18. Thus, 20 variables must be specified. Since only 16 variables are specified by the problem, four additional variables must be arbitrarily chosen. In the solution illustrated, the velocities VR1, V12, V2D, and V3B were set to the desired value of 5.0 m/s. The input column of the

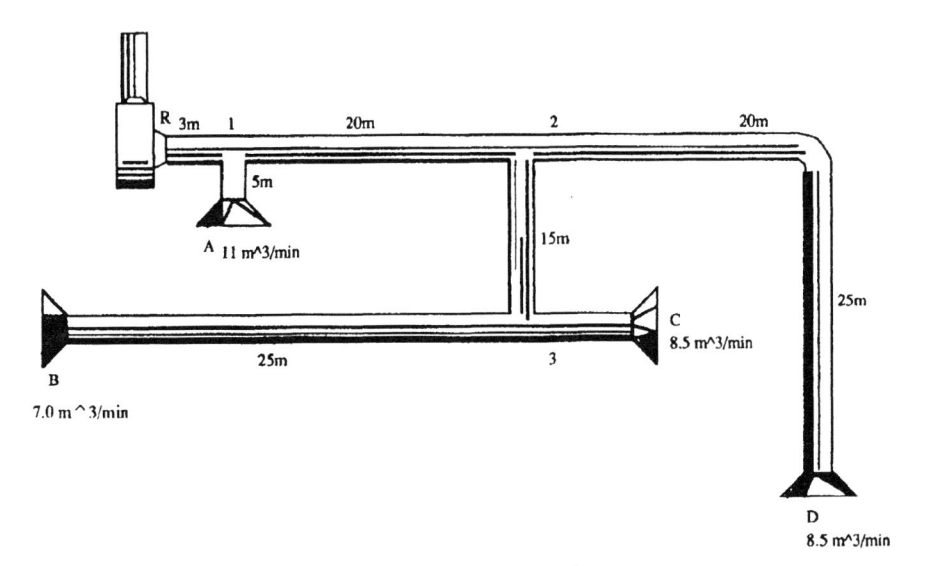

Figure 2.16. Schematic drawing of an exhaust system.

Principles of Process Engineering

```
============================= RULE SHEET ==============================
S Rule─────────────────────────────────────────────────────────────────
* fR1=fricfac(ReR1,rrR1)
* f12=fricfac(Re12,rr12)
* f2D=fricfac(Re2D,rr2D)
* f23=fricfac(Re23,rr23)
* f3B=fricfac(Re3B,rr3B)
* f3C=fricfac(Re3C,rr3C)
* f1A=fricfac(Re1A,rr1A)
* ReR1=Reynolds(VR1,DR1,rho,mu)
* Re12=Reynolds(V12,D12,rho,mu)
* Re2D=Reynolds(V2D,D2D,rho,mu)
* Re23=Reynolds(V23,D23,rho,mu)
* Re3B=Reynolds(V3B,D3B,rho,mu)
* Re3C=Reynolds(V3C,D3C,rho,mu)
* Re1A=Reynolds(V1A,D1A,rho,mu)
* rrR1=RelRough(eps,DR1)
* rr12=RelRough(eps,D12)
* rr2D=RelRough(eps,D2D)
* rr23=RelRough(eps,D23)
* rr3B=RelRough(eps,D3B)
* rr3C=RelRough(eps,D3C)
* rr1A=RelRough(eps,D1A)
* FR1=Darcy(fR1,LR1,DR1,VR1,g)
* F12=Darcy(f12,L12,D12,V12,g)
* F2D=Darcy(f2D,L2D,D2D,V2D,g)+Entrance(f2D,V2D,g)+Elbow(Ke,V2D,g)
* F23=Darcy(f23,L23,D23,V23,g)+Elbow(Ke,V23,g)
* F3B=Darcy(f3B,L3B,D3B,V3B,g)+Entrance(f3B,V3B,g)+Elbow(Ke,V3B,g)
* F3C=Darcy(f3C,L3C,D3C,V3C,g)+Entrance(f3C,V3C,g)+Elbow(Ke,V3C,g)
* F1A=Darcy(f1A,L1A,D1A,V1A,g)+Entrance(f1A,V1A,g)+Elbow(Ke,V1A,g)
* F=FR1+F12+F2D
* F=FR1+F12+F23+F3B
* F=FR1+F12+F23+F3C
* F=FR1+F1A
* D2D=sqrt(4*QD/(pi()*V2D))
* D3C=sqrt(4*QC/(pi()*V3C))
* D3B=sqrt(4*QB/(pi()*V3B))
* D1A=sqrt(4*QA/(pi()*V1A))
* D23=sqrt(4*(QB+QC)/(pi()*V23))
* D12=sqrt(4*(QB+QC+QD)/(pi()*V12))
* DR1=sqrt(4*(QA+QB+QC+QD)/(pi()*VR1))


========================== FUNCTION SHEET =============================
Name─────────────── Type ──── Arguments── Comment─────────────────────
fricfac             Rule       2;1
Reynolds            Rule       4;1
RelRough            Rule       2;1
Darcy               Rule       5;1
Entrance            Rule       3;1
Elbow               Rule       3;1
```

Figure 2.17. TK Solver plus rule and function sheets for solving example 2.7.

TK Solver variables sheet (fig. 2.18) gives the values for the 20 variables which were either specified by the problem or arbitrarily set. The resulting solution is unique for this set of inputs but would be different for a different set of input variables.

In order to iteratively solve the set of 39 equations, first guesses of 0.025 were used for the seven friction factors and first guesses of 5.0 m/s were used for the three unknown velocities. The output column of figure 2.18 gives the resulting solution for the 39 variables which were unknown. These 39 variables include the 7 diameter values which will result in equal friction losses through all paths of the system.

```
================ RULE FUNCTION: fricfac ================
S Rule
  if Re<2130 then f=64/Re else f=0.25/(log(rr/(3.7)+2.51/(Re*f^0.5)))^2

================ RULE FUNCTION: Reynolds ================
S Rule
  Re=v*d*rho/mu

================ RULE FUNCTION: RelRough ================
S Rule
  rr=eps/d

================ RULE FUNCTION: Darcy ================
S Rule
  F=f*(1/d)*v^2/(2*g)

================ RULE FUNCTION: Entrance ================
S Rule
  Fent=20*f*v^2/(2*g)

================ RULE FUNCTION: Elbow ================
S Rule
  Felb=K*v^2/(2*g)
```

Figure 2.17. Continued.

The student should recognize that though the computer software package, TK Solver, is successfully used here, any method of solution (including manual) which can simultaneously solve the set of equations can be used to obtain the desired values for the pipe diameters.

2.19 Compressibility Error

Air that is subjected to a pressure to force it through a series of pipes, a mass of grain, or a heat exchanger is compressed so that γ changes through the system. In most calculations where drying and ventilation problems are being considered, air is assumed to be incompressible to simplify calculations, specific weight at atmospheric pressure being used throughout. Pressures under these conditions seldom exceed 2.5 kPa and are usually in the order of 1 kPa or less. The error resulting by neglecting compression is dependent upon the absolute pressures. If atmospheric pressure is 101.35 kPa and the pressure varies within the system by 2.5 kPa, then the error from assuming incompressibility will be approximately 2.5%.

2.20 Optimum Rates of Flow

The question frequently arises as to whether one should have a small pipe or conduit with high velocity or a large pipe with low velocity. Although each installation should be analyzed carefully from the standpoint of initial cost, power requirement, noise level, and operating costs, the following suggestions may be used as a guide. Velocities of 1.2 to 1.8 m/s are usually best for water. Three meters per second may be used if the system resistance is low. Where noise is not a problem, air systems may be designed for velocities of 5.0 to 7.5 m/s. Velocities up to 10.0 m/s may be used in large pipes.

```
=================================== VARIABLE SHEET ===================================
St Input------ Name------ Output----Unit------ Comment-----------------------------
               fR1        .01922033            Friction factor between R and 1
               f12        .02006347            Friction factor between 1 and 2
               f2D        .02268734            Friction factor between 2 and D
               f23        .02114768            Friction factor between 2 and 3
               f3B        .02323422            Friction factor between 3 and B
               f3C        .02267               Friction factor between 3 and C
               f1A        .02205523            Friction factor between 1 and A
               FR1        .19063149    m       Friction loss between R and 1
               F12        1.6020536    m       Friction loss between 1 and 2
               F2D        9.0200262    m       Friction loss between 2 and D
               F23        2.5411223    m       Friction loss between 2 and 3
               F3B        6.478904     m       Friction loss between 3 and B
               F3C        6.478904     m       Friction loss between 3 and C
               F1A        10.62208     m       Friction loss between 1 and A
               ReR1       127271.61            Reynolds No. between R and 1
               Re12       105390.91            Reynolds No. between 1 and 2
               Re2D       62720.134            Reynolds No. between 2 and D
               Re23       80634.497            Reynolds No. between 2 and 3
               Re3B       56917.594            Reynolds No. between 3 and B
               Re3C       75498.581            Reynolds No. between 3 and C
               Re1A       97520.595            Reynolds No. between 1 and A
               rr12       .00046999            Relative roughness between 1 and 2
               rrR1       .00038919            Relative roughness between R and 1
               rr2D       .00078975            Relative roughness between 2 and D
               rr23       .00055679            Relative roughness between 2 and 3
               rr3B       .00087026            Relative roughness between 3 and B
               rr3C       .00095065            Relative roughness between 3 and C
               rr1A       .00094886            Relative roughness between 1 and A
    5          VR1                      m/s    Average velocity between R and 1
    5          V12                      m/s    Average velocity between 1 and 2
    5          V2D                      m/s    Average velocity between 2 and D
               V23        4.5319492    m/s     Average velocity between 2 and 3
    5          V3B                      m/s    Average velocity between 3 and B
               V3C        7.2449205    m/s     Average velocity between 3 and C
               V1A        9.3406042    m/s     Average velocity between 1 and A
               DR1        .38541486    m       Diameter between R and 1
               D12        .31915382    m       Diameter between 1 and 2
               D2D        .18993452    m       Diameter between 2 and D
               D23        .26940321    m       Diameter between 2 and 3
               D3B        .17236276    m       Diameter between 3 and B
               D3C        .15778731    m       Diameter between 3 and C
               D1A        .15808414    m       Diameter between 1 and A
    1.202      rho                      kg/m^3 Density of air
    .0000182   mu                       Pa*s   Viscosity of air
    .00015     eps                      m      Roughness factor for ducts
    3          LR1                      m      Distance between R and 1
    20         L12                      m      Distance between 1 and 2
    45         L2D                      m      Distance between 2 and D
    15         L23                      m      Distance between 2 and 3
    25         L3B                      m      Distance between 3 and B
    5          L3C                      m      Distance between 3 and C
    5          L1A                      m      Distance between 1 and A
    9.81       g                        m/s^2  Gravitational acceleration
    1.25       Ke                              K-value for elbow
               F          10.812711            Total friction loss
    .18333333  QA                       m^3/s  Volume flow rate at A
    .11666667  QB                       m^3/s  Volume flow rate at B
    .14166667  QC                       m^3/s  Volume flow rate at C
    .14166667  QD                       m^3/s  Volume flow rate at D
```

Figure 2.18. TK Solver variables sheet for example 2.7.

Fluid Mechanics

The reader should realize that these values are general and that frequently values above or below these should or could be used.

Flow of Granular Materials

Grain, ground feed, and other similar materials flow in an entirely different manner than liquids.

2.21 Rate of Flow

Leva (1959) reviewed a number of investigations of the flow rate from orifices. The depth of the solids and solids carrying duct diameter had very little effect on the flow rate. If the diameter of the granular particles is greater than 0.04 times the diameter of the orifice, a flow rate reduction may be experienced. The angle of approach of the floor to the orifice and the coefficient of friction of the granular material on the floor may, in combination, affect the flow rate. The flow rate varied nearly as the cube of the orifice diameter, the exponent range being from 2.50 to 2.96, with 2.75 seeming to be the average.

2.22 Angle of Repose

When a granular material is permitted to flow from a point into a pile as shown in figure 2.19, the shape of the pile is characteristic of the material. The angle ϕ which the side of the pile makes with a horizontal is called the angle of repose. For any material, it varies with the moisture content and amount of foreign material present, increasing with an increase in either. The tangent of the angle of repose is recognized as the coefficient of friction of the material on itself.

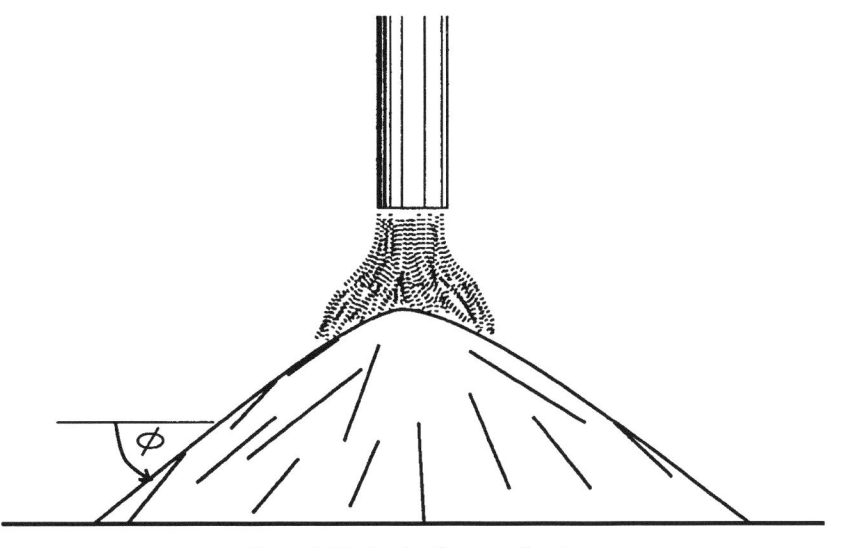

Figure 2.19. Angle of repose of grain.

Table 2-6. Coefficients of friction of various kinds of grains

	Bulk Density (kg/m³)	Grain on Grain	Grain on Rough Board	Grain on Smooth Board	Grain on Iron	Grain on Cement
Grain						
Wheat	785	0.466	0.412	0.361	0.414	0.444
Barley	625	0.507	0.424	0.325	0.376	0.452
Oats	450	0.532	0.450	0.369	0.412	0.466
Corn	705	0.521	0.344	0.308	0.374	0.423
Beans	735	0.616	0.435	0.322	0.366	0.442
Peas	800	0.472	0.287	0.268	0.263	0.296
Tares	785	0.554	0.424	0.359	0.364	0.394
Flax Seed	655	0.456	0.407	0.308	0.339	0.414
Peanuts (Unshelled)		0.649				

This property is important in material transfer since it affects the capacity of belt conveyors and other bulk transfer devices and partially determines the minimum slope of floors in self-emptying bins; coefficient of friction of grain on the bin material being another factor. Some materials, particularly those that have been produced by grinding, have such steep angles of repose that they are not completely self-flowing. Agitation is usually necessary to maintain flow.

2.23 Coefficient of Friction

Granular materials will not flow through pipes or chutes unless the pitch is sufficient to overcome the coefficient of friction of the material upon the conduit. This characteristic determines the minimum pitch of conduit suitable for moving materials by gravity. Grain or other granular materials will flow in a conduit at a flatter angle if it is moving when introduced into the conduit. If a system is designed on this basis, however, trouble may arise from accidental stoppage since starting flow may be difficult if a minimum pitch is used. Coefficients of friction for a few grains are listed in table 2-6.

Nomenclature

A	area (m²)
C_f	correction factor for pressure drop through beds (dimensionless)
D	diameter (m)
D_h	hydraulic diameter (m)
D_p	diameter of particles (m)
e	efficiency of pump, fan, etc. (dimensionless decimal)
F	energy loss to friction per unit weight of fluid (J/N)
f	friction factor (dimensionless)
g	acceleration due to gravity (9.81 m/s²)
h	elevation above a reference plane (m)
K	friction loss factor (dimensionless)

L	length (m)
$\dot{m}$	mass flow rate (kg/s)
O_f	percent opening in floor (decimal)
p	pressure (Pa)
p_{atm}	atmospheric pressure (Pa)
P	force (N)
P_r	power required by pump, fan, etc. (W)
P_s	power supplied to the fluid system (W)
Q	volumetric flow rate (m^3/s)
r	radius (m)
R_h	hydraulic radius (m)
Re	Reynolds number (dimensionless)
s_p	particle surface area (m^2)
t	time (s)
V	velocity (m/s)
v_p	particle volume (m^3)
W	work done on fluid per unit weight of fluid (J/N)
y	separation distance (m)
ε	roughness factor (m)
ε_p	voidage fraction (dimensionless)
γ	specific weight of fluid (N/m^3)
μ	dynamic viscosity of fluid (Pa s)
ν	kinematic viscosity (m^2/s)
ϕ	angle of repose (rad)
ρ	fluid density (kg/m^3)

References

Abrams, C. F. and J. D. Fish Jr. 1982. Air flow resistance characteristics of bulk piled sweet potatoes. *Transactions of ASAE* 25(4):1103–1106.

ASAE Standards, 32nd Ed. 1996. Resistance to Airflow of Grains, Seeds, Other Agricultural Products, and Perforated Metal Sheets. D272.2. St. Joseph, Mich.: ASAE.

Bakker-Arkema, F. W., R. J. Patterson and W. G. Bickert. 1969. Static pressure-airflow relationships in packed beds of granular biological materials such as cherry pits. *Transactions of ASAE* 12(1):134–140.

Baumeister, T., E. A. Avallone and T. Baumeister III. 1978. *Marks' Standard Handbook for Mechanical Engineers*. New York: McGraw-Hill Book Co.

Bern, C. J. and L. F. Charity. 1975. Airflow resistance characteristics of corn as influenced by bulk density. ASAE Paper No. 75-3510. St. Joseph, Mich.: ASAE.

Bird, R. B., W. E. Stewart and E. N. Lightfoot. 1960. *Transport Phenomena*. New York: John Wiley & Sons, Inc.

Calderwood, D. L. 1973. Resistance to airflow of rough, brown and milled rice. *Transactions of ASAE* 16(3):525–527, 532.

Crane Co. 1935. Engineering data on flow of fluids in pipes and heat transmission. Chicago, Ill.

Dahlberg, A. C. and J. C. Hening. 1925. Viscosity, surface tension, and whipping properties of milk and cream. New York Expt. Sta. Tech. Bull. 113.
Ergun, S. 1952. Fluid flow through packed columns. *Chemical Engineering Progress.* 48(2):89–94.
Gaffney, J. J. and C. D. Baird. 1975. Forced air cooling of bell peppers in bulk. ASAE Paper No. 75–6525. St. Joseph, Mich.: ASAE.
Haque, E., G. H. Foster, D. S. Chung, and F. S. Lai. 1978. Static pressure across a corn bed mixed with fines. *Transactions of ASAE* 21(5):997–1000.
Harper, J. C. 1960. Viscometric behaviour in relation to evaporation of fruit purees. *Food Technology* 14:557–561.
Hart, J. A., K. R. Lawther, and E. Szomanski. 1965. Fluid flow in packed beds. *Inst. of Engineers, Australia. 2nd Australian Conference on Hydraulics and Fluid Mechanics Proceedings* B263–B286.
Henderson, S. M. 1943. Resistance of shelled corn and bin walls to airflow. *Agricultural Engineering* 24(11):367–369.
Hicks, R. E. 1970. Pressure drop in packed beds of spheres. *Ind. Eng. Chem. Fundam.* 9(3):500–502.
Hukil, W. V. and N. C. Ives. 1955. Radial air flow resistance of grain. *Agricultural Engineering* 36(5):332–335.
Kratz, A. P. and J. R. Fellows. 1938. Pressure losses resulting from changes in cross-sectional area in air ducts. Illinois Eng. Expt. Sta. Bull. 300.
Leva, M. 1959. *Fluidization.* New York: McGraw-Hill Book Co.
Mohsenin, N. N. 1963. A testing machine for determination of mechanical and rheological properties of agricultural products. Pennsylvania Agr. Expt. Sta. Bull. 701.
Moody, L. F. 1944. Friction factors for pipe flow. *Transactions of ASME.* 66:671–684.
Nguyen, V. T. 1981. Airflow resistance of sunflower seed. Unpublished term project for AE568 at Iowa State University, Ames, IA. Under the direction of Carl J. Bern.
Schuler, R. T. 1974. Drying-related properties of sunflower seeds. ASAE Paper No. 74–3534, ASAE, St. Joseph, MI 49085.
Shedd, C. K. 1953. Resistance of grains and seeds to air flow. *Agricultural Engineering* 34(9):616–619.
Sheldon, W. H., C. W. Hall, and J. K. Wang. 1960. Resistance of shelled corn and wheat to low airflows. *Transactions of ASAE* 3(2):92–94.
Staley, L. M. and E. L. Watson. 1961. Some design aspects of refrigerated potato storages. *Canadian Agricultural Engineering* 3(1):20–22.
Steele, J. L. 1974. Resistance of peanuts to airflow. *Transactions of ASAE* 17(3):573–577.
Szomanski, E. 1968. Packed beds—Their dynamics, structure, fluid flow, heat transfer and other characteristics. *Inst. of Engineers, Australia.* In *Conference on Hydraulics and Fluid Mechanics* 1968, 117–123.
Vennard, J. K. 1961. *Elementary Fluid Mechanics.* New York: John Wiley & Sons, Inc.
Wilkinson, W. L. 1960. *Non-Newtonian Fluids.* London: Pergamon Press.
Zenz, F. A. and D. F. Othmer. 1960. *Fluidization and Fluid-particle Systems.* New York: Van Nostrand Reinhold Co.

Fluid Mechanics

Problems

1. Find Reynolds number for milk at 20.2 C flowing at 0.075 m³/min in sanitary (smooth) tubing with a 4.0 cm inside diameter. Milk has a density of 1 030 kg/m³. What would be the diameter of a tube in which laminar flow could be expected for this volumetric flow rate?

2. Milk at 20.2 C is to be lifted 3.6 m through 10 m of sanitary pipe that contains two Type A elbows. Milk in the lower reservoir enters the pipe through a type A entrance. Assuming a pump efficiency of 80%, plot the power required by the pump versus volumetric flow rate if 2-cm-inside-diameter pipe is used. Consider volumetric flow rates between 0 and 0.3 m³/min (Hint: Power requirement at 0.225 m³/min is approximately 4 033.6 W). Repeat the problem for a 4-cm-inside diameter pipe.

3. Plot the power required to pump molasses at 21 C (specific gravity = 1.43) through the system of problem 2. Consider flow rates from 0 to 0.06 m³/min and assume a pump efficiency of 80%.

4. Air (21 C) at the rate of 0.1 m³/m²s is to be moved vertically through a crib of shelled corn 1.6 m deep. The area of the floor is 12 m² with an opening percentage of 10% and the connecting galvanized iron pipe is 0.3 m in diameter and 12 m long. What is the power requirement, assuming fan efficiency to be 70%? If the diameter of the connecting pipe is increased to 0.45 m, how much power will be required?

5. Air is flowing through a conduit system at 1 m³/s. A 20-cm galvanized iron pipe enlarges abruptly to 40 cm. The 40 cm section is 7 m long. It decreases abruptly at the end of the section to 20 cm in diameter. Would the power requirement increase or decrease, and by how much, if the central section were reduced in diameter to 20 cm?

6. Determine the size of the pipes required for the installation in the accompanying figure. Assume that the hoods have a friction loss equivalent to a length of pipe 20 times the diameter of the connecting pipe and that the optimum velocity in the pipe is 5 m/s.

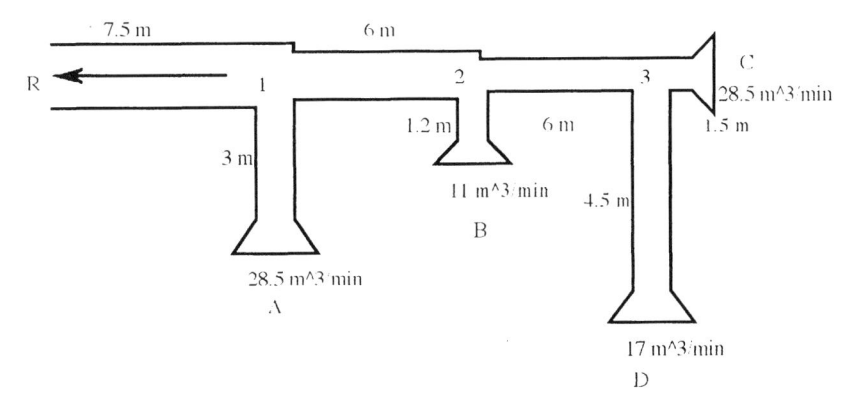

7. Develop a system characteristic curve (plot both total pressure and static pressure curves) for a system similar to that of example 2.6, but with an entrance pipe diameter of 0.4 m, pipe length of 2 m, and a 2 m depth of oats. Use other characteristics as given in the example problem. Plot the system characteristic curve for flow rates from 0 to 4.0 m^3/s.

8. Use Darcy's law and the friction factor equation for laminar flow to show that in laminar flow the pressure drop due to friction is linearly proportional to the fluid flow rate through the system.

9. At very high flow rates the friction factor approaches a constant value. Use Darcy's law to show that under these conditions the pressure drop is proportional to the square of fluid flow rate.

3 Fluid-Flow Measurements

Many devices for measurement of fluid flow rates depend upon the accurate measurement of pressure. These pressure measurements are then used to calculate fluid velocities and/or volumetric flow rates. Thus, we will begin this chapter with a review of methods for pressure measurement. We will then introduce devices and procedures for fluid flow measurement which depend on differential pressure measurement and will then discuss other flow meters which do not depend on pressure measurement. Due to space limitations and to the type of fluid systems generally encountered by the processing engineer, we will not discuss those flow measuring devices such as weirs and flumes which are widely used in open channel flow measurement. For additional information on the devices introduced here as well as others not discussed here, the reader is referred to reviews by Replogle and Birth (1983) and by Cheremisinoff (1979).

Pressure Measurement

3.1 Units of Pressure

Pressure has been previously defined in Chapter 1 as a force per unit area. Within the SI system of measurement it is given units of pascals (Pa) or newtons per square meter (N/m^2). Pressure is also often expressed in terms of the height of a column of fluid which would produce the specified force per unit area (i.e., centimeters of water). This alternate designation of pressure is really a measure of pressure head in the Bernoulli equation (eq. 2.13). It is mandatory that the fluid be specified if the pressure head designation is used. The true pressure term is obtained from the pressure head by multiplying by the specific weight of the fluid as follows:

$$p = H\gamma \tag{3.1}$$

where

p = pressure (Pa)
H = pressure head or column height of fluid (m)
γ = specific weight of fluid (N/m^3)

Equation 3.1 can be written as:

$$p = H\rho g \tag{3.2}$$

where

ρ = density of fluid (kg/m^3)
g = acceleration due to gravity (m/s^2)

Example 3.1. The pressure head in a fluid system is expressed as 2 cm of water at 20 C (2.0 cm H$_2$O). Determine the true pressure in pascals.

Solution
The density of water at 20 C is 998.207 kg/m^3.
Then:

$$p = H\rho g$$
$$p = (2.0 \text{ cm H}_2\text{O}) (1 \text{ m}/100 \text{ cm}) (998.207 \text{ kg/m}^3) (9.81 \text{ m/s}^2)$$
$$p = 195.8 \text{ Pa}$$

Example 3.2. The pressure head in a fluid system is expressed as two millimeters of mercury at 0 C (2.0 mm Hg). Determine the true pressure in pascals.

Solution
The density of mercury at 0 C is 13 595.1 kg/m^3.
Then:

$$p = H\rho g$$
$$p = (2.0 \text{ mm Hg}) (1 \text{ m}/1\,000 \text{ mm}) (13\,595.1 \text{ kg/m}^3) (9.81 \text{ m/s}^2)$$
$$p = 266.7 \text{ Pa}$$

Table 3-1 gives some conversion factors between various pressure and pressure head units. Some units in the English system are included in table 3-1 due to their continued usage.

3.2 Types of Pressure Measurements

Pressure may be expressed as *absolute, differential*, or *gage*. *Absolute* pressure is the total pressure exerted by a fluid. Its value will always be greater than zero (zero for a

Table 3-1. Pressure conversion factors

1 atmosphere (atm)	= 101325.05 Pa	= 101.32505 kPa
1 millibar (mb)	= 100.00 Pa	= 0.10000 kPa
1 mm mercury (mm Hg)	= 133.32 Pa	= 0.13332 kPa
1 in. mercury (in. Hg)	= 3386.36 Pa	= 3.38636 kPa
1 lb$_f$ per sq in. (psi)	= 6894.76 Pa	= 6.89476 kPa
1 lb$_f$ per sq ft	= 47.88 Pa	= 0.04788 kPa
1 cm water 20 C	= 97.89 Pa	= 0.09789 kPa
1 in. water 20 C	= 248.65 Pa	= 0.24865 kPa
1 dyne/cm^2	= 0.10000 Pa	= 0.00010 kPa
1 torr (tor)	= 133.32 Pa	= 0.13332 kPa
1 N/m^2	= 1.0000 Pa	= 0.00100 kPa

Fluid-Flow Measurements

perfect vacuum). *Differential* pressure represents the algebraic difference between two pressures. *Gage* pressure is a special case of *differential* pressure since it is the difference between *absolute* pressure and *atmospheric* pressure. Thus,

$$p_g = p_a - p_{atm} \tag{3.3}$$

where

p_a = absolute pressure (Pa)
p_{atm} = atmospheric pressure (Pa)
p_g = gage pressure (Pa)

At standard atmospheric pressure, gage pressures will range upward from -101.325 kPa. Values lower than -101.325 kPa are not possible since they would indicate a negative absolute pressure. Fluid system calculations which result in lower values indicate that the physical system is not possible.

Vacuum is also a differential pressure obtained by subtracting the absolute pressure from atmospheric pressure. Thus,

$$p_{vac} = p_{atm} - p_a \tag{3.4}$$

where p_{vac} = vacuum (Pa)

Comparison of equations 3.3 and 3.4 indicates that a negative gage pressure and a positive vacuum have the same meaning. Thus, vacuums above 101.325 kPa are impossible under standard atmospheric conditions.

Pressures are also referred to as either *static* or *dynamic*. *Static* pressures are those resulting from pressure and elevation and indicate forces perpendicular to the walls of the container. *Dynamic* pressures (also called velocity pressures) which result from the force due to a change in velocity can be used to measure the velocity head in the Bernoulli equation. Care must be exercised to differentiate between these in setting up equipment, making readings, and analyzing data.

In general, pressures taken normal to the direction of fluid motion are static pressures. Static pressures can be observed in one of two ways: (1) by making the observation through a small hole in the container wall or (2) by using a disc or tube-type static head device which is inserted into the stream (fig. 3.1). For observations requiring a high degree of accuracy, particularly if there is a question as to the characteristics of flow, a number of small holes can be spaced evenly around the conduit in question and connected together by a manifold. This arrangement is called a piezometer ring. The disc and tube shown in figure 3.1 can be used if installation of the other types are inadvisable or impractical.

The hole in the conduit must be perpendicular to the conduit wall and free of burrs. The conduit wall must be uniform in contour and smooth in the region of the hole. A hole 0.3 cm in diameter may be used for pipes of 6 cm diameter or under. For pipes up to 40 cm in diameter, a 0.6- to 1.3-cm-diameter hole may be used. In any case, the smallest practical hole should be used; the smaller the hole, the greater the accuracy. Although single holes give reliable results if the velocity pattern in the conduit is symmetrical

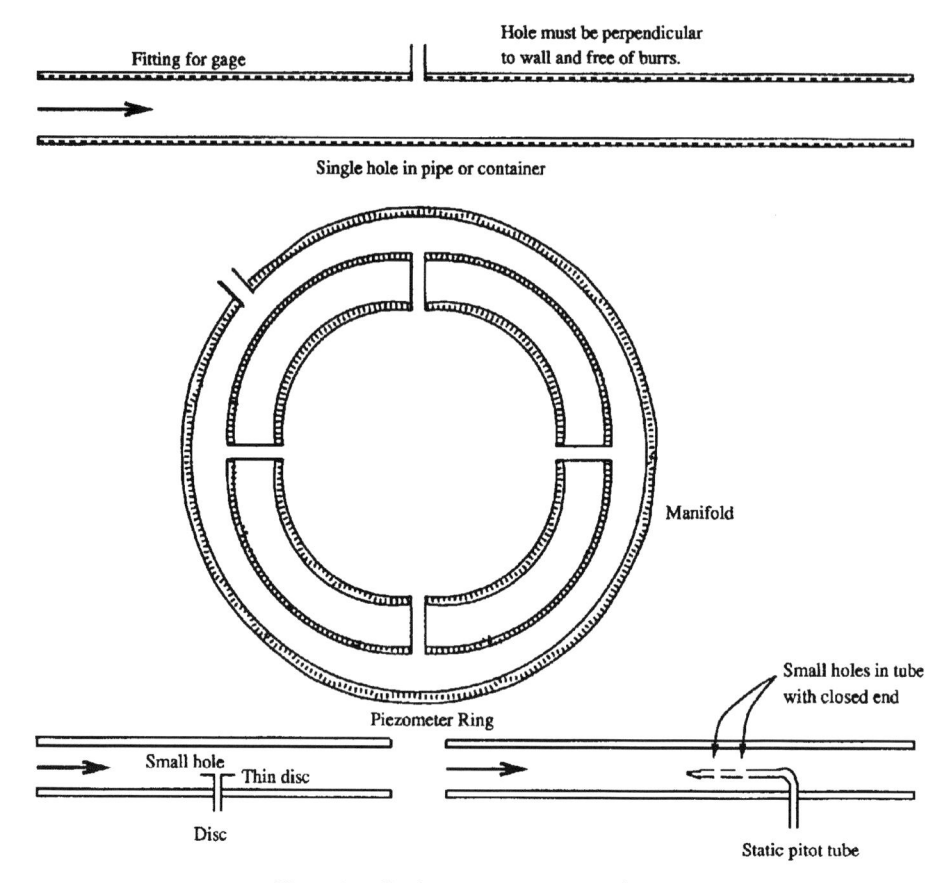

Figure 3.1. Static pressure measuring devices.

and the inside surface is uniform and smooth, the piezometer-ring type of connection assures more reliable observations if flow properties or conduit characteristics are not ideal.

A simple but reliable piezometer for low-pressure measurement can be constructed as shown in figure 3.2. The holes should be spaced evenly around the tube. Four holes are recommended as minimum; six or eight would be more reliable. A large number of small holes are better than a small number of large holes.

3.3 Manometers

The simplest type of pressure gage is the manometer, which may take many forms in practice. Figure 3.3 shows three types which may be useful for the processing engineer. The U-tube is the simplest. The pressure head is indicated by the difference in height of the liquid columns, h, and is expressed in meters or centimeters of the fluid contained in the manometer.

Fluid-Flow Measurements

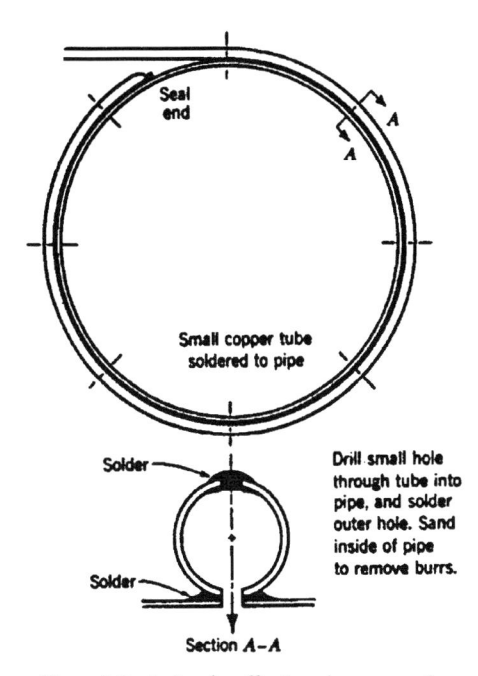

Figure 3.2. A simple, effective piezometer ring.

In general, any pressure gage should be located level with the desired point of pressure observation. If this is not done a significant error may result from the column of fluid in the connecting pipe.

The differential ΔC of figure 3.4A is a true measure of pressure head since the pressures at points C_1, C_2, and C_3 are equal. The pressure in system A is then,

$$p_A = \Delta C \rho_m g \tag{3.5}$$

where $\rho_m =$ density of manometer fluid (kg/m^3).

However, as shown in system B, the lower meniscus of the manometer fluid is seldom level with the pipe axis. The pressure is equal at points d_2, d_3, and d_4 and is given by:

$$p_4 = \Delta d \rho_m g \tag{3.6}$$

where $p_4 =$ pressure at elevation of lower manometer meniscus (Pa).

The pressure at the axis of the pipe of system B is greater than at d_4 by the height of the fluid column a. It is then given by:

$$p_B = \Delta d \rho_m g + a \rho_f g \tag{3.7}$$

where

$p_B =$ pressure at pipe axis of system B (Pa)
$\rho_m =$ density of manometer fluid (kg/m^3)
$\rho_f =$ density of fluid in system (kg/m^3)

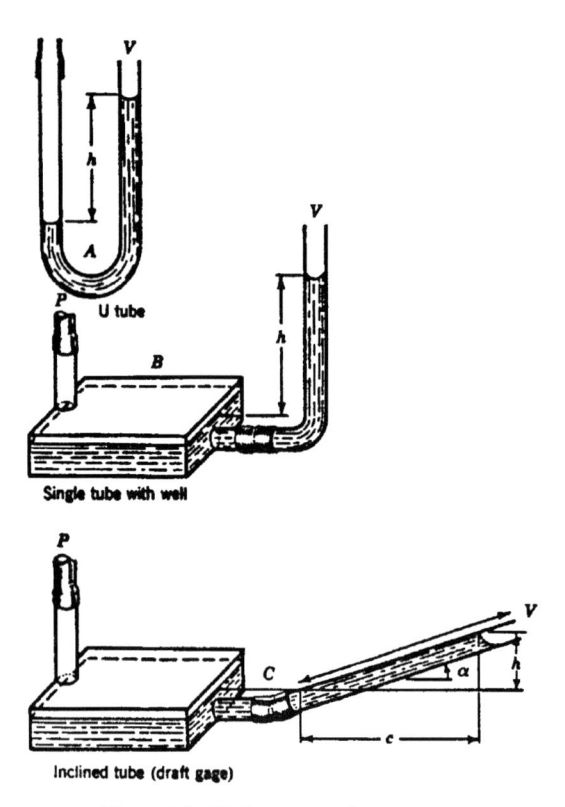

Figure 3.3. Various types of manometers.

Since a changes with each change in pressure, it is often convenient to calculate the pressure at the average level of the manometer fluid, d_5, which is less than d_3 by the height of the column $0.5\ \Delta d$. Then the pressure at d_5 in the fluid of the system (*Note: Not in the manometer fluid.*) is:

$$p_5 = \Delta d \rho_m g - 0.5 \Delta d \rho_f g \tag{3.8}$$

and the pressure at the axis of the pipe of system B becomes:

$$p_B = \Delta d \rho_m g - 0.5 \Delta d \rho_f g + b \rho_f g$$

or

$$p_B = \Delta d(\rho_m - 0.5\rho_f)g + b\rho_f g \tag{3.9}$$

Equation 3.9 is valid only if the connecting tube is completely filled with the fluid of the system. If another fluid such as air is trapped in the tube, the pressure calculations must be corrected for the presence of the additional material.

When the fluid flowing within the system such as B is a gas, the density of the gas becomes insignificant compared with the manometer fluid and equation 3.9 becomes

Fluid-Flow Measurements

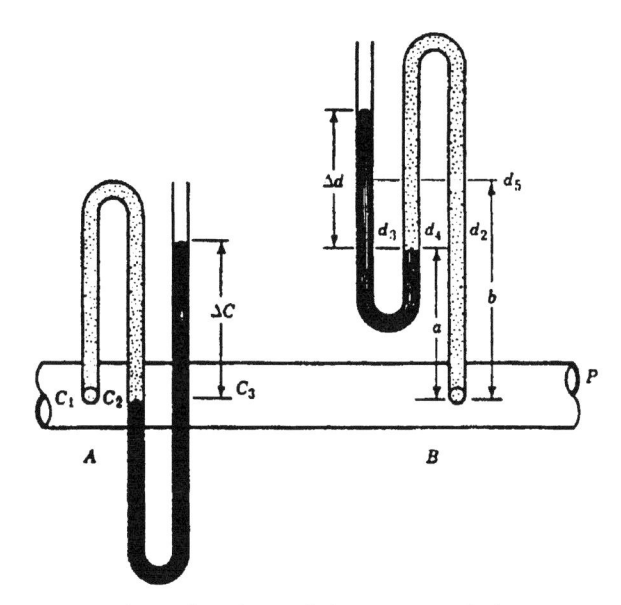

Figure 3.4. U tube in line and above point at which pressure is to be measured.

approximately:

$$p_B = p_4 = \Delta d \rho_m g \tag{3.10}$$

Differential pressure between two points of the same elevation in a fluid system may be measured by connecting the two sides of a U-tube manometer to the two points. The differential pressure is then given by:

$$\Delta p = \Delta d (\rho_m - \rho_f) g \tag{3.11}$$

where

Δp = differential pressure (Pa)
Δd = difference in elevation of manometer fluid in two legs of U-tube (m)
ρ_m = density of manometer fluid (kg/m^3)
ρ_f = density of fluid in system (kg/m^3)
g = acceleration due to gravity (m/s^2)

The most common liquid used in manometers is mercury. With a density of approximately 13.6 times that of water, it provides a convenient scale length for barometers (absolute pressure gages) and other rather large differential pressure gages. In addition, mercury has a low freezing point of approximately −39 C. Other liquids used in manometry include water, organic liquids with densities less than water, and bromide compounds with densities of about three times that of water.

The single tube B (fig. 3.3) has most of the advantages of the U tube without its disadvantages. The well is large as compared to the tube so that the change in fluid level in the well is not significant for a fluid elevation in the column. Consequently, the manometer difference h can be read directly from a scale, no significant corrections being necessary. If desirable, the scale can be adjusted to compensate for variations in the level of the well fluid, thus giving a true reading.

The inclined tube C, also called a draft gage because of its use for observing chimney draft on furnaces, is a convenient means of increasing sensitivity. The scale multiplication or increase in sensitivity varies according to the cosecant of the angle α. The inclination of the tube is limited by the surface tension characteristics of the fluid meniscus. When the inclination is too great, the meniscus has a tendency to "stick" and accurate readings are difficult. It is usually inadvisable to attempt multiplications of more than 20 with this type of gage. Because of irregularities in bore and straightness, these gages must be calibrated individually if a high degree of accuracy is desired.

A micromanometer such as that of figure 3.5 can be used to read to 0.0025 cm (0.001 in.) of the manometer fluid. The micromanometer consists of a well-type manometer with the well movable in a vertical plane by means of a precision ground lead screw. A uniformly graduated scale and micrometer graduated wheel are attached directly to the lead screw, providing direct reading indication of the well travel. The glass indicating tube consists of two vertical legs with an inclined portion in the center. The angle of this incline tube is such to provide sensitivity for accurate indication. On the inclined portion a reference hairline has been etched into the glass and serves as the zero, or fixed reference, for the instrument.

3.4 Bourdon Tube

The Bourdon tube is one of the most common pressure measuring elements. The tube is formed in an oval or flattened shape which tends to become circular when internal pressure is applied. The Bourdon tube may be bent to various shapes. The most prevalent is the C type, with the helical, the spiral, and the twisted tube sections also used. Figure 3.6 shows a schematic of a C-type Bourdon tube. The tube is mechanically linked through gears to a pointer such that as the tube unwinds the pointer rotates giving an indication of the internal pressure.

The Bourdon tube is a secondary instrument since it must be calibrated against a known primary standard. Many precision gages have accuracies of 0.1 to 0.5 % of full-scale reading. Some precision gages incorporate automatic temperature correction by means of a bimetallic bar forming part of the linkage in the pointer mechanism.

When hot vapors or steam pressure is being observed with a Bourdon-type gage, a loop or "pigtail" filled with liquid is used in the connecting pipe to protect the gage from excessive temperatures or corrosive vapors. To prevent solids or corrosive fluids from entering the Bourdon tube, seals are often used.

The most common cause of failure in Bourdon-tube gages is rapid undamped pressure fluctuation which wears the linkage parts. A shut-off valve or pulsation dampers may be used to reduce the effects of pressure pulsations.

Although Bourdon-type gages normally measure gage pressure, they can be constructed for absolute pressure or for differential pressure.

Fluid-Flow Measurements

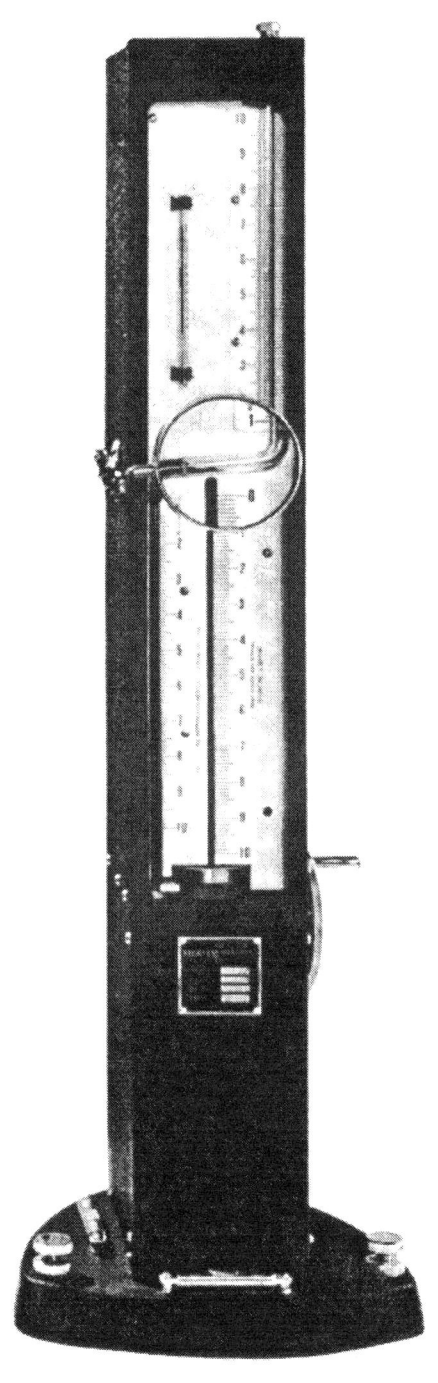

Figure 3.5. A precision micromanometer.

56   Principles of Process Engineering

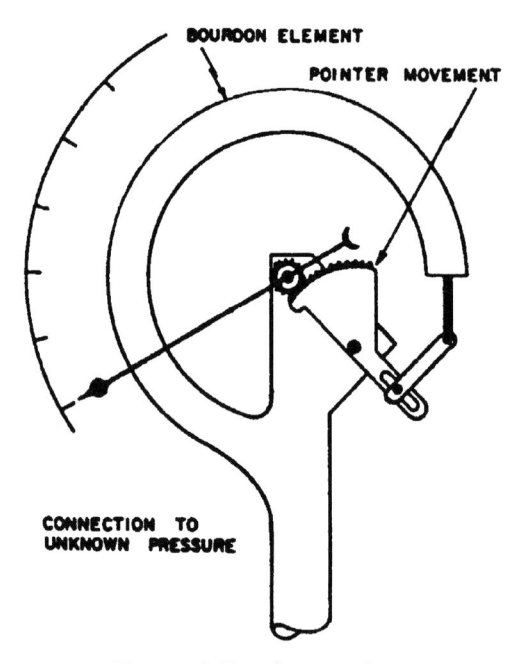

Figure 3.6. Bourdon gage, C type.

3.5 Diaphragm Gages

The diaphragm type of gage (fig. 3.7) consists of a spring-loaded diaphragm which actuates a series of levers attached to the indicating hand. Gages of this type are designed for relatively low-pressure operation (up to 35 kPa). The diaphragm may be metallic or nonmetallic. Nonmetallic diaphragms may be constructed of flexible materials such as leather, neoprene, polyethylene, teflon, or silk.

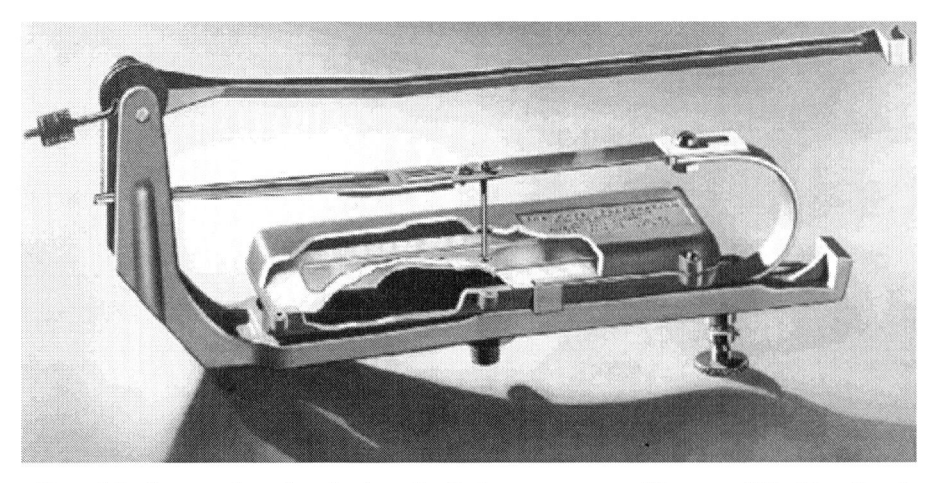

Figure 3.7. Cutaway view of mechanism of a diaphragm-type gage. (Courtesy of The Hays Corp.)

3.6 Bellows Gages

A metallic bellows is constructed from a thin seamless tube into an expansible or collapsible member. Movement of the bellows is in an axial direction. The thermostat valve of an automotive cooling system is actuated by a bellows. The advantage of the bellows unit over a diaphragm or Bourdon tube is the larger force which may be developed to operate mechanical linkages. This force may be increased for a given pressure by increasing the bellows diameter. The axial movement of the bellows is approximately 5 to 10 % of its length. Hysteresis and zero shift problems are more common with the bellows unit than with Bourdon tubes or diaphragms.

The bellows gage may be used for measuring either gage, absolute, or differential pressure. For absolute pressure measurement the bellows is enclosed within an evacuated housing.

Bellows may be constructed of brass, phosphor bronze, stainless steel, Monel, or beryllium copper.

3.7 Electrical Pressure Transducers

Electrical pressure transducers combine the pressure sensing devices of the previous sections with a means for transforming the movement of these elements to an electrical output signal. Zoerb (1983) gives a thorough review of different techniques used for generating the electrical output. They include: (1) strain gages mounted on diaphragms, Bourdon tubes, or bellows; (2) capacitance units in which a diaphragm moves one of the plates; (3) piezoelectric units consisting of crystal or ceramic elements used with a flat diaphragm; (4) linear variable differential transformer units which measure the displacement of a diaphragm, Bourdon tube, or bellows; (5) magnetic reluctance units used with a diaphragm or Bourdon tube; and (6) moving-contact potentiometer units used with a bellows or Bourdon tube.

Electrical pressure transducers often incorporate methods for temperature compensation when the gages are used at temperatures other than that for which the primary sensor is calibrated. Thus, while electrical pressure measuring devices depend upon previously discussed primary sensors, they convert signals to a more convenient form for observation and recording.

Differential Pressure Flow Meters

Differential pressure flow measuring devices are generally termed head or rate meters. The primary device creates a differential head which is a function of the fluid velocity and density. This pressure differential may be due to a conversion of energy from kinetic to potential as is the case in a pitot tube, to a conversion of energy from potential to kinetic as is the case in the venturi meter, or to a decrease in energy due to friction in the device as is the case in the laminar flow meter.

3.8 Pitot Tube

The pitot tube is essentially an open tube pointing into the stream of fluid flow as shown in figure 3.8. It finds important applications in research laboratories although its

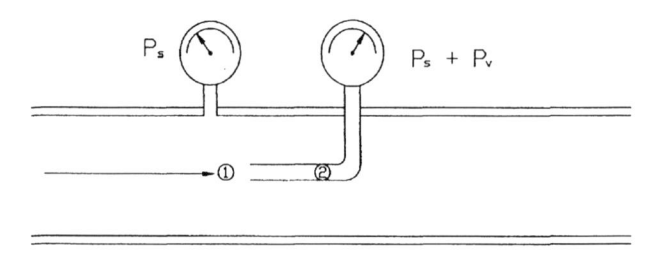

Figure 3.8. Elementary pitot tube.

commercial application is limited because the range of velocities that can be measured, using standard differential measuring units, is quite narrow and the variations in velocity profiles in pipes are quite wide, requiring a complete velocity transverse to obtain an accurate estimate of average velocity for volumetric flow rate determination.

If the Bernoulli equation 2.13 is written for points 1 and 2 in the pitot tube setup of figure 3.8, the following results:

$$h_1 + p_1/\gamma + V_1^2/(2g) + W - F = h_2 + p_2/\gamma + V_2^2/(2g)$$

Since $h_1 = h_2$ and $V_2 = W = 0$, the Bernoulli equation reduces to:

$$p_1/\gamma + V_1^2/(2g) - F = p_2/\gamma \qquad (3.12)$$

where

p_1 = fluid pressure in stream prior to pitot tube (Pa)
p_2 = fluid stagnation pressure in pitot tube (Pa)
V_1 = velocity of fluid prior to pitot tube (m/s)
F = energy loss due to friction (J/N or m)

If we assume that the friction loss is negligible, then equation 3.12 becomes:

$$p_1/\gamma + V_1^2/(2g) = p_2/\gamma \qquad (3.13)$$

or

$$V_1 = \sqrt{2g(p_2 - p_1)/\gamma} \qquad (3.14)$$

The pressure at point 1 is the static pressure within the fluid stream while the pressure at point 2 is the sum of the static pressure and the pressure resulting from the change in velocity and is referred to as the total pressure. The difference between total and static pressure is known as the velocity pressure. Then,

$$p_v = p_t - p_s \qquad (3.15)$$

where

p_v = velocity pressure (Pa)
p_t = total pressure (Pa)
p_s = static pressure (Pa)

Fluid-Flow Measurements

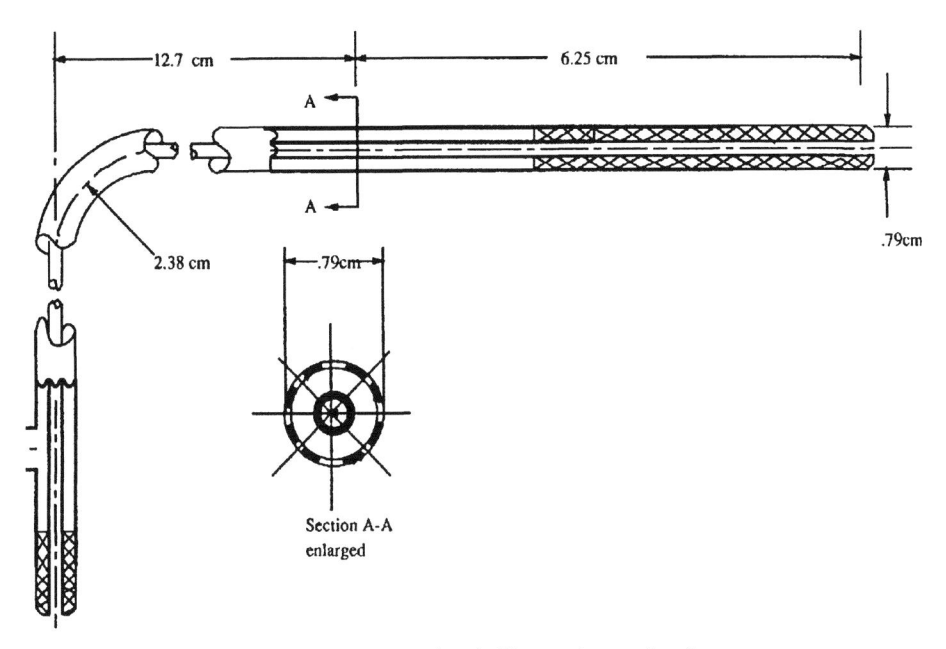

Figure 3.9. A practical and efficient pitot-static tube.

Equation 3.14 can then be written in terms of the velocity pressure as:

$$V_1 = \sqrt{2gp_v/\gamma} \qquad (3.16)$$

In order to correct for possible errors due to the assumption of zero friction loss, a correction factor K may be added to equation 3.16 as follows:

$$V_1 = K\sqrt{2gp_v/\gamma} \qquad (3.17)$$

The value of K may vary from 1.00 to about 0.965 depending on the gradient of the velocity profile into which the pitot tube is placed (ASME, 1971). It is common practice to assume a value of 1.00 for K.

Usually the static and total pressure elements are unified into a combined pitot tube as shown in figure 3.9. In general, the elementary type yields more reliable static pressure readings than the combined type because eddy currents may exist in the region of the static holes in the combined type. However, combined tubes designed and constructed on the basis of exhaustive tests will give results well within accepted engineering tolerances.

The static and total pressure ports of the pitot tube are usually connected to a differential pressure gage so that the velocity pressure is measured directly. This single reading then is used to calculate the velocity at the tip of the gage.

The velocity of the fluid will usually be a maximum at the center of a conduit and decrease toward the walls. Average velocities can be determined by dividing the conduit into a number of small equal concentric areas, observing the velocity at the center of each area, and finding the average of these. A system for doing this is shown in figure 3.10. If

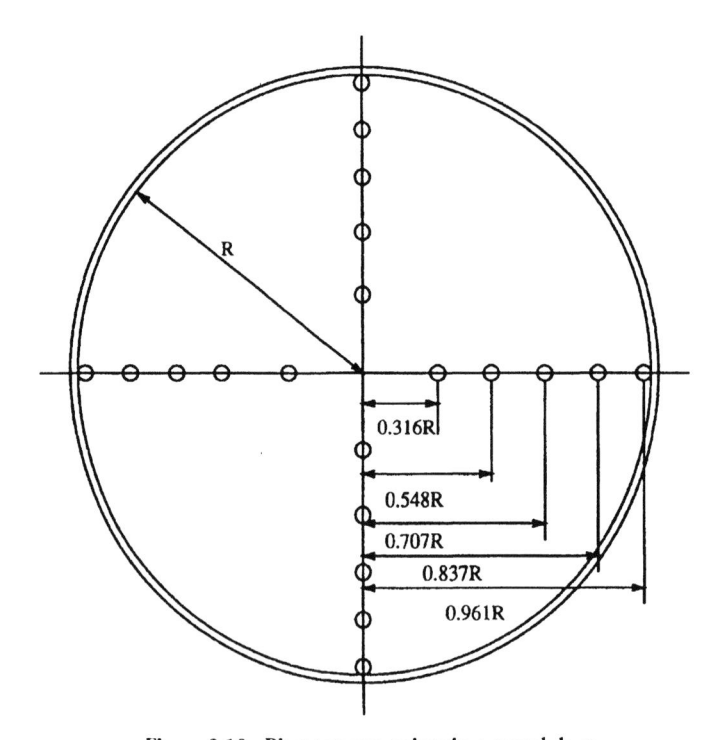

Figure 3.10. Pitot traverse points in a round duct.

the installation is a permanent one, the average velocity can be determined for a number of volumetric flow rates through the normal operating range and these velocities can be referred to the maximum velocity at the center by a factor so that the single center reading will indicate the true average. For laminar flow, the average velocity will be 0.5 times the center velocity while the factor will usually vary from 0.85 to 0.97 for turbulent flow.

Observations should be made in the middle of a long straight uniform section of pipe. For best results, the length should be at least 20 times the diameter. Observations close to bends, outlets, blowers, grills, and other restrictions will probably give erroneous results.

Although the pitot tube performs properly at very low velocities, the sensitivity of the manometer gage with which it is usually used is insufficient to measure air velocities of under 2.0 m/s with the requisite accuracy. In order to increase the differential pressure, the pitot-venturi design, shown in figure 3.11, has evolved. Single- or double-venturi sections may be added to a pitot tube to increase the pressure differential. The venturi sections essentially decrease the static pressure, thus creating a magnified differential. Experimental calibrations must be provided for the pitot-venturi meter.

Example 3.3. A combination pitot tube is used to measure the velocity of water at the center of a pipe. The static and total pressure ports of the tube are connected to a mercury-filled U-tube manometer and a difference of 4 cm is observed in the elevation

Fluid-Flow Measurements

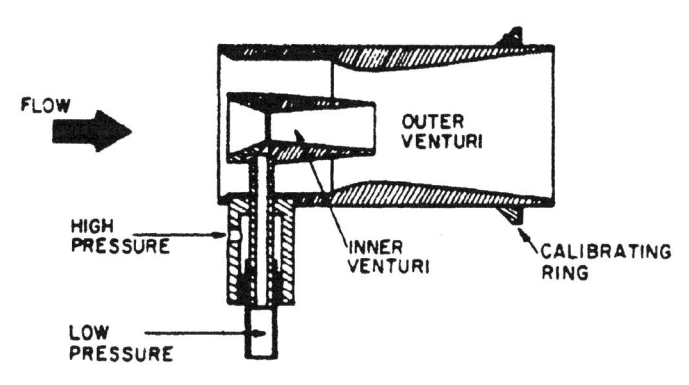

Figure 3.11. Pitot-venturi design.

of mercury in the two sides of the manometer. The point midway between the upper and lower meniscus of the U-tube manometer is 1 m above the axis of the pipe. Determine the velocity of water at the center of the pipe.

Solution
The velocity pressure from the U-tube reading is:

$$p_v = \Delta d(\rho_{Hg} - \rho_{H_2O})g$$
$$p_v = (.04 \text{ m})(13\,595.1 \text{ kg/m}^3 - 1\,000.0 \text{ kg/m}^3)(9.81 \text{ m/s}^2)$$
$$p_v = 4\,942.3 \text{ Pa}$$

(Note that the 1 m difference in elevation of the manometer and the pipe is not important in the differential pressure measurement since it has equal effects on the static and total pressures.)

The water velocity is then:

$$V = \sqrt{2gp_v/\gamma_{H_2O}} = \sqrt{2p_v/\rho_{H_2O}}$$
$$V = \sqrt{(2)(4\,942.3 \text{ Pa})/(1\,000.0 \text{ kg/m}^3)} = 3.14 \text{ m/s}$$

3.9 Venturi Meter

The venturi meter shown in figure 3.12 is preferable to the pitot tube when average cross-sectional velocities are desired. The velocity indicated is a true average, and the pressure difference can be magnified by increasing the diameter ratios so that more accurate readings can be obtained. It is an excellent measuring device for permanent installations, but, because of its bulk and the fact that it is an integral section of the conduit system, it is not readily mobile.

A typical venturi meter consists of a cylindrical inlet, convergent entrance, throat, and divergent outlet. The convergent entrance has an included angle of about 21° and the divergent cone 7 to 8°. The effect of the divergent cone is to reduce the overall pressure loss of the meter and its removal will have no effect on the discharge coefficient.

![Figure 3.12. Venturi tube.]

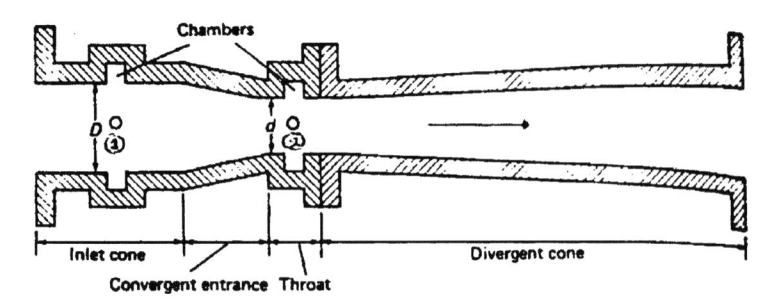

Figure 3.12. Venturi tube.

Considering the Bernoulli equation from point 1 to 2, figure 3.12, and neglecting the friction term gives:

$$p_1/\gamma + V_1^2/(2g) = p_2/\gamma + V_2^2/(2g) \qquad (3.18)$$

The conservation of mass equation (equation of continuity) for the incompressible fluid is:

$$A_1 V_1 = A_2 V_2 \qquad (3.19)$$

and

$$V_1 = (A_2 V_2)/A_1 = (D_2/D_1)^2 V_2 = \beta^2 V_2 \qquad (3.20)$$

where

$$\beta = D_2/D_1. \qquad (3.21)$$

Substituting this value of V_1 into equation 3.18 and solving for V_2 yields:

$$V_2 = \left(C/\sqrt{1-\beta^4}\right)\sqrt{2g(p_1-p_2)/\gamma} \qquad (3.22)$$

or

$$V_2 = \left(C/\sqrt{1-\beta^4}\right)\sqrt{2(p_1-p_2)/\rho} \qquad (3.23)$$

where C is a discharge coefficient that corrects for energy loss due to friction in the meter. Discharge coefficients for venturi meters constructed according to the procedures of ASME (1971) are given in table 3-2. For throat Reynolds numbers outside the tabulated limits or for non-standard venturi dimensions, the meter must be calibrated to obtain the

Table 3-2. ASME coefficients for venturi meters

Type of Inlet Cone	Machined	Rough Welded Sheet Metal	Rough Cast
Throat Re Range	5×10^5–1×10^6	5×10^5–2×10^6	5×10^5–2×10^6
Inlet Dia. Range (cm)	5–25	20–120	10–80
Range of β	0.4–0.75	0.4–0.7	0.3–0.75
C	0.995	0.985	0.984
Tolerance (%)	±1.0	±1.5	±0.7

Fluid-Flow Measurements

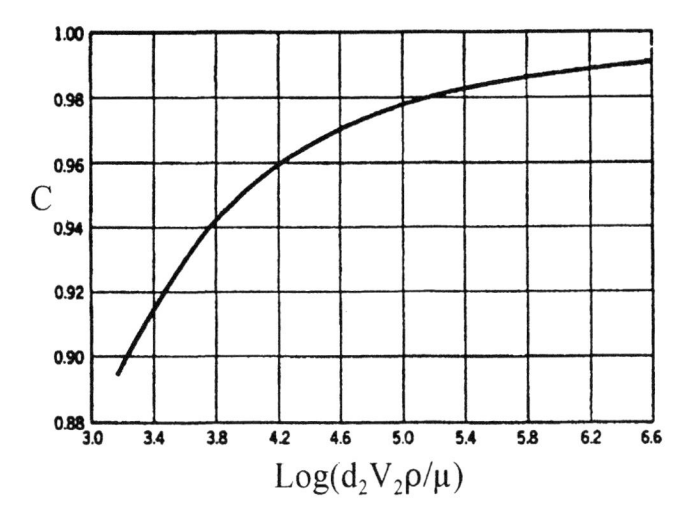

Figure 3.13. Values of the venturi coefficient C referred to the log of Reynolds number.

numerical value of the discharge coefficient. Figure 3.13 gives an example calibration curve for a venturi meter with throat Reynolds numbers dropping to approximately 1500. Applications of the agricultural process engineer seldom have Reynolds numbers high enough to assume the constant C-values given in table 3-2.

Note from equation 3.23 that an increase in V_2 or an increase in β will cause a decrease in p_2, assuming that p_1 remains constant. With liquids, if p_2 drops as low as the vapor pressure of the fluid, vaporization will occur at any slight irregularity. The formation and subsequent collapse of vapor bubbles promotes erosion of metal. This process, called cavitation, not only limits the venturi as a measuring device, but also causes an increase in energy loss and erosion or pitting of the tube itself.

Although cavitation must be avoided when a venturi is used as a measuring device, the phenomenon which produces it is used in certain types of pumps. If V_1 and n are sufficient to produce a p_2 less than atmospheric, this rarefied pressure, vacuum, can be used for evacuating or pumping. The laboratory suction pump that is fastened to a water faucet and basement sump pumps operated off the house water systems are good examples of venturi pumps.

The previous discussion applies to liquids which are incompressible ($\gamma_1 = \gamma_2$). If γ_2/γ_1 is nearly equal to 1, gases may be considered incompressible and the error resulting is negligible. Ratios of 0.95 and 0.90 produce errors of 4 and 6%, respectively. Since it is advisable to calibrate an individual tube against a known standard for most accurate results, preliminary calculations for gases can be made with equation 3.23 without serious error resulting. Ventilation, drying, and air-conditioning pressures with which the processing engineer is active seldom will exceed a gage pressure of 1 kPa. The ratio of specific weights at this pressure referred to atmospheric at 101.325 kPa is approximately 0.99, the resulting error being 0.5%.

Example 3.4. Water at 21 C is being measured by a venturi meter having the discharge coefficient relationship of figure 3.13 with diameter $D_1 = 15$ cm and $D_2 = 10$ cm. Gages attached at points 1 and 2, figure 3.12, read 10 and 5 cm of water gage pressure head, respectively. Determine the average velocity of the water at diameter D_1 and the volumetric flow rate of the water.

Solution

$$\beta = D_2/D_1 = (10\,\text{cm})/(15\,\text{cm}) = .6667$$
$$\sqrt{1 - \beta^4} = \sqrt{1 - .1975} = \sqrt{0.8025} = .896$$
$$p_1 - p_2 = (10 - 5)\,\text{cm H}_2\text{O}(1\,\text{m}/100\,\text{cm})\,(998\,\text{kg/m}^3)\,(9.81\,\text{m/s}^2)$$
$$= 489.5\,\text{Pa}$$
$$V_2 = (C/.896)\,\sqrt{2(489.5\,\text{Pa})/(998\,\text{kg/m}^3)}$$
$$= (C/.896)\,\sqrt{0.981\,\text{m}^2/\text{s}^2} = 1.105\,C\,\text{m/s}$$

Since C is a function of V_2, the solution must be by trial and error. Assume $C = 0.95$. Then,

$$V_2 = 0.95(1.105\,\text{m/s}) = 1.050\,\text{m/s}$$
$$Re = \rho V_2 d_2/\mu$$
$$= (998\,\text{kg/m}^3)\,(1.050\,\text{m/s})\,(0.10\,\text{m})/(0.000984\,\text{Pa s})$$
$$= 106\,494$$
$$\log Re = 5.02$$

Then from figure 3.13 the next estimate of C is 0.978. Then,

$$V_2 = 0.978\,(1.105\,\text{m/s}) = 1.081\,\text{m/s}$$
$$Re = (998\,\text{kg/m}^3)\,(1.081\,\text{m/s})\,(0.10\,\text{m})/(0.000984\,\text{Pa s})$$
$$= 109\,638$$
$$\log Re = 5.04$$

Referring again to figure 3.13, the C estimate is still 0.978 and thus the correct value of V_2 is 1.081 m/s.

The volumetric flow rate of the water is given by,

$$Q = A_2 V_2 = \left(\pi d_2^2/4\right) V_2$$
$$= (\pi (0.1\,\text{m})^2/4)\,(1.081\,\text{m/s})$$
$$= 0.00849\,\text{m}^3/\text{s}$$

The velocity of water at d_1 is obtained from conservation of mass:

$$Q = A_1 V_1$$
$$V_1 = Q/A_1$$
$$= 4Q/\pi d_1^2 = 4(.00849\,\text{m}^3/\text{s})/(\pi\,(0.15\,\text{m})^2) = 0.480\,\text{m/s}$$

Fluid-Flow Measurements

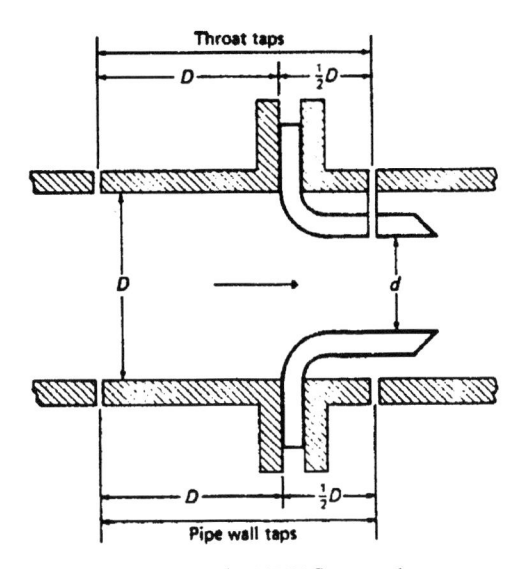

Figure 3.14. ASME flow nozzle.

3.10 Nozzle Meter

Figure 3.14 shows an ASME (1971) flow nozzle. The pressure differential may be sensed using either throat or pipe wall taps. In either case, one tap is located one pipe diameter upstream and the other one-half pipe diameter downstream from the nozzle inlet. Equation 3.23 is used for nozzles with the discharge coefficient of nozzles constructed according to ASME specifications being given by:

$$C = 0.9975 - 0.00653(10^6/Re)^a \qquad (3.24)$$

where

Re = Reynolds number for the fluid in the throat of the nozzle (dimensionless)
a = 0.5 for $Re < 10^6$ and 0.2 for $Re > 10^6$

Most of the data upon which equation 3.24 is based were obtained for pipe diameters between 5 and 40 cm, Re between 10^4 and 10^6, and β between 0.15 and 0.75. For values in these ranges the value of C determined by equation 3.24 are accurate to within 2%.

Example 3.5. An ASME nozzle is to be designed to measure the flow of 1.5 m³/min of water at 21 C in a 15-cm-inside-diameter pipe. The pressure differential across the nozzle is not to exceed 20 kPa. What should be the throat diameter of the nozzle?

Solution
A trial and error procedure is necessary since both C and β of equation 3.23 are dependent on the nozzle diameter. A first estimate may be obtained by assuming:

$$C/\sqrt{1-\beta^4} = 1$$

Then,
$$V_2 = \sqrt{2(20\,000 \text{ Pa})/(998 \text{ kg/m}^3)} = 6.33 \text{ m/s}$$
and
$$A = Q/V_2 = (1.5 \text{ m}^3/\text{min})(1 \text{ min}/60 \text{ s})/(6.33 \text{ m/s})$$
$$= 3.949 \times 10^{-3} \text{ m}^2$$

Then,
$$D_2 = \sqrt{4A/\pi} = .07091 \text{ m} = 70.91 \text{ mm}$$

Check the trial values:
$$\beta = 70.91 \text{ mm}/150 \text{ mm} = 0.4727$$
$$Re = \rho V_2 D_2/\mu$$
$$= (998 \text{ kg/m}^3)(6.33 \text{ m/s})(.07091 \text{ m})/(0.000984 \text{ Pa s})$$
$$= 455\,247$$
$$C = 0.9975 - 0.00653\,(10^6/Re)^a$$
$$Re < 10^6 \text{ therefore } a = 0.5$$
$$C = 0.9975 - 0.00653(10^6/455\,247)0.5$$
$$= 0.9878$$
$$V_2 = ((0.9878)/\sqrt{1-\beta^4})\sqrt{2(20\,000 \text{ Pa})/(998 \text{ kg/m}^3)}$$
$$= (1.01345)(6.33 \text{ m/s}) = 6.415 \text{ m/s}$$
$$A = Q/V_2 = 3.897 \times 10^{-3} \text{ m}^2$$
$$D_2 = \sqrt{4A/\pi} = 70.44 \text{ mm}$$

Further trials are not necessary.

3.11 Orifice Meter

Orifice meters, figure 3.15, are convenient devices for measuring rates of flow because they are simply constructed, easily installed, and occupy little space as compared to the venturi. Although commercial units are available, shop-made meters give reliable results. Orifices are subject to considerable friction loss and are inferior to venturi in this respect. These meters are convenient for measuring or calculating discharges into the air and into or out of large bodies such as storage tanks since a hole or valve is essentially an orifice. Equation 3.23 is basically applicable to orifice meters. But with an orifice, figure 3.15, the fluid is unable to make an abrupt turn as it passes the edge of the orifice and the minimum diameter of the fluid stream (called the vena contracta) is at a point downstream from the orifice. The location of the vena contracta is a function of β as shown in figure 3.16.

Fluid-Flow Measurements

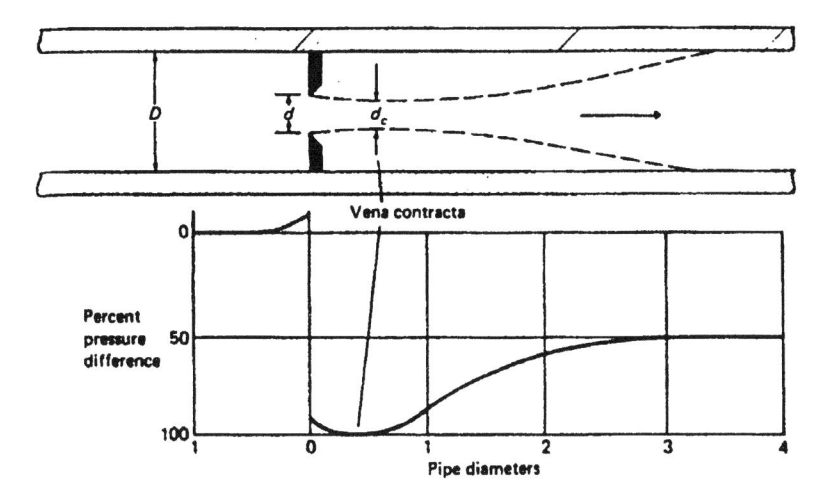

Figure 3.15. Relative pressure changes due to flow through an orifice.

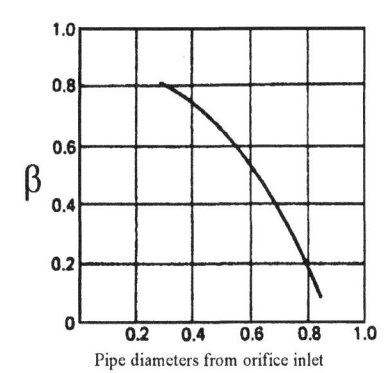

Figure 3.16. Location of vena contracta.

Since the location of the pressure taps is critical, it is necessary to specify the exact position of the downstream pressure tap. The jet contracts to about 60% of the orifice area so that discharge coefficients are on the order of 0.6 compared to near unity for venturi and nozzle meters.

Three tap locations are specified by ASME (1971) for measuring pressure differential. These are flange, vena contracta, and the 1 D and 0.5 D methods. In the flange tap, the location is always 2.5 cm (1 in.) from either face of the orifice plate regardless of the size of the pipe. In the vena contracta tap, the upstream tap is located one pipe diameter from the inlet face of the orifice plate and the downstream tap is located at the vena contracta as determined from figure 3.16. In the 1 D and 0.5 D tap, the upstream tap is located one pipe diameter and the downstream tap is located one-half pipe diameter from the inlet face of the orifice plate.

Table 3-3. Values of C_o, ΔC, and a for use in equation 3.25

β	$D = 2$ in. $= 50$ mm C_o	ΔC	$D = 4$ in. $= 100$ mm C_o	ΔC	$D = 8$ in. $= 200$ mm C_o	ΔC	$D = 16$ in. $= 400$ mm C_o	ΔC
				Flange Taps $a = 1$				
0.20	0.5972	127	0.5946	200	0.5951	327	0.5955	551
0.30	0.5978	144	0.5977	209	0.5978	307	0.5980	457
0.40	0.6014	181	0.6005	256	0.6002	362	0.6001	514
0.50	0.6050	260	0.6034	386	0.6026	584	0.6022	903
0.60	0.6078	392	0.6055	622	0.6040	1015	0.6032	1710
0.70	0.6068	573	0.6030	953	0.6006	1637	0.5991	2998
				Vena Contracta Taps $a = 1/2$				
0.20	0.5938	1.61	0.5928	1.61	0.5925	1.61	0.5924	1.61
0.30	0.5939	1.78	0.5934	1.78	0.5933	1.78	0.5932	1.78
0.40	0.5970	2.01	0.5954	2.01	0.5953	2.01	0.5953	2.01
0.50	0.5994	2.29	0.5992	2.29	0.5992	2.29	0.5991	2.29
0.60	0.6042	2.68	0.6041	2.68	0.6041	2.69	0.6041	2.70
0.70	0.6069	3.34	0.6068	3.37	0.6067	3.44	0.6068	3.57
				$1D$ and $1/2 D$ Taps $a = 1/2$				
0.20	0.5909	2.03	0.5922	1.41	0.5936	1.10	0.5948	0.94
0.30	0.5915	2.02	0.5930	1.50	0.5944	1.24	0.5956	1.12
0.40	0.5936	2.17	0.5951	1.72	0.5963	1.49	0.5974	1.38
0.50	0.5979	2.40	0.5978	1.99	0.5999	1.79	0.6007	1.69
0.60	0.6036	2.67	0.6040	2.31	0.6044	2.12	0.6048	2.11
0.70	0.6078	3.19	0.6072	2.98	0.6068	3.07	0.6064	3.51

The discharge coefficient, C, for orifice meters may be determined from:

$$C = C_o + \Delta C/Re^a \tag{3.25}$$

where

Re = Reynolds number for fluid based on orifice diameter
$C_o, \Delta C, a$ = constants determined from table 3-3.

Example 3.6. Air at 21 C flows at a rate of 1 m³/s through a pipe having an inside diameter of 40 cm. It is proposed to measure this flow with a 24-cm-diameter ASME orifice equipped with vena contracta taps. What differential pressure may be expected?

Solution
$\beta = D_2/D_1 = (24 \text{ cm})/(40 \text{ cm}) = 0.6$
$V_2 = Q/A = 4Q/(\pi d_2^2)$
$\quad = 4 (1 \text{ m}^3/\text{s})/(\pi (0.24 \text{ m})^2) = 22.10 \text{ m/s}$
$Re = \rho V_2 D_2 / \mu$
$Re = (1.202 \text{ kg/m}^3)(22.10 \text{ m/s})(0.24 \text{ m})/(0.0000182 \text{ Pa s})$
$\quad = 350\,297$

Fluid-Flow Measurements

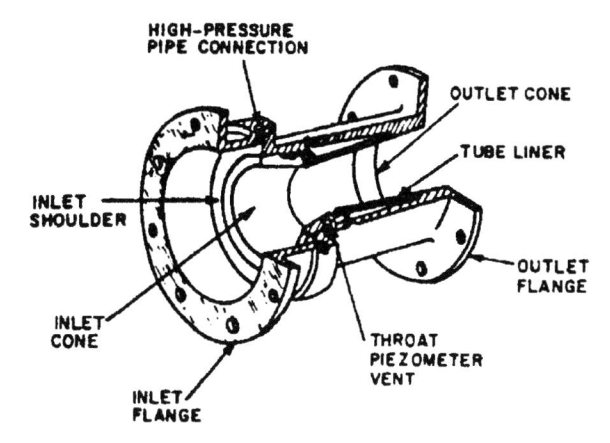

Figure 3.17. The Dall tube.

From table 3-3,

$$C_o = 0.6041, \quad \Delta C = 2.70, \text{ and } a = 0.5$$

Then,

$$C = 0.6041 + 2.70/(350\,297)^{0.5} = 0.6087$$

Equation 3.23 may be rearranged to give:

$$\Delta p = \rho V_2^2 (1 - \beta^4)/(2C^2)$$
$$= (1.202 \text{ kg/m}^3)(22.10 \text{ m/s})^2 (1 - 0.6^4)/2 (0.6087)^2$$
$$= 689.3 \text{ Pa}$$

3.12 Dall Flow Tube

The Dall flow tube is a modified venturi. As shown in figure 3.17, the device has a short, straight inlet section, the end of which decreases in diameter to the inlet shoulder. It has a converging cone section, a narrow annular gap or slotted throat annulus and a diverging outlet cone.

Despite abrupt fluid direction changes in the meter, there is no separation of the jet and thus there is high pressure recovery. The Dall tube is not recommended for dirty fluids, because deposits of foreign materials will change the sharp contours which govern its coefficient of discharge.

3.13 Laminar Flow Meter

The primary element of the laminar flow meter is a matrix of very small passages which produce laminar flow through the meter. Figure 3.18 illustrates a laminar flow meter installation and figure 3.19 represents an end view of a matrix composed of triangular passages. The effective diameter of the passages is only a few thousandths of a centimeter, while the length of the passages is normally several centimeters.

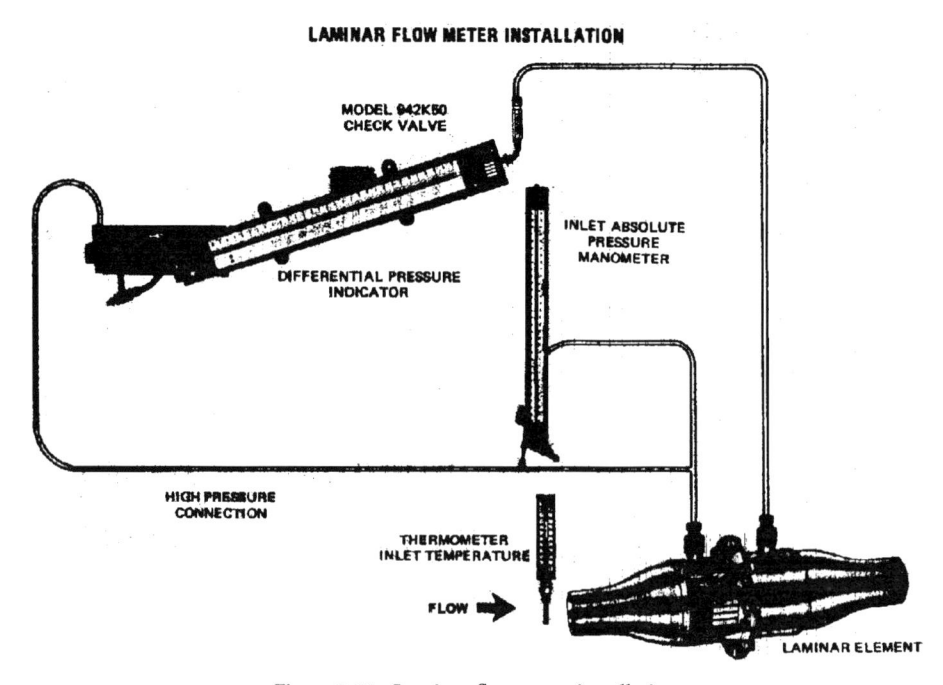

Figure 3.18. Laminar flow meter installation.

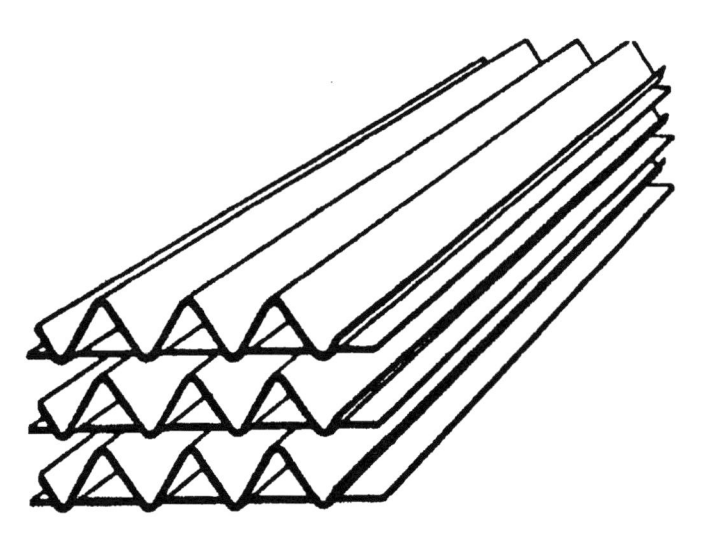

Figure 3.19. Laminar flow element matrix.

Fluid-Flow Measurements

The laminar flow meter is usually designed so that within its recommended operating range the Reynolds number for the fluid will be less than 1000 to insure laminar flow. Under these conditions, equations 2.26 and 2.27 combine to give:

$$F = (64/Re)(L/D)(V^2/(2g)) \qquad (3.26)$$

or

$$F = 32\mu L V/(\gamma D^2) \qquad (3.27)$$

where

F = energy loss due to friction per unit weight of fluid in a given passage (J/N)
Re = Reynold's number (dimensionless)
L = length of passages (m)
D = effective diameter of matrix passages (m)
V = velocity of fluid through passages (m/s)
g = acceleration due to gravity (m/s^2)
μ = dynamic viscosity of fluid (Pa s)
γ = specific weight of fluid (N/m^3)

Inspection of equation 3.27 reveals that the friction loss across the laminar flow element is linearly proportional to the fluid velocity through the passages. If passage walls are sufficiently thin, then the average fluid velocity in the matrix is approximately equal to the average velocity in the entrance duct and the velocity is given by:

$$V = 4Q/(\pi D_m^2) \qquad (3.28)$$

where

Q = volumetric flow rate (m^3/s)
D_m = total diameter of meter entrance duct (m)

The diameter of the entrance duct may be expressed as a multiple of the effective passage diameter as:

$$D_m = nD \qquad (3.29)$$

where n = dimensionless ratio.

Also, since the velocity and elevation heads do not change in the meter and there is no work done on the system, the friction head is equal to the change in the pressure head:

$$F = \Delta p/\gamma \qquad (3.30)$$

Substitution of equations 3.28, 3.29, and 3.30 into equation 3.27 yields:

$$\Delta p/\gamma = 128n^2 \mu L Q/(\pi \gamma D_m^4) \qquad (3.31)$$

or

$$Q = \pi D_m^4 \Delta p/(128n^2 \mu L) \qquad (3.32)$$

Thus, the volumetric flow rate is linearly proportional to the pressure drop through the meter. Note that the density of the fluid does not enter directly in equation 3.32. However, dynamic viscosity appears in the equation and it may vary with density as well as with temperature of the fluid.

For a given laminar flow meter, equation 3.32 can be written as:

$$Q = C_m \Delta p / \mu \tag{3.33}$$

where C_m = meter calibration constant (m^3).

Volumetric Flow Meters

Volumetric flow meters function by dividing the fluid into separate increments of volume and then counting or totalizing these volumes. One or more moving parts are positioned in the flow stream to physically separate the fluid into volumetric increments. Devices in this class of meters include nutating disks, gear or lobed impeller, sliding vane, rotating vane, reciprocating piston, oscillating or rotary piston, and diaphragm or bellows meters.

3.14 Nutating Disk Meters

The nutating disk is a popular quantity meter used for volumetric measurements of liquids, figure 3.20. The meter consists of a disk supported at the center by a ball. Each

Figure 3.20. Cutaway view of a nutating disk meter. (Courtesy of Hersey Products Inc., Spartanburg, S.C.)

Fluid-Flow Measurements

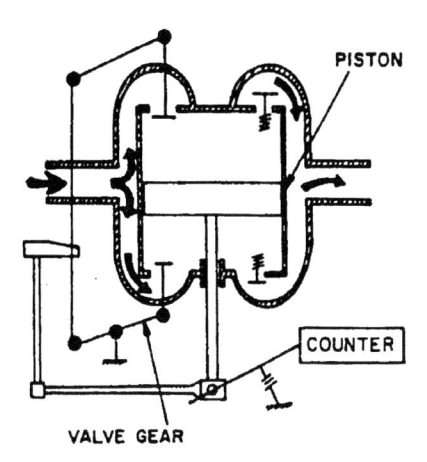

Figure 3.21. Operating principle of the reciprocating piston meter.

cycle of the measuring disk displaces a fixed volume of liquid. The nutating disk is the only moving part in the meter. Its movement is used to rotate a shaft which drives a counter to indicate the quantity of liquid passing through the meter. A variety of disk and chamber materials are used depending upon the type of material to be measured. Cheremisinoff (1979) gives a good review of available materials and situations in which they are used.

3.15 Reciprocating Piston Meters

The operating principle of the reciprocating piston meter is shown in figure 3.21. The meter is essentially a piston pump operated backwards. The piston moves back and forth due to the flow of the fluid. As the piston reaches the end of its stroke, it shifts the intake and discharge valves. This in turn operates a counter which adds a fixed volume increment of fluid with each cycle or stroke of the piston.

3.16 Oscillating or Rotary Piston Meters

Figure 3.22 shows and describes the operation of an oscillating piston meter. It is similar in operation to the nutating disc meter but superior in performance because of less frictional drag and better balance.

3.17 Bellows Meters

The familiar household gas meter, figure 3.23, consists of two bellows inner connected by valves. As one bellows is being filled from the supply line, the other is emptying into the service line. Valves shift the direction of flow at the end of the stroke, and the emptied bellows fills from the supply line. The oscillation of the mechanism activates a volumetric indicator.

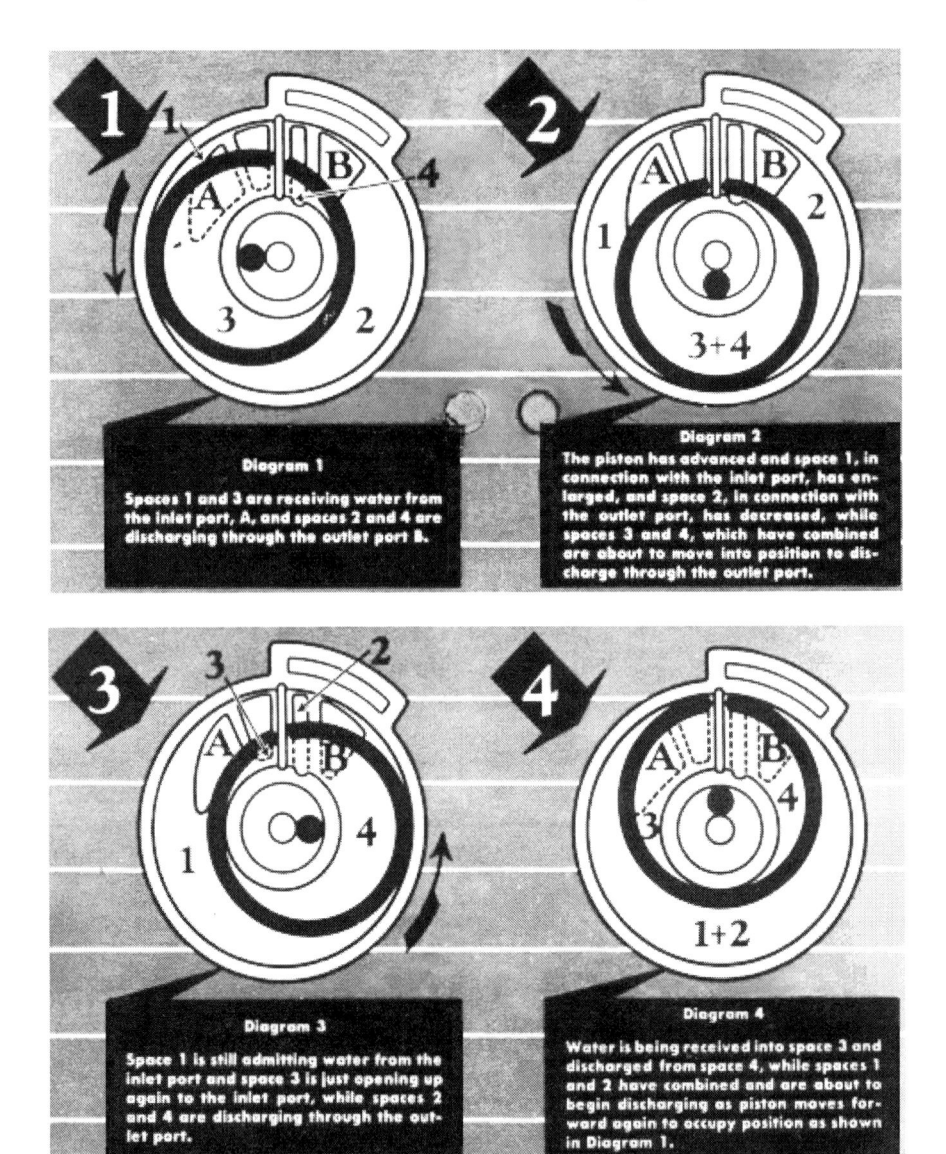

Figure 3.22. Oscillating-piston meter and its operation. (Courtesy of the Pittsburgh Equitable Meter Division, Rockwell Manufacturing Co.)

Force-Displacement Meters

Meters in this class operate by a force displacing an obstruction immersed in the flowing fluid. The force, caused by drag effects of the fluid, causes a deflection of the primary element. A measure of this displacement is a function of the differential

Fluid-Flow Measurements

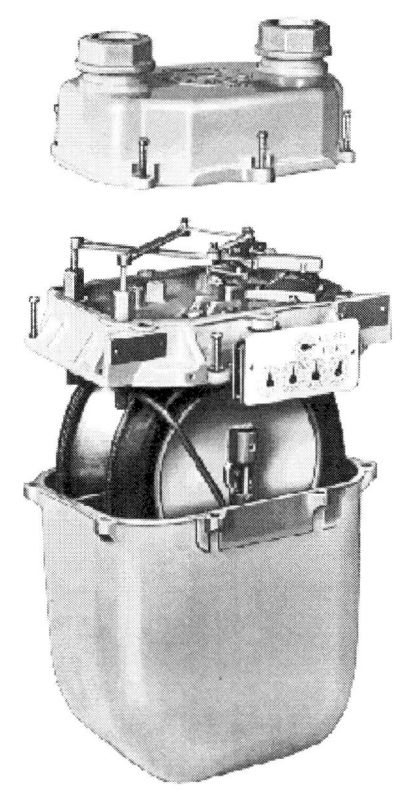

Figure 3.23. Bellows gas meter. (Courtesy of the Pittsburgh Equitable Meter Division, Rockwell Manufacturing Co.)

pressure and serves to indicate the flow rate. Meters of this type include rotameters, turbine or propeller meters, rotating vane anemometers, swinging vane anemometers, and cup anemometers.

3.18 Rotameters

A commercial rotameter and its schematic elements are shown in figure 3.24. The rotor is supported by the upward motion of the fluid, and its position in the tube indicates the rate of flow.

The rotor is stationary when the drag force on the rotor plus the weight of the fluid displaced by the rotor is equal to the weight of the rotor:

$$F_D + \rho_f g \phi = \rho_r g \phi \qquad (3.34)$$

where

F_D = drag force (N)

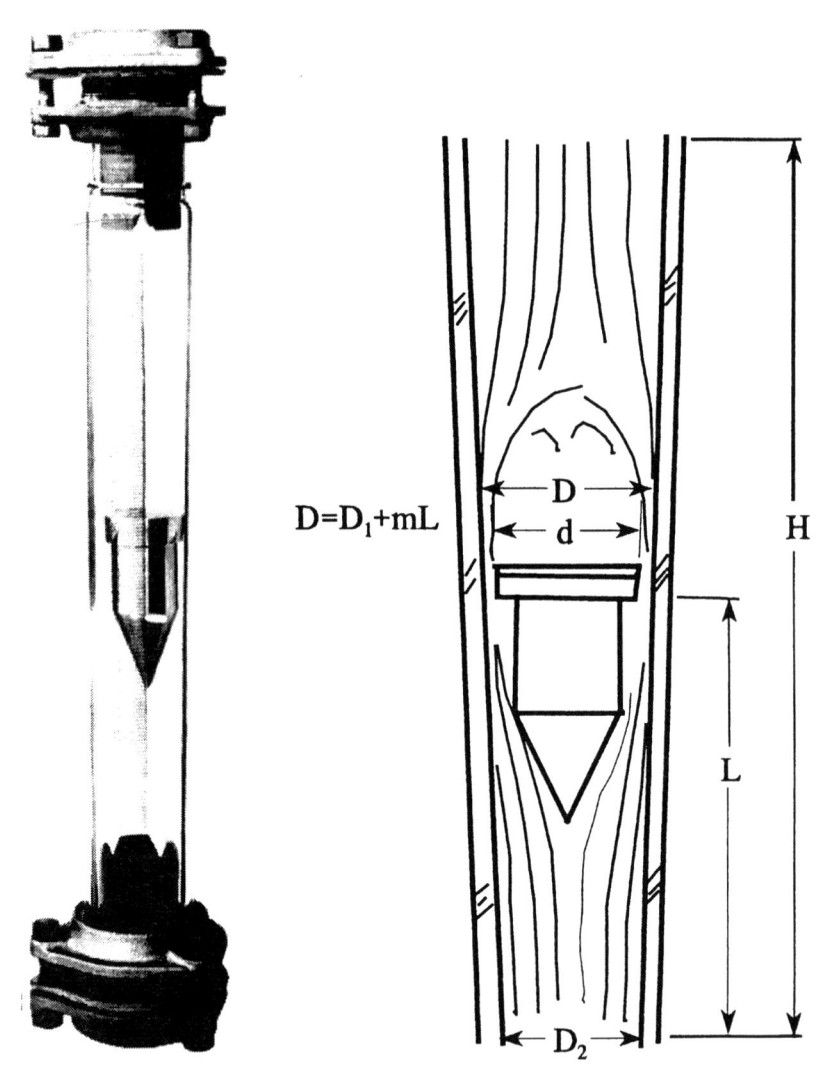

Figure 3.24. A commercial rotameter with important details shown. (Courtesy of Schutte and Koerting Co.)

$D = D_1 + mL$

ρ_f = density of fluid (kg/m^3)
ρ_r = density of rotor material (kg/m^3)
g = acceleration due to gravity (m/s^2)
ϕ = volume of rotor (m^3)

The drag force is given by the following relationship:

$$F_D = C_D A_B \gamma_f (V^2/2g) \tag{3.35}$$

Fluid-Flow Measurements

where

C_D = drag coefficient (dimensionless)
A_B = projected area of rotor in direction of fluid flow (m²)
γ_f = specific weight of fluid (N/m³)
V = velocity of fluid (m/s)

If equation 3.35 is substituted into equation 3.34 and then solved for the velocity, the following relationship for velocity results:

$$V = \sqrt{2g\phi(\rho_r - \rho_f)/(C_D A_B \rho_f)} \tag{3.36}$$

Note that the velocity will be fixed by the above relationship so that an increase in flow rate through the meter does not result in a change in velocity but results in a change in the area through which flow occurs. The drag coefficient, C_D, is in general a function of the Reynolds number and the shape of the object around which flow occurs. However, since the velocity is fixed in the case of the flow of a given fluid in a rotameter, the drag coefficient is also constant.

From figure 3.24, the area through which flow occurs may be expressed as:

$$A = \pi(D^2 - d^2)/4 \tag{3.37}$$

where

A = area of annulus through which flow occurs (m²)
D = diameter of tube at location of rotor (m)
d = diameter of rotor (m)

or

$$A = \pi((D_1 + mL)^2 - d^2)/4 \tag{3.38}$$

where

D_1 = diameter of tube at bottom (m)
m = taper of tube (m/m)
L = elevation of rotor from bottom of tube (m)

The volumetric flow rate of the fluid is the product of the velocity through the annulus and the area of the annulus. Then,

$$Q = (\pi/4)((D_1 + mL)^2 - d^2)\sqrt{2g\phi(\rho_r - \rho_f)/(C_D A_B \rho_f)} \tag{3.39}$$

The mass flow rate is the product of the volume flow rate and the density of the fluid. Thus,

$$\dot{m} = (\pi/4)((D_1 + mL)^2 - d^2)\sqrt{2g\phi_f(\rho_r - \rho_f)/(C_D A_B)} \tag{3.40}$$

If D_1 is equal to d and m is small, equation 3.40 becomes approximately:

$$\dot{m} = C'L\sqrt{\rho_f(\rho_r - \rho_f)} \tag{3.41}$$

where C' = meter constant (m²/s).

For maximum accuracy, the meter should be calibrated through its entire range against a known standard.

Rotameters can be used for liquids or gases. Chemicals, oils, food products, and fluids carrying suspended solid material can be metered. Large or small quantity rates can be handled. The pressure drop through the meter is nominal.

An operational problem frequently encountered with rotameters is a tendency of the rotor to stick to the tube wall upon contact during unstable flow. This problem may be eliminated by the use of a guide wire which passes through the center of the rotor and tube or by the use of glass ribs on the inside of the tube.

3.19 Turbine or Propeller Meters and Rotating Vane or Cup-Type Anemometers

The operating principle of each of these devices is the rotation of a "propeller" which is placed in the fluid stream. Turbine or propeller meters are generally designed for liquid flow measurement. The propeller may directly drive a register for flow indication. Figure 3.25 illustrates a propeller meter.

Rotating vane and cup-type anemometers are designed for measurement of gas flow rates (especially air). A rotating vane anemometer is essentially a small windmill which indicates the linear air travel through the meter. The average velocity is then determined by dividing the distance by the time as observed by a stop watch. Each instrument must be calibrated individually and usually operates satisfactorily in the range of 1.5 to 15 m/s. Special instruments are available for lower velocities, but extreme care must be used in handling and maintaining them to insure continued accuracy. When making a test, the

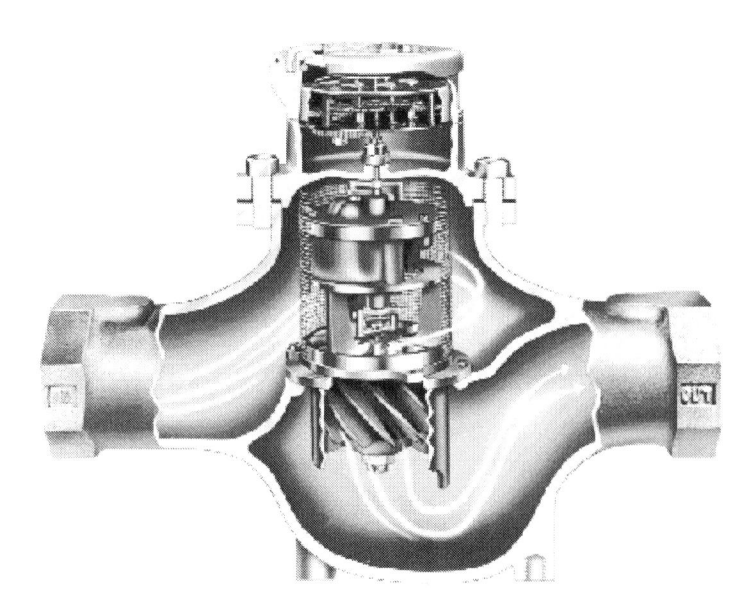

Figure 3.25. A propeller meter. (Courtesy of the Pittsburgh Equitable Meter Division, Rockwell Manufacturing Co.)

Fluid-Flow Measurements

instrument must be reasonably well aligned in the direction of air motion. No individual reading should be made for a time of less than 1 min or a linear reading of less than 30 m.

Cup-type anemometers for wind velocity measurement have been the standard method used by the Weather Bureau for many years. The cup anemometer (fig. 3.26) usually consists of three or four conical cups which rotate about a vertical axis. There is a considerable

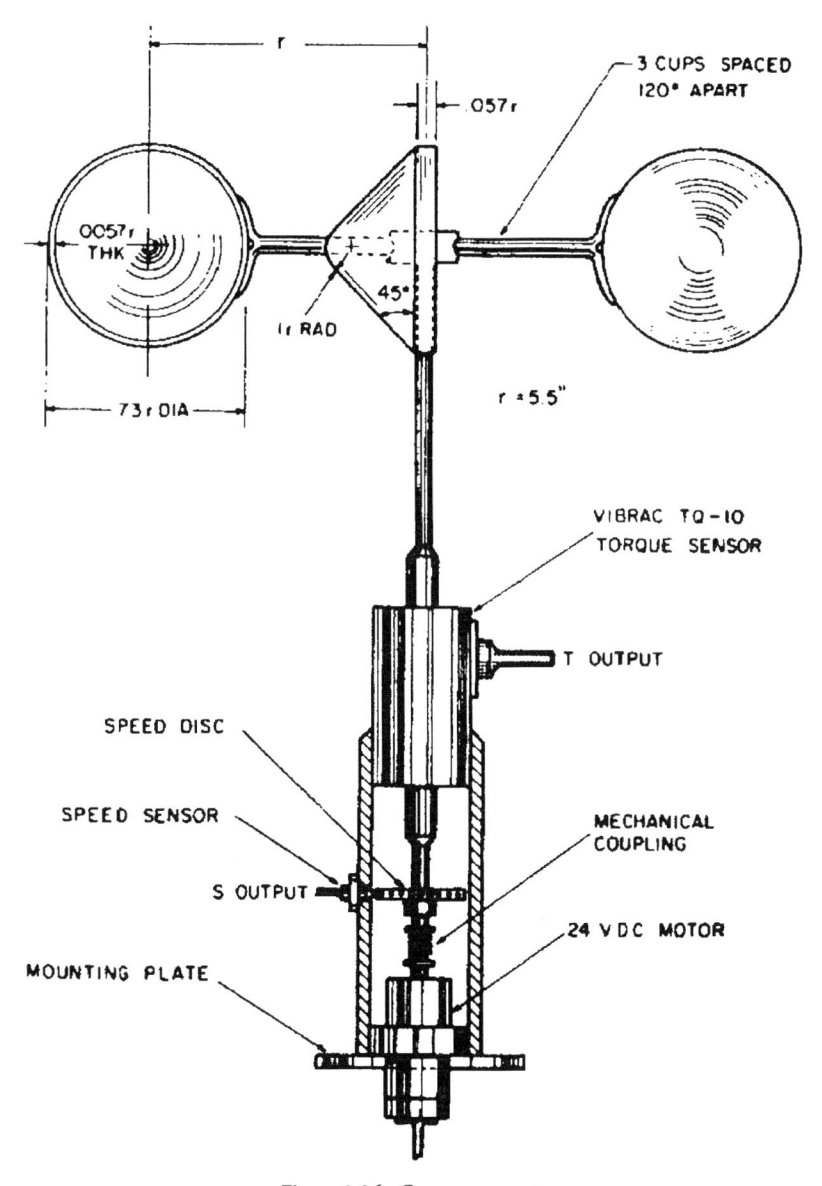

Figure 3.26. Cup anemometer.

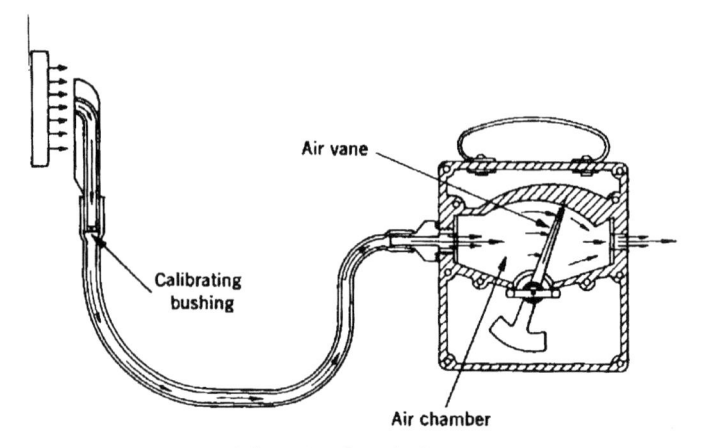

Figure 3.27a. Schematic of a swinging vane anemometer.

Figure 3.27b. A commercial swinging vane meter for measuring air flow.

Fluid-Flow Measurements

range in size of the various cup anemometers. The larger are generally more rugged, but the larger mass and friction of the larger rotating elements may make the starting velocities excessively high. A high moment of inertia causes the anemometer to lag when the velocity is increasing and coast when the velocity is decreasing. Wind velocity is proportional to the rate of rotation of the element which may be electronically sensed.

3.20 Swinging Vane Anemometers

The swinging vane meter is essentially a spring- or gravity-loaded gate which is moved by the impact of the flowing fluid, shown in figure 27a. Analysis of the flow past the vane and the resulting vane deflection is complicated. Consequently, meters of this type are calibrated experimentally. A commercial swinging-vane meter is shown in figure 3.27b. It is fitted so that high and low air velocities and static pressures can be observed both directly and remotely. Static pressure observations are possible since a certain static pressure operating through a definite resistance will produce a definite rate of flow through the instrument. The pressure-rate-of-flow relationship is determined, and the instrument is also calibrated in terms of pressure.

The degree of accuracy of the instrument depends upon the precision of manufacture, calibration accuracy, and care in operation and handling.

Thermal Meters

Thermal flow meters depend upon the addition of energy to the fluid and measurement of either the change in temperature of the fluid or of the energy source. This temperature change is then related to the flow rate.

3.21 Thomas Meter

This measuring system is based upon the rise in temperature of the fluid that results from the introduction of energy into a confined stream of flowing fluid that can be either liquid or gaseous. An electrical heating element is placed in the stream of flowing fluid and raises the temperature of the fluid.

The rate at which energy is given off by the heater must equal the rate at which energy is absorbed by the fluid. Thus,

$$P = EI = \dot{m} c_p (t_2 - t_1) \tag{3.42}$$

where

$P =$ power (W)
$E =$ electrical potential (volts)
$I =$ current (amperes)
$\dot{m} =$ mass flow rate (kg/s)
$c_p =$ specific heat of fluid (J/kg K)
$t_2 =$ downstream or hot temperature (C)
$t_1 =$ upstream or cool temperature (C)

If the relationship for $\dot{m}$ in terms of average fluid velocity, fluid density, and cross-sectional area of conduit is substituted into equation 3.42, the equation for average velocity results:

$$V = EI/(A\rho c_p(t_2 - t_1)) \qquad (3.43)$$

where

V = average velocity of fluid (m/s)
A = cross-sectional area of conduit (m^2)
ρ = density of fluid (kg/m^3)

Performance of the Thomas meter is highly dependent on the ability to accurately measure small temperature differences and electrical power consumption. It is also important that heat losses by radiation or conduction be eliminated. Small temperature differences are advisable in order to minimize heat losses. Losses by radiation can be eliminated by shielding the heating element while conduction losses can be eliminated by insulating the conduit containing the metering elements. The heating element current is usually adjusted to maintain a constant temperature differential of 1 to 3 C.

3.22 Hot-Wire Anemometer

The hot-wire anemometer, figure 3.28 (left), is based on the rate of heat transfer between a heated wire and the fluid stream flowing perpendicular to the wire. A small platinum wire 0.01 cm or less in diameter is heated by an electric current to a high temperature. Since the resistance of the wire varies with temperature, the amount of current flowing will vary with the velocity of fluid past the heated wire. An increase in velocity will permit an increase in the current flowing since the cooled wire will offer less resistance to electrical flow.

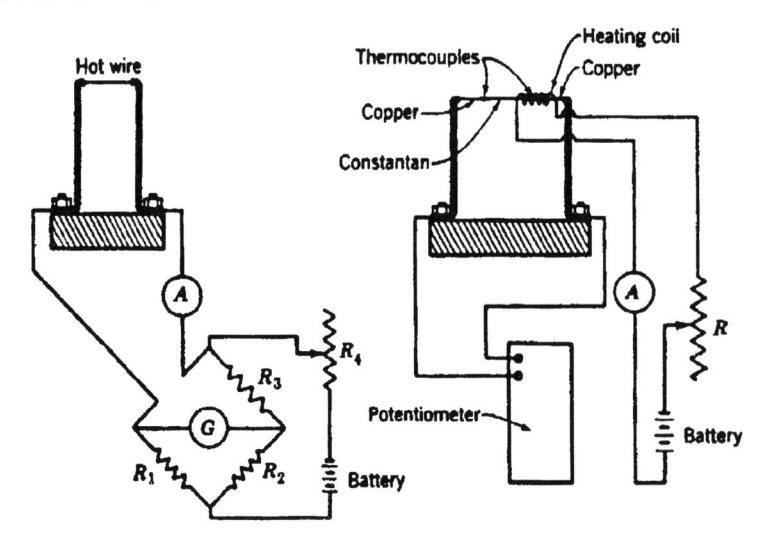

Figure 3.28. Hot-wire anemometer, left, and thermocouple anemometer, right.

Fluid-Flow Measurements

The basic equation describing the performance of the hot-wire anemometer equates the electrical power supplied to the wire and the rate of heat transfer to the fluid:

$$I^2 R = q = hS(t_s - t_f) \tag{3.44}$$

where

I = current (amperes)
R = electrical resistance (ohms)
q = heat transfer from wire to fluid (W)
h = convective heat transfer coefficient (W/m^2 K)
S = surface area of wire (m^2)
t_s = surface temperature of wire (C)
t_f = temperature of fluid (C)

The convective heat transfer coefficient will be discussed in greater detail in Chapter 7. However, for purposes of discussion of the operating characteristics of the hot-wire anemometer, we can estimate h as follows:

$$h = c_1 + c_2(\rho V)^n \tag{3.45}$$

where

c_1 and c_2 = coefficients which depend on thermal and viscous properties of the fluid and include appropriate dimensions
ρ = density of fluid (kg/m^3)
V = velocity of fluid (m/s)
n = empirical constant which has a value similar to 0.5.

Thus, the convective heat transfer coefficient is a function of the mass flow rate of the fluid. When equation 3.45 is substituted into equation 3.44 the following results:

$$I^2 R = (\alpha + \beta(\rho V)^n)(t_s - t_f) \tag{3.46}$$

where α and β = coefficients with appropriate dimensions.

Several methods of operation of the hot-wire anemometer can be used. First, a constant wire surface temperature (constant resistance) method may be used. In this case, if the sensor temperature is much higher than the fluid temperature, the signal (voltage or amperage) will be insensitive to temperature of the fluid but very sensitive to mass flow rate.

Second, if the fluid temperature is varying, controls may be used which maintain a constant temperature difference between the fluid and sensor.

Third, a constant current method in which the resistance and temperature of the sensor will vary with velocity while current is maintained constant.

Although the hot-wire anemometer can be used at low-wire temperatures for measuring very low velocities, it performs best under conditions of moderate to hot temperatures for measuring moderate to high velocities.

The direction of air motion must be known so that the hot wire can be located perpendicular to it. If the wire is placed at an angle to flow, low-velocity indications result.

Corrections cannot be made under this condition since the relationship between the adjustment factor and angle is not known.

Fluctuating velocity is difficult to observe since the equipment must be balanced for each velocity. Automatic adjusting and recording equipment can be used under fluctuating temperatures if the high cost can be justified.

This measuring device can be used in a small space and operated and observed from a remote location.

3.23 Thermocouple Anemometer

The thermocouple anemometer shown in figure 3.28 (right) operates on essentially the same principle as the hot-wire anemometer. A predetermined standard current is passed through the heating coil. This raises the temperature of the enclosed thermocouple. Air moving past the heated coil cools it. The cooling effect is reflected in the difference in temperature between the thermocouples. Consequently, the air velocity is related to the potential between the thermocouples. The potential across the thermocouples is nearly a linear function of the temperature difference and can be observed by a potentiometer. However, since it operates on the Wheatstone-bridge principle, no current is flowing when a reading is made. Consequently, the size and length of leads, if reasonable, do not affect performance.

The basic equations for the hot-wire anemometer apply also to the thermocouple anemometer, except that the heat is not supplied directly to the sensor where the heated thermocouple is located. The relationship between supplied current and the temperature difference and mass flow rate is essentially the same except for different calibration constants.

The hot junction of the thermocouple anemometer is usually maintained at approximately 10 C above air temperature, whereas hot-wire anemometer sensors operate at temperatures 30 to 100 C above air temperature. A normal change in air temperature does not affect the performance of the thermocouple anemometer since the heating element is made of Manganin, which has a constant resistance through a wide temperature range, and the two thermocouples are referred to ambient air temperature.

The thermocouple anemometer is affected by direction of approach of the air. Fluctuations in velocity may create a reading difficulty if an automatic recorder is not used.

The thermocouple anemometer, like the hot-wire anemometer, can be placed in a small space and operated from a remote location.

Special Metering Methods

There are a number of other metering methods for which operating principles are beyond the scope of this book. The reader is referred to the reviews by Replogle and Birth (1983) and Cheremisinoff (1979) for more information on these meters which exploit other properties of flowing fluids, such as the propagation of various sonic and electromagnetic waves. Other metering methods include chemical dilution techniques, electromagnetic meters, ultrasonic meters, nuclear magnetic resonance flowmeters, optical-ring laser flowmeters, laser Doppler meters, and vortex shedding meters.

Nomenclature

A	area (m²)
A_B	projected area of rotameter rotor in direction of fluid flow (m²)
C	discharge coefficient (dimensionless)
C'	rotameter constant (m²/s)
C_D	drag coefficient (dimensionless)
c_p	specific heat (J/kg K)
D	diameter (m)
E	electrical potential (volts)
F	energy loss due to friction per unit weight of fluid (J/N or m)
F_D	drag force (N)
H	head (m)
I	current (amperes)
K	meter coefficient (dimensionless)
L	elevation of rotameter rotor (m)
m	taper of tube in a rotameter (m/m)
$\dot{m}$	mass flow rate (kg/s)
p	pressure (Pa)
p_s	static pressure (Pa)
p_t	total pressure (Pa)
p_v	velocity pressure (Pa)
Q	volumetric flow rate (m³/s)
V	velocity of fluid (m/s)
W	energy added to fluid per unit weight of fluid (J/N or m)
β	ratio of venturi, nozzle, or orifice diameter to inlet diameter (dimensionless)
γ	specific weight of fluid (N/m³)
μ	dynamic viscosity of fluid (Pa s)
ϕ	volume of rotor in a rotameter (m³)
ρ	density of fluid (kg/m³)

References

Am. Soc. Mech. Engrs. 1962. In *Symposium on Measurement in Unsteady Flow*. New York: ASME.

Am. Soc. Mech. Engrs. 1971. *Fluid Meters, Their Theory, and Application*, 6th Ed., ed. H. S. Bean. New York: ASME.

Am. Soc. Mech. Engrs. 1979. *Glossary of Terms Used in the Measurement of Fluid Flow in Pipes*. ASME MFC-1M-1979. New York: ASME.

Am. Soc. Mech. Engrs. 1983. *Measurement Uncertainty for Fluid Flow in Closed Conduits*. ASME MFC-2M-1983. New York: ASME.

Baumeister, T., E. A. Avallone and T. Baumeister III. 1978. *Marks' Standard Handbook for Mechanical Engineers*. New York: McGraw-Hill Book Co.

Cheremisinoff, N. P. 1979. *Applied Fluid Flow Measurement. Fundamentals and Technology*. New York: Marcel Dekker, Inc.

Cox, W. F. 1959. *Flow Measurement and Control*. London: Heywood & Co., Ltd.
Dwyer Instruments, Inc. 1985. The 1985 Dwyer Catalog. Michigan City, Ind.: Dwyer Instruments, Inc.
Iversen, H. W. 1956. Orifice coefficients for Reynolds numbers from 4 to 50000. *Transactions of ASME* 78: 359–364.
Linford, A. 1961. *Flow Measurement and Meters*, 2nd Ed. London: E. and F. N. Spon, Ltd.
Meriam Instrument Co. Undated. Laminar Flow Meters. File No. 501: 215–1. Cleveland, Ohio: Meriam Instrument Co.
Murdock, J. W. 1964. Tables for the Interpolation and Extrapolation of ASME Coefficients for Square-Edged Concentric Orifices. ASME 64-WA/FM-6. New York: ASME.
———. 1976. *Fluid Mechanics and Its Applications*. Boston, Mass.: Houghton Mifflin Co.
Omega Engineering, Inc. 1984. Pressure and Strain Measurement Handbook. Stamford Conn.: Omega Engineering, Inc.
Replogle, J. A. and G. S. Birth. 1991. Flow. Chapter 5. In *Instrumentation and Measurement for Environmental Sciences*, 2nd Ed, ed. B. W. Mitchell Jr., Ch. 5. St. Joseph, Mich.: ASAE.
Seban, R. A., W. H. Hillendahl, E. J. Gallagher, and A. L. London. 1943. A thermal anemometer for low velocity flow. *Transactions of ASME* 65: 843–846.
Smith, H. W. 1936. Improved micromanometer. *Ind. Eng. Chem. Anal.*, 8: 151–152.
Spink, L. K. 1958. *Principles and Practice of Flow Meter Engineering*, 8th Ed. Foxboro, Mass.: Foxboro Co.
Thermo-Systems, Inc. Undated. General System Information for 1050 Series Anemometry. St. Paul, Minn.: TSI Inc.
Zoerb, G. C. 1991. Pressure and Vacuum. *Instrumentation and Measurement for Environmental Sciences*. 2nd Ed., ed. B. W. Mitchell Jr., Ch. 4. St. Joseph, Mich.: ASAE.

Problems

1. The pressure head in a fluid system is expressed as 5 cm of milk at 20.2 C. Determine the true pressure in pascals.

2. Pressure in a stream of water is being measured with a mercury filled U-tube manometer with one leg of the manometer exposed to atmospheric pressure and the other connected to a pressure tap at the center line of the water stream. The connecting tube is filled with water. The average level of the manometer fluid is 1 m above the pressure tap. Determine the gage pressure in the water stream if the manometer indicates a difference of 1 cm of mercury.

3. A water pressure gage located 1.5 m above a pressure source reads 1 kPa. The connecting tube is full of water at 20 C. What is the actual pressure at the source?

Fluid-Flow Measurements

4. The system of problem 3 has air filling the connecting tube. Assume that the compression of the air is negligible and estimate the true pressure at the source.

5. A combination pitot tube and a mercury filled U-tube manometer are used for measuring the velocity pressure in a stream of water at 21 C. If the difference in mercury level in the two legs of the manometer is 6 cm, determine: (a) the velocity pressure and (b) the velocity of water.

6. A combination pitot tube located at the center of a cylindrical air tube produces a velocity pressure head of 6.6 cm H_2O. What is the velocity at the impact end of the pitot tube? The gauge reads 6.6, 6.6, 6.4, 5.3, and 4.6 cm H_2O when placed at points 1 through 5 in figure 3.10. What is the average velocity in the tube? What factor would be applied to the center reading to indicate a true velocity? *Note:* In actual practice it would be necessary to check this factor through the entire range of velocities to be encountered, because variation may be expected.

7. Air at 21 C is being measured by a venturi meter with basic diameters of 35 and 25 cm, respectively. Gauges connected at points 1 and 2 of figure 3.12 indicate a pressure head difference of 9 cm H_2O. Assume the calibration curve of figure 3.13 to hold and determine the quantity of air flowing.

8. A nozzle meter is to be designed to fit into a 45-cm-diameter pipe. Air at room temperature flows through the pipe at velocities varying from 16.7 to 26.7 m/s. If a 5-cm inclined manometer containing alcohol with a specific gravity of 0.89 is to be used with the nozzle, what should be the diameter of the nozzle?

9. The differential pressure gauge on an orifice meter reads 35 kPa. What is the velocity of water if the inside pipe diameter is 30 cm and the sharp-edged orifice is 21 cm in diameter with taps located at the vena contracta?

10. A laminar flow element produces a pressure head difference of 8 cm H_2O when air at 21 C flows through the element at a rate of 15 m^3/min. Determine the meter calibration constant.

11. A rotameter has a rotor with a diameter of 2 cm, a volume of 5 cm^3, and a density of 2000 kg/m^3. The rotameter tube has a diameter of 2 cm at the bottom and a taper of 0.08 m/m. If the drag coefficient is 100 and the density of the fluid is 1000 kg/m^3, plot the fluid mass flow rate versus the elevation of the rotor for elevations from zero to 50 cm.

12. A Thomas meter is located in an air duct of 0.2 m^2 cross-sectional area. The density of the air is 1.2 kg/m^3 and its specific heat is 1.0048 kJ/kg K. Assume a controlled temperature differential of 2.5 C and a heater potential of 110 volts. Plot the velocity as abscissa and amperage as ordinate for velocities from 0 to 1.5 m/s. Assume a constant current of 3 amps and plot temperature difference against velocity.

4 Pumps

Pumps are generally considered as devices for elevating or moving liquids. Although this is a satisfactory conception in certain regards, it would be more exact to state that they increase the work head W in the Bernoulli equation 2.13. This statement implies that the pumping effect upon the fluid might be to elevate the fluid, change its internal pressure, change its velocity, or a combination of these.

The processing engineer is concerned with the performance of pumps on particular systems of interest. Consequently, the selection, installation, and to a lesser extent the design of pumps have to be dealt with by this person. There are many different types of pumps to choose from depending on the particular head-flow rate combination desired. This plethora of pump types can be classified into two general categories according to the method in which energy is imparted to the fluid: displacement (or positive displacement) and dynamic. Displacement pumps impart energy to the liquid in a pulsating fashion while dynamic pumps impart energy in a steady fashion. Displacement pumps generally produce a relatively low but constant flow rate against widely varying and often quite high heads. Dynamic pumps are more likely to be used in cases where high flow rates are desired with relatively low heads.

4.1 Evaluating Performance

The mechanical efficiency of pumps is the ratio of the power output to the input. The power output is the product of the total energy head W provided by the pump to the fluid and the weight flow rate of the fluid. Then,

$$P_o = W Q \rho g \tag{4.1}$$

where

P_o = power output of pump (W)
W = work done on fluid per unit weight of fluid (J/N)
Q = volumetric flow rate of fluid (m³/s)
ρ = density of fluid (kg/m³)
g = acceleration due to gravity (m/s²)

The mechanical efficiency is then given by:

$$e_m = P_o/P_i = W Q \rho g / P_i \tag{4.2}$$

where

e_m = mechanical efficiency (decimal)
P_i = power input to the pump (W)

Example 4.1. A pump provides 0.009 m³/s of water and a total head of 10.6 m. Determine the power output of the pump. If the power input is 1310 W, determine the mechanical efficiency of the pump.

Solution

(a) $P_o = WQ\rho g$
$= (10.6 \text{ m})(0.009 \text{ m}^3/\text{s})(1\,000 \text{ kg/m}^3)(9.81 \text{ m/s}^2)$
$= 935.9 \text{ W}$
(b) $e_m = P_o/P_i = (935.9 \text{ W})/(1310 \text{ W}) = 0.714 = 71.4\%$

The volumetric efficiency which applies to displacement pumps only is the ratio of the volume of fluid moved per unit of time to the displacement volume of the pump per unit time. Then,

$$e_v = Q/Q_d \qquad (4.3)$$

where

e_v = volumetric efficiency (decimal)
Q = volumetric flow rate of fluid (m³/s)
Q_d = displacement volume of pump per unit time (m³/s)

Displacement Pumps

Displacement pumps may be further subdivided into reciprocating or rotary-type pumps. Each of these types are often referred to as positive displacement pumps and are used in situations where constant and/or highly predictable flow rates are desired. They may also be capable of providing flow against relatively high heads.

4.2 Reciprocating or Piston Pumps

Reciprocating pumps (also called piston pumps) consist of a crank and connecting rod which are driven by a motor to move a piston or plunger as shown in figure 4.1. The piston or plunger may be used to directly move the fluid by providing appropriate

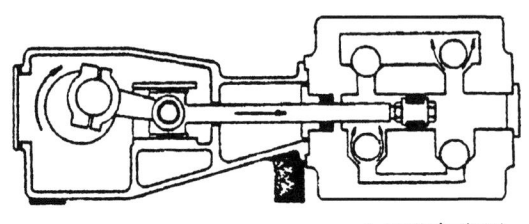

Hydraulic Institute).

Figure 4.1. Piston-packed pump.

Pumps

intake and exhaust valves or it may be connected to a diaphragm with a known volumetric displacement for each stroke of the piston. Volumetric efficiencies above 97% and overall mechanical efficiencies around 90% are common for these pumps.

Piston pumps manufactured for reliable, heavy-duty service at pressures of 35 MPa or more are quite common. However, when the pressures of operation exceed 105 MPa, the piston velocities have to be considerably reduced to get satisfactory packing life and to decrease the number of pressure reversals and cyclic stresses that are involved in fatigue failures. Consequently, as a result of this piston velocity reduction the piston sizes that will be required and the subsequent loads on them will make the more common crank-driven piston pumps impractical.

In an operation which involves continuous full flow operation at pressures of 10 MPa or more, the energy savings that can result from the relatively high overall efficiency of a piston pump can be considerable. The dairy industry takes advantage of the high efficiency of piston pumps when using them with two-stage valves attached as homogenizers in the pasteurization of milk. Also, piston pumps do not need priming, and because of the relatively slow speed of operation the parts that are subjected to the most wear, like gaskets, valves, pistons, etc., are simple and easy to maintain.

The two strokes that comprise a complete cycle of the piston pump are demonstrated in figure 4.2. Each cylinder of this type pump requires the operation of two valves; one for the suction stroke and one for the discharge stroke with only the discharge stroke being productive as far as product flow through the system is concerned. Due to this lack of productive flow during the suction stroke of an individual cylinder, multiplexing or coupling of two or more liquid pumping cylinders is the general rule in piston pump design. The most inexpensive way to accomplish this coupling is to use a double acting piston. This results in a pump which discharges for the full 360 degree revolution of

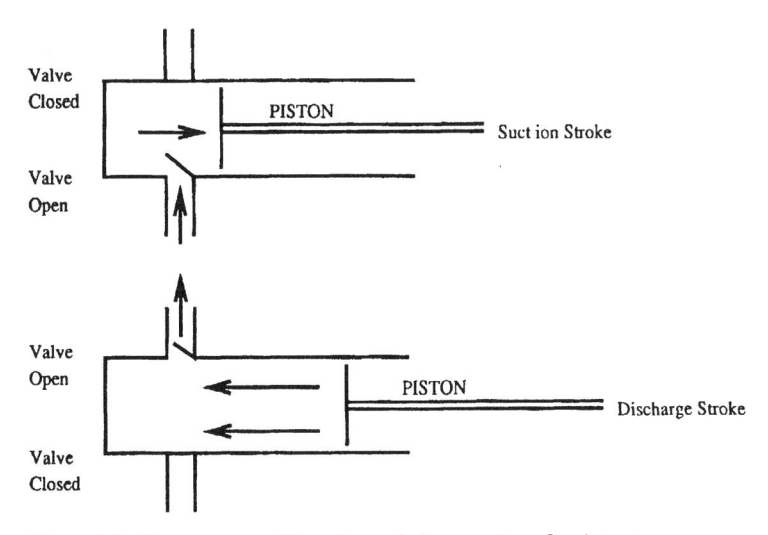

Figure 4.2. Plunger pump. The valve and piston action of a piston-type pump or both the suction and discharge strokes.

Principles of Process Engineering

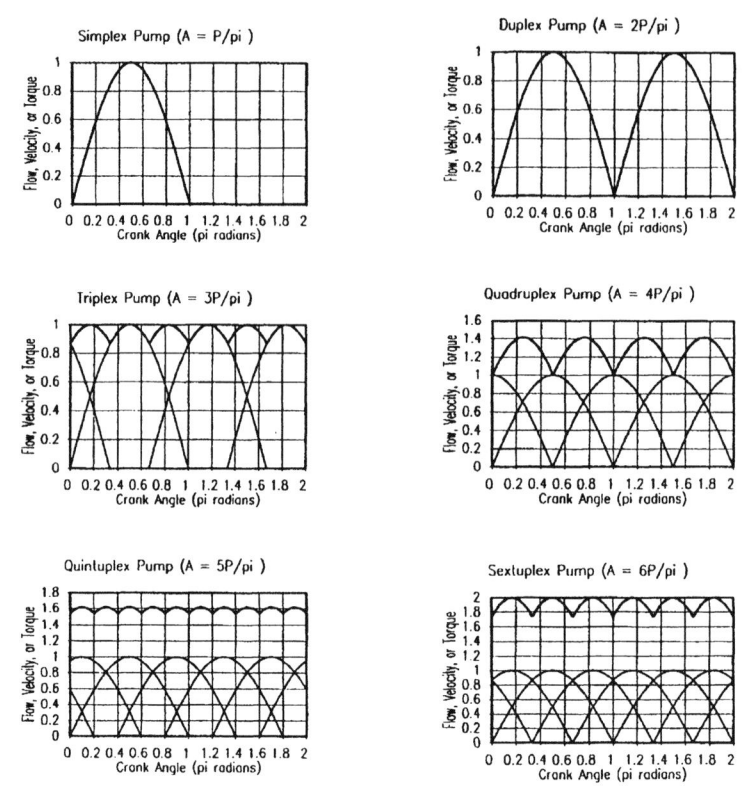

Figure 4.3. Flow-pressure-torque curves on multiplex pumps (A = average for multiplex pump, P = peak for single cylinder).

the crankshaft. The major limitation of this configuration is that duplexing is all that can be accomplished with it; therefore, multiplexing of single acting pistons is the most common configuration for piston pumps. Duplexing individual cylinders in this manner, so that one piston is 180 degrees out of phase with the other, allows the doubling of efficiency and flow rate without requiring an increase in torque from the crankshaft, since the available power from the crankshaft is used for productive forward flow in the system 100% of the time rather than 50% of the time. With duplexing there is still considerable fluctuation in flow rate and pressure as indicated in figure 4.3. Thus, the addition of more individual cylinders working in concert is beneficial to the stability of flow obtained from a piston pump. It is important to note that if stability of flow is of prime importance, the total number of individual cylinders in the pump should be an odd number. However, the addition of an odd cylinder to a piston pump entails an increase in the torque required from the crankshaft, as opposed to the lack of an increase in crankshaft torque for the duplex arrangement. So, the odd cylinder stabilizes flow but requires additional input energy. This trade off situation requires consideration when one is deciding which piston pump is appropriate for a particular application.

Pumps

4.3 Rotary Pumps

Rotary pumps are inexpensive and simple to construct. If constructed and maintained with very close tolerances, the volumetric efficiency is high (e.g., 95%) and high pressures can be produced. Mechanical efficiencies may be 90% or more under the best conditions.

There are several types of rotary pumps. Each one is used for a different need or purpose. These types include gear, lobe, screw, and vane.

The gear pump is probably the most common of the rotary pumps. It works with either external or internal gears as shown in figures 4.4 and 4.5. One gear is driven while

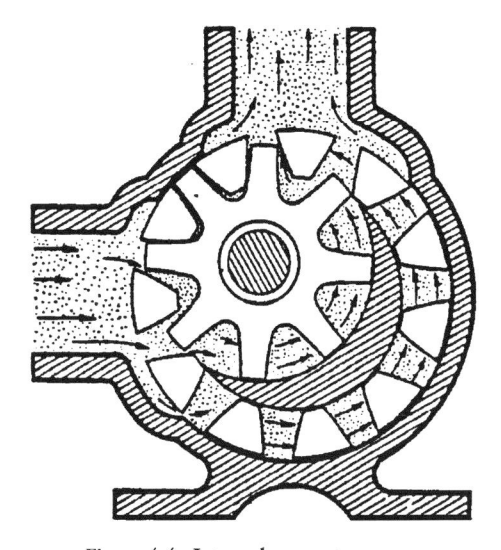

Figure 4.4. Internal-gear rotary pump.

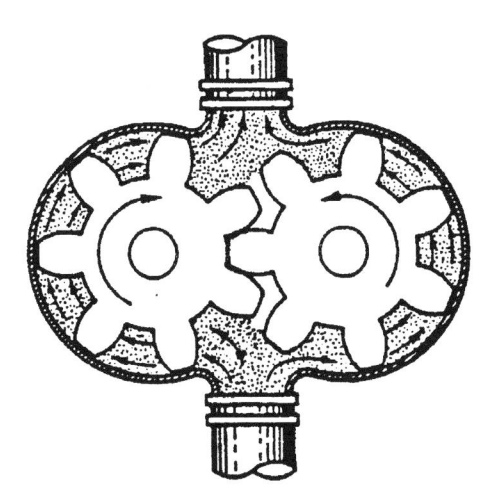

Figure 4.5. External-gear rotary pump.

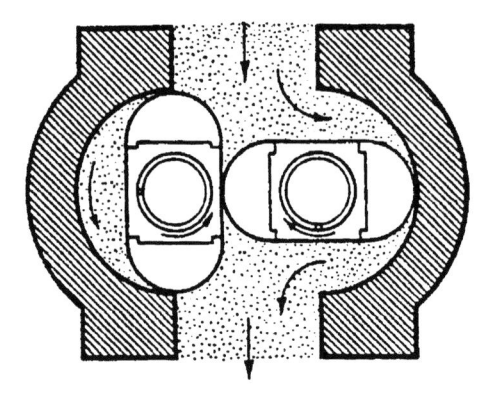

Figure 4.6. Lobe pump, used for both gases and liquids.

the other rotates freely. A partial vacuum is created by the unmeshing of the gears near the inflow of the fluid. This vacuum draws the fluid into the pump. As the gears mesh together, the fluid is pushed out the other side. The gears need constant lubrication, therefore viscous substances are better for gear pumps. They are especially suited for viscous substances such as ice-cream mix, molasses, and oils. Only liquid or liquid gas mixtures can be used since the gear pump is subject to clogging. Though classified as displacement pumps which implies a pulsating flow, gear pumps provide relatively constant outputs.

Figure 4.6 shows a lobe pump. This pump type uses two, three, or four lobes in place of gears. Otherwise, the principles behind the operation of the lobe pump are basically

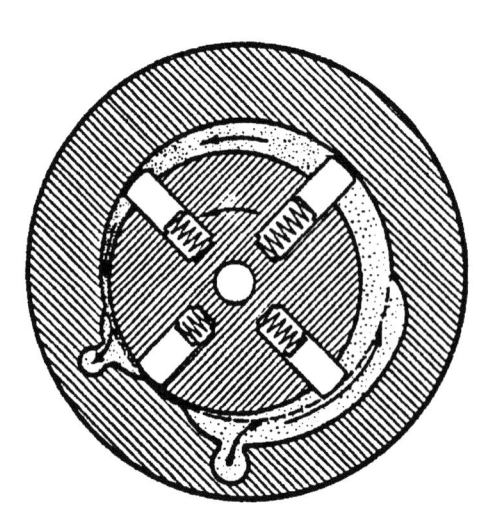

Figure 4.7. Vane pump, used for both gases and liquids

the same as for the gear pump. The two rotors on the lobe pump are driven—none of the lobes rotate freely. The lobe pump produces more of a pulsating output than the gear pump. It can pump both liquids and gases.

The screw pump consists of a helical screw which rotates within a fixed casing and transfers fluid from the inlet to the outlet. Since the screw pump does not operate with any kind of rotating gears, it can pump liquids with solid particles intermixed. The screw pump also produces a very steady output.

Figure 4.7 shows a vane pump. This pump operates with sliding vanes and hinged parts to maintain the seal that is needed. The vane is used in evacuating processes where a vacuuming effect is needed, e.g., a milking machine. The vane pump moves vapor and liquids only—no solid mixtures. It also provides a relatively steady output.

Dynamic Pumps

Dynamic pumps may be further subdivided into centrifugal or special effect pumps. These pumps are usually used in situations where high volumetric flow rates are required with relatively low heads.

4.4 Centrifugal Pumps

The centrifugal pump is widely used for pumping fluid materials such as water, milk, lubricants, chemical solutions, and materials being processed. Its popularity is due to relative simplicity, mechanical efficiencies as high as 90% under favorable conditions, and ability to handle fluids containing solids in suspension. Centrifugal pumps can be designed for high-pressure operation where necessary. Because of simplicity and ease of disassembling, which facilitate cleaning, washing, and sterilizing, they are satisfactory for food products.

Centrifugal pumps consist of two basic parts: (1) an impeller, which forces the liquid into a rotary motion, and (2) the pump casing, which directs the liquid to the impeller and leads it away under a higher pressure (fig. 4.8). The designation "centrifugal pump" is used for a family of pumps having various inlet conditions and various fluid flow directions at the exit of the impeller. Centrifugal pumps which have the inlet fluid directed to the center of the impeller may further be divided into radial, mixed, and axial flow

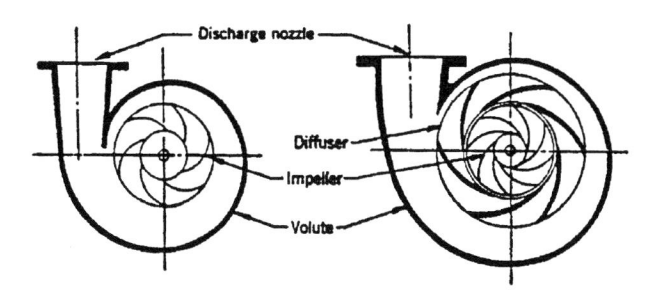

Figure 4.8. Volute and diffusion casing pumps.

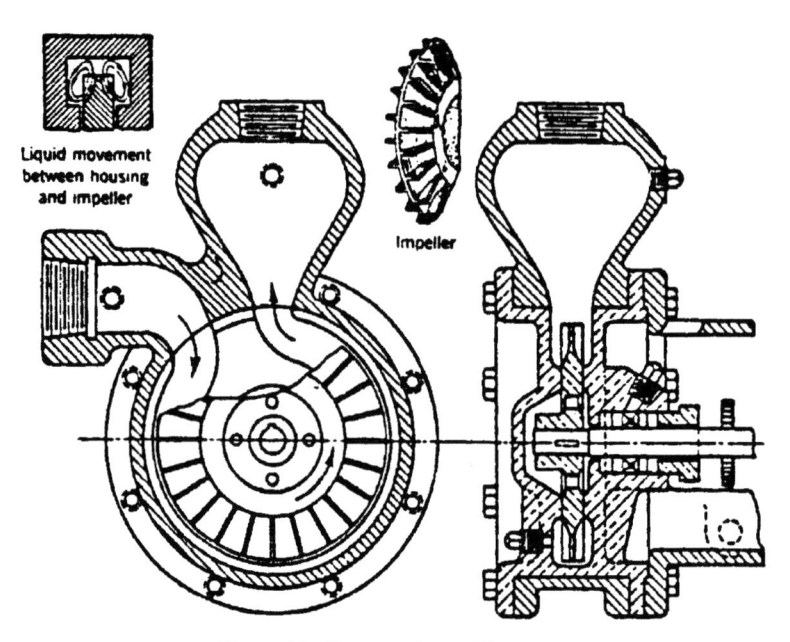

Figure 4.9. Regenerative turbine pump.

pumps depending on the impeller design. In the radial flow centrifugal pump, the fluid enters the pump parallel to the axis of rotation of the impeller and is then directed outward from the impeller in a radial direction. In an axial flow centrifugal pump (sometimes called propeller pumps), the fluid enters parallel to the axis of rotation of the impeller and also exits parallel to the axis of rotation. In mixed flow centrifugal pumps the exiting fluid has both a radial and an axial component. In a fourth type of centrifugal pump the fluid enters the impeller peripherally rather than at its center. The regenerative turbine pump of figure 4.9 is an example of this type pump.

The basic principles of design of centrifugal pumps also apply to fans, blowers, and compressors and are important to an understanding of performance and proper selection. Considering figure 4.10 we assume that (1) the vane thickness is negligible, (2) friction losses are negligible, and (3) the peripheral velocity at the inlet is zero.

The velocities of a fluid moving along a vane of the impeller or between vanes can be described vectorially as noted in figure 4.10. The vane velocities, v, are related to the vane configuration by:

$$v_2/v_1 = r_2/r_1 \tag{4.4}$$

where

v_2 = peripheral vane velocity at the outer radius of the impeller (m/s)
v_1 = peripheral vane velocity at the inner radius of the impeller (m/s)
r_2 = outer radius of the impeller (m)
r_1 = inner radius of the impeller (m)

Pumps

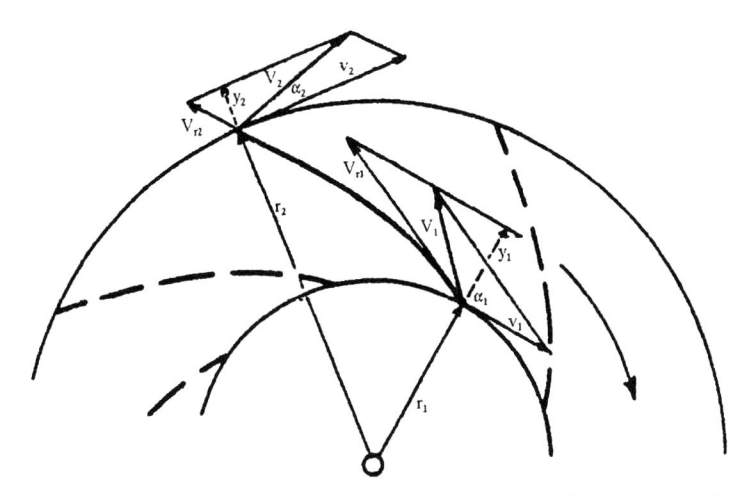

Figure 4.10. Velocity vectors for a centrifugal pump impeller or wheel with impeller widths of W_1 and W_2 at the inner and outer peripheries.

The vane velocities are fixed for any design and wheel speed. The velocities of the fluid relative to the vane, V_{r1} and V_{r2}, are dependent upon the mass fluid rate and the vane configuration. The absolute velocities, V_1 and V_2, are graphic resolutions of v_1, v_2, V_{r1}, and V_{r2}. The fluid velocities V_{r1}, V_{r2}, V_1, and V_2 are functions of the fluid volumetric flow rate through the radial components of velocity y_1 and y_2. For an incompressible fluid the volumetric flow rate at the inner radius is equal to that at the outer radius of the impeller as follows:

$$Q = 2\pi r_1 w_1 y_1 = 2\pi r_2 w_2 y_2 \tag{4.5}$$

where

Q = volumetric flow rate (m³/s)
w_1 = vane width at r_1 (m)
w_2 = vane width at r_2 (m)
y_1 = radial fluid velocity at r_1 (m/s)
y_2 = radial fluid velocity at r_2 (m/s)

An expression for the theoretical head (work done on fluid per unit weight of fluid) of a centrifugal pump may be obtained by applying the principle of angular momentum to the mass of fluid going through the impeller. This principle states that the rate of change of angular momentum of a body with respect to the axis of rotation is equal to the torque of the resultant force on the body with respect to the same axis.

The angular momentum of a body is given by:

$$J = I\omega \tag{4.6}$$

where

J = angular momentum (kg m²/s)

I = moment of inertia with respect to axis of rotation (kg m^2)
ω = angular velocity (s^{-1})

The moment of inertia of an element of fluid about its axis of rotation is given by:

$$I = dmr^2 \tag{4.7}$$

where

dm = mass of fluid element (kg)
r = distance from axis of rotation (m)

The angular velocity of the fluid element is given by:

$$\omega = V \cos\alpha / r \tag{4.8}$$

where

V = fluid velocity (m/s)
α = angle between fluid velocity vector and vane velocity vector (radians)

The torque exerted by the vane may then be computed as the difference in angular momentum of a differential mass of fluid entering the impeller at r_1 and an equal mass of fluid exiting at r_2:

$$T = (J_2 - J_1)/dt \tag{4.9}$$

where

T = torque exerted by vanes (N m)
J_2 = angular momentum of fluid element at r_2 (kg m^2/s)
J_1 = angular momentum of fluid element at r_1 (kg m^2/s)
dt = differential time during which fluid element passes (s)

Substitution of equations 4.6, 4.7, and 4.8 yields:

$$T = \dot{m}(r_2 V_2 \cos\alpha_2 - r_1 V_1 \cos\alpha_1) \tag{4.10}$$

or

$$T = Q\rho(r_2 V_2 \cos\alpha_2 - r_1 V_1 \cos\alpha_1) \tag{4.11}$$

The input power is the product of the torque and the angular velocity of the impeller:

$$P_i = T\omega \tag{4.12}$$

Then,

$$P_i = Q\rho\omega(r_2 V_2 \cos\alpha_2 - r_1 V_1 \cos\alpha_1) \tag{4.13}$$

or substituting v for ωr,

$$P_i = Q\rho(v_2 V_2 \cos\alpha_2 - v_1 V_1 \cos\alpha_1) \tag{4.14}$$

Pumps

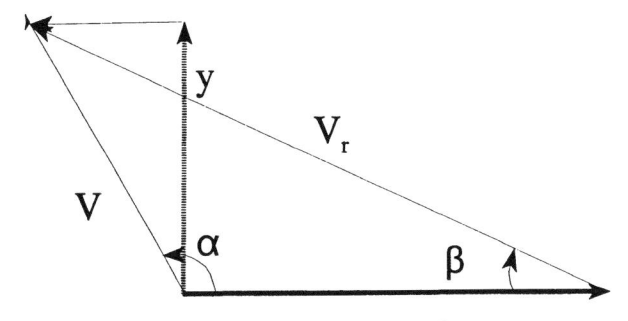

Figure 4.11. Velocity triangle.

If the mechanism is an idealized one, the power input is equivalent to the power output given by equation 4.1. Then,

$$W Q \rho g = Q\rho(v_2 V_2 \cos\alpha_2 - v_1 V_1 \cos\alpha_1) \tag{4.15}$$

and

$$W = (v_2 V_2 \cos\alpha_2 - v_1 V_1 \cos\alpha_1)/g \tag{4.16}$$

This equation which assumes no hydraulic losses relates the impeller configuration and fluid flow through it to the theoretical total head and is known as *Euler's* equation. If the fluid enters the impeller with no tangential component, if α_1 is 90°, Euler's equation reduces to:

$$W = v_2 V_2 \cos\alpha_2/g \tag{4.17}$$

Geometric consideration of the velocity triangles in the impeller of the centrifugal pump as illustrated in figure 4.11 result in the following equalities:

$$V_r = Q/(2\pi r w \sin\beta) \tag{4.18}$$

and

$$V \cos\alpha = v - V_r \cos\beta \tag{4.19}$$

where β = vane pitch angle of the impeller (radians).

Substitution of equations 4.18 and 4.19 and the relationships $v = \omega r$ and $\omega = 2\pi N$ into equation 4.16 yields:

$$W = 4\pi^2 N^2 (r_2^2 - r_1^2)/g + (NQ/g)(1/(w_1 \tan\beta_1) - 1/(w_2 \tan\beta_2)) \tag{4.20}$$

where N = rotational speed of impeller (rev/s)

Equation 4.20 relates the total head, W, produced by the pump to the dimensions of the pump impeller, the rotational speed, and the volumetric flow rate. The relationship is for an ideal situation in which the flow through the impeller is frictionless and the fluid velocity vectors can be determined from the vane angles. In practice the flow is not frictionless and fluid velocities cannot be accurately predicted from vane angles.

However, equation 4.20 and the ideal fluid assumptions used in its development allow us to make some very useful inferences concerning the operation of a real pump. Note the following significant characteristics:

1. The total theoretical head at complete shut-off is given by the first term on the right side of equation 4.20. Thus, the total head at shut-off is proportional to the square of the rotational speed and to the difference in the squares of the inner and outer radii of the impeller vanes.
2. For a given vane pitch angle at the inner radius of the impeller, the free delivery flow rate (i.e., volumetric flow rate at zero total head) increases with the vane pitch angle at the outer radius of the impeller. Note that from equation 4.20 it is possible to choose values of the pitch angles and vane widths such that the total head would increase rather than decrease with an increase in volumetric flow rate. However, this is not possible to achieve in a real pump due to relative circulation within the impeller channel and due to the fact that the tangential velocity of the fluid cannot exceed the tangential velocity of the impeller vane.
3. The total head delivered by the pump is a function of the volumetric flow rate as opposed to the mass flow rate. This means that approximately the same total head should be provided to different fluids. Differences in fluid viscosity and thus frictional effects will prevent this relationship from holding exactly.
4. For a specific pump with speed varying:
    a. *The volumetric capacity varies directly as the speed.* This is true since y is proportional to v in figure 4.10 and v is proportional to speed.
    $$Q_1/Q_2 = N_1/N_2 \qquad (4.21)$$
    b. *The total head varies as the square of speed.* This can be seen from equation 4.20 since the square of speed occurs in the first term on the right-hand side of the equation and the product of speed and volumetric capacity (which varies directly with speed) occurs in the second term of the equation.
    $$W_1/W_2 = N_1^2/N_2^2 \qquad (4.22)$$
    c. *The power required varies as the cube of the speed.* Since power is a product of the total head and weight flow rate, it follows from (a) and (b) that,
    $$P_{o1}/P_{o2} = N_1^3/N_2^3 \qquad (4.23)$$
5. For geometrically similar pumps with speed constant and diameter varying:
    a. *The volumetric capacity varies as the cube of the diameter.* The capacity is a direct function of periphery speed and periphery area. Since the periphery speed varies directly as the diameter and the area as the square of the diameter, it follows that,
    $$Q_1/Q_2 = D_1^3/D_2^3 \qquad (4.24)$$
    b. *The total head varies as the square of the diameter.* The first term of the right side of equation 4.20 varies with the square of the diameter while the second term varies with the flow rate divided by the vane width. Since the flow rate varies

Pumps

with the cube of diameter and the vane width is proportional to the diameter for geometrically similar pumps, then both terms vary with the square of diameter. Thus,

$$W_1/W_2 = D_1^2/D_2^2 \qquad (4.25)$$

c. *The power varies as the fifth power of the diameter.* Since power is the product of total head and weight flow rate, (a) and (b) combine to give,

$$P_{o1}/P_{o2} = D_1^5/D_2^5 \qquad (4.26)$$

Example 4.2. Determine the theoretical total head at complete shut-off developed by a pump having an inner impeller radius of 7 cm and an outer impeller radius of 15 cm if the pump operates at 1 760 rpm.

Solution
$Q = 0$ at complete shut-off.
Therefore:

$$W = 4\pi^2 N^2 (r_2^2 - r_1^2)/g$$
$$W = 4\pi^2 (1\ 760\ \text{rpm})^2 (1\ \text{min}/60\ \text{s})^2 ((.15\ \text{m})^2 - (.07\ \text{m})^2)/(9.81\ \text{m/s}^2)$$
$$W = 61\ \text{m}$$

The total head produced by a pump is the sum of a velocity head and a pressure head. Since pumps are used to move a quantity of fluid against a resistance which may be attributable to elevation or friction of conduits, nozzles, and other fittings; it is usually desirable that most of the velocity head produced by the pump be converted into a pressure head. This conversion is attempted by gradual reduction of the velocity in one of two ways.

Diffuser or guide vanes may conduct the fluid away from the impeller and gradually lower its velocity by increasing the conduit area. The reduction in velocity effects an increase in pressure head as a result of the conservation of energy as described by the Bernoulli equation. The vanes are so bent that the water is turned gradually and is finally discharged into a manifold.

A second method, which is simpler and less expensive, is the volute manifold or casing outlined by the dotted line in figure 4.10. The casing is so designed that the average velocity is constant at all cross sections and is approximately $V/2$ in figure 4.10. When properly designed, each fluid element is gradually turned toward the discharge outlet so that turbulence losses are at a minimum.

A further reduction in velocity effect may be obtained by gradually expanding the diameter of the discharge pipe.

4.5 Jet Pumps

The jet pump is a special effect pump shown elementarily in figure 4.12 which operates on the velocity energy of a jet of fluid. Water (or other fluid, either compressible or

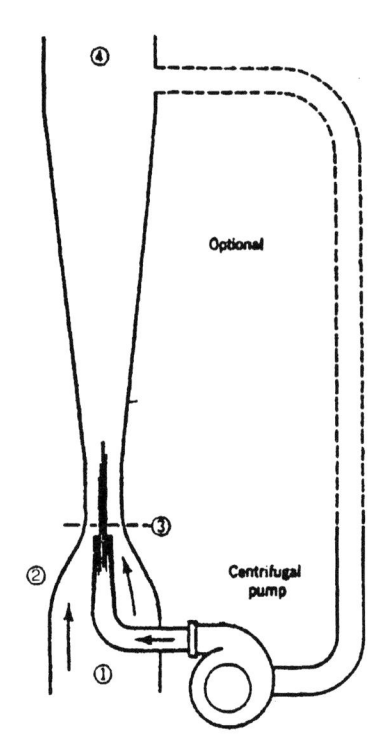

Figure 4.12. Schematic elements of a jet pump.

incompressible) is forced through a jet or nozzle of such dimensions that all or nearly all the energy involved is converted into velocity energy. This energy, which is directional, is applied to the fluid to be moved. The jet is produced by recirculating a portion of the liquid or gas in those cases where the material will not damage the pump. Jet pumps are frequently used for pumping sumps or processing residues that contain solid matter or chemically active materials that would not pass through a mechanical pump satisfactorily. For example, if the material is a gas, water or air is provided from an external source to supply the jet energy. The diluted mixture is discarded.

Jet pump theory (Stepanoff, 1957) is rational and straightforward but too involved to include in this text. However, the basic consideration is the conservation of energy, which may be expressed as:

$$H_1 V_1 A_1 \gamma_1 + H_2 V_2 A_2 \gamma_2 = H_4 V_4 A_4 \gamma_4 + P_f \qquad (4.27)$$

where

H = total fluid hydraulic head (m)
V = fluid velocity (m/s)
A = cross-sectional area (m^2)
γ = specific weight of fluid (N/m^3)
P_f = power loss due to friction and turbulence in the mixing process (W)

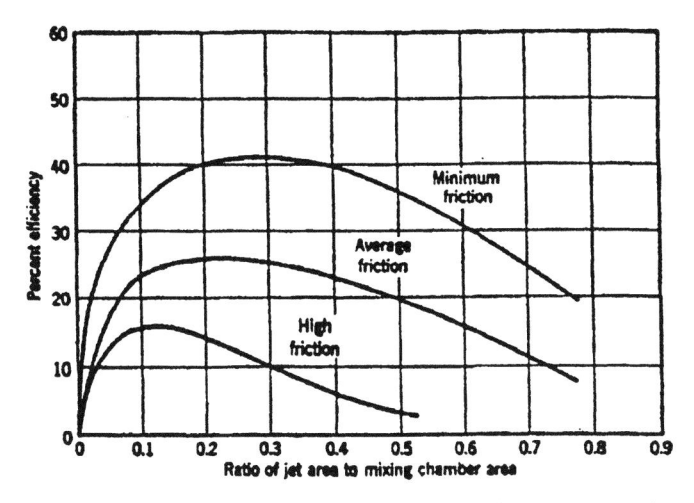

Figure 4.13. Jet-pump efficiency related to the relative jet area and friction in the jet unit.

The total hydraulic head of the fluid streams can be made up of any combination of elevation, pressure, and velocity heads. However, H_2, the head of the recirculated or external fluid, must be predominately velocity head in order to transfer the energy to the fluid entering at 1 in figure 4.12. The efficiency may be expressed as:

$$e_j = (H_4 V_4 A_4 \gamma_4 - H_1 V_1 A_1 \gamma_1)/(H_2 V_2 A_2 \gamma_2) \qquad (4.28)$$

where e_j = efficiency of the jet portion of pump (decimal).

The efficiency is closely related to the ratio of the nozzle area to the mixing-cylinder area and the friction in the system. The relationship of these factors is shown in figure 4.13. These are theoretical curves that have been substantiated by observation.

In spite of its low efficiency, the simplicity of the jet pump, its freedom from moving parts, its ability to pump materials of sludge consistency, and its low initial cost fit it for use in situations where other pumping devices would be impractical.

4.6 Air Lift Pumps

The air lift pump is a special effect pump having the arrangement illustrated in figure 4.14. Compressed air is released through a "foot-piece", or air diffuser, at the bottom of the vertical pipe called the eductor pipe. A mixture of air bubbles and liquid to be lifted is created which has a density less than that of the liquid alone. This lower density mixture rises in the pipe and is discharged at a point above the level of the liquid. The weight of the column of air and liquid, $h_f + h_s$, is equal to the weight of the column of liquid, h_s. For maximum efficiency of the airlift pump the air bubbles should be small. The small bubbles exert no appreciable lifting force on the liquid.

Air lifts are the most simple and foolproof type of pump and, in operation, give the least trouble because there are no remote or submerged moving parts. They can be

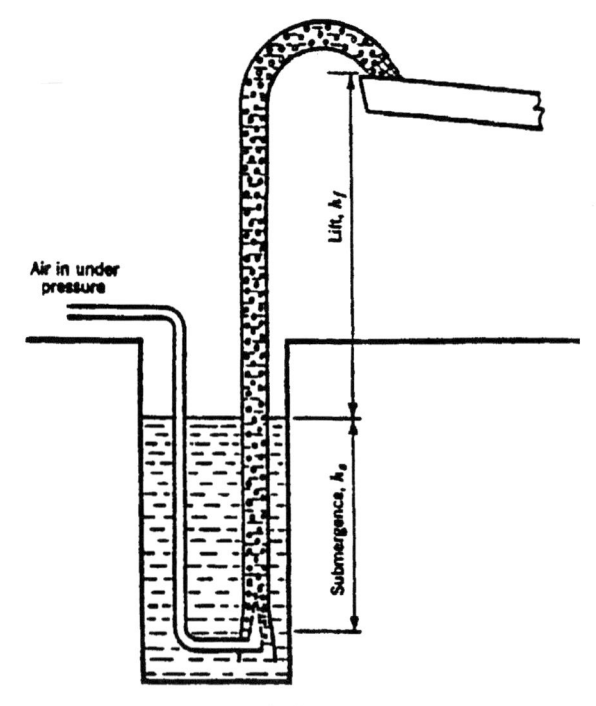

Figure 4.14. The air lift.

operated successfully in holes of any practicable size, and they can be used in crooked holes not suited to any other type of pump.

The principal disadvantages of air lifts are the necessity for making a deeper well than is required for other pump types, the intermittent nature of the flow, and the relatively low efficiencies obtained. Babbitt et al. (1967) give approximate efficiencies to be expected for air lift pumps while pumping water. Table 4-1 summarizes the expected efficiencies

Table 4-1. Effect of submergence on efficiencies of air lift at Hattiesburg, Miss.

Ratio, h_s/h_f	Submergence Ratio, $h_s/(h_s + h_f)$	Percent Efficiency
8.70	0.896	26.5
5.46	0.845	31.0
3.86	0.795	35.0
2.91	0.745	36.6
2.25	0.693	37.7
1.86	0.650	36.8
1.45	0.592	34.5
1.19	0.544	31.0
0.96	0.490	26.5

Table 4-2. Some recommended submergence percentages for air lifts

Lift, m	Submergence Percentage
Up to 15	70–66
15–30	66–55
30–60	55–50
60–90	50–43
90–120	43–40
120–150	40–33

as a function of the ratio, h_s/h_f, and the submergence ratio, $h_s/(h_s + h_f)$. Table 4-2 gives submergence percentages recommended for air lifts.

The volume flow rate of air which is required to give the desired liquid flow rate may be estimated from the following empirical equation:

$$Q_a = Q_w(h_f + h_v)/(22.9E \log((h_s + 10.4)/10.4) \tag{4.29}$$

where

Q_a = volumetric flow rate of air (m³/s)
Q_w = volumetric flow rate of liquid (m³/s)
h_f = lift required (m)
h_v = velocity head of liquid at discharge (m)
E = pump efficiency estimated from table 4-1 (decimal)
h_s = submergence (m)

The air pressure which must be provided at the submerged end of the air pipe or foot piece is equal to the pressure exerted by a column of the fluid equal to h_s.

Some excess air capacity should be provided, because if the free-liquid surface in the well should fall more than anticipated after prolonged pumping, more air will be required to maintain the discharge.

Performance

Performance characteristics of pumps are specified in three different ways. All three should be considered in the selection of a pump for a specific job. They are: (1) performance curves, (2) specific speed, and (3) net positive suction head.

4.7 Performance Curves

The American Society of Mechanical Engineers (1965) and the Hydraulic Institute (1965) have developed standard methods for testing centrifugal and rotary pumps. These test code series should be studied if formal tests are to be made or performance data subjected to a critical analysis.

Tests are made by operating the pump at a constant speed and varying the capacity by throttling the outlet. The total head, the velocity head, and perhaps the static head are plotted against the volumetric discharge rate. The power input and efficiency are also

106   Principles of Process Engineering

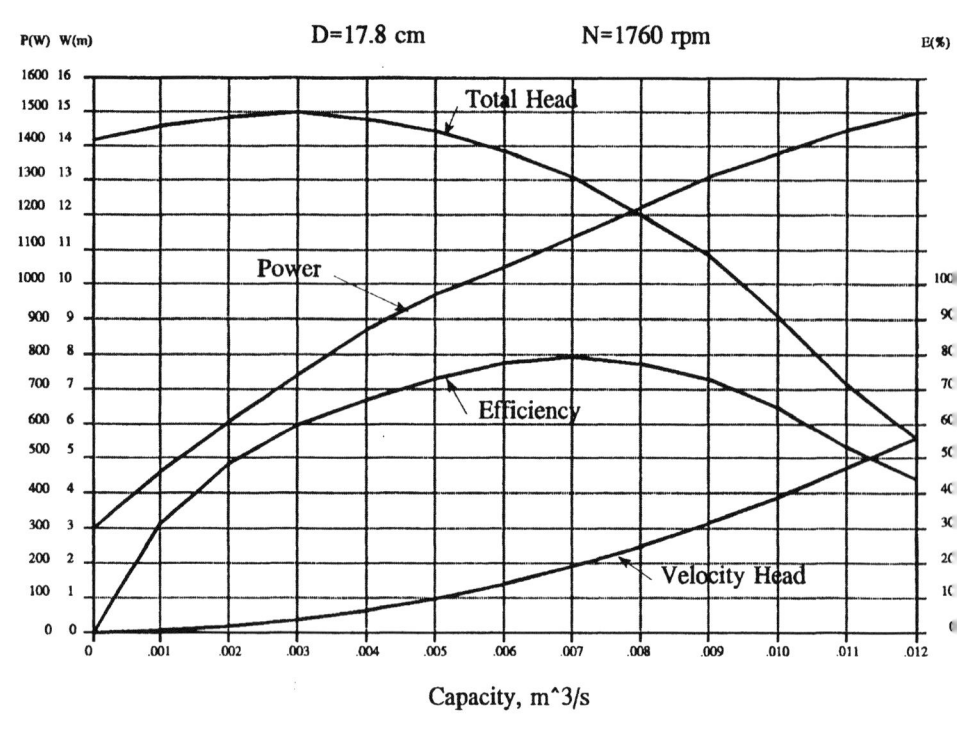

Figure 4.15. Representative centrifugal-pump test plot (D = 17.8 cm and N = 1760 rpm).

determined and plotted against the discharge rate. The power input to the pump shaft is determined by any of the accepted procedures. Calibrated electric motors are included. The efficiency is calculated as discussed in section 4.1.

A representative pump test plot for a centrifugal pump is shown in figure 4.15. These curves represent the performance of the pump connected directly to an electric motor that operates at 1760 rpm. A complete performance study would include a series of tests made at different pump speeds. Note that some power is required at the no-discharge position.

Performance of centrifugal pumps while pumping water is used as a standard for comparison of pumps. As previously discussed in section 4.4, the total head versus capacity curves should not vary greatly for different fluids.

Centrifugal pump performance at a speed other than that at which tests were conducted or performance curves for geometrically similar pumps may be estimated from the affinity laws which may be derived from the relationships of section 4.4. The centrifugal pump affinity laws may be stated as:

$$Q_1/Q_2 = (N_1/N_2)(D_1/D_2)^3 \quad (4.30)$$

$$W_1/W_2 = (N_1/N_2)^2(D_1/D_2)^2 \quad (4.31)$$

$$P_{o1}/P_{o2} = (N_1/N_2)^3(D_1/D_2)^5(\rho_1/\rho_2) \quad (4.32)$$

Pumps

[Figure 4.16: Graph titled "Representative Regenerative Turbine-Pump" showing P(W), H(m) on left axes and E(%) on right axis vs Capacity in m³/s. Curves shown for POWER, TOTAL HEAD, and EFFICIENCY.]

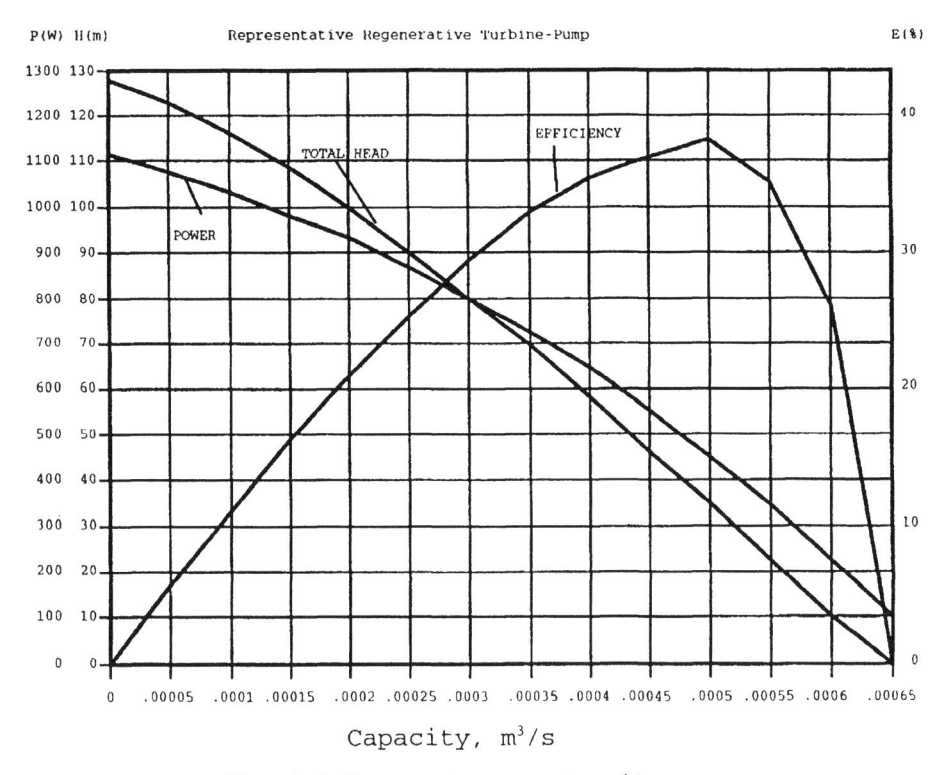

Figure 4.16. Representative regenerative turbine-pump.

Figures 4.16 and 4.17 are representative performance curves for a regenerative turbine pump and a gear pump, respectively.

Example 4.3. A pump, operating at 1760 rpm, delivering 0.008 m³/s at 12 m of head, and requiring 1180 W, is speeded up to 2100 rpm. What are the corresponding operating parameters at the new speed?

Solution

$$Q_1/Q_2 = (N_1/N_2)(D_1/D_2)^3$$
$$Q_2 = Q_1(N_2/N_1)$$
$$= (0.008 \text{ m}^3/\text{s})(2100 \text{ rpm}/1760 \text{ rpm}) = 0.00955 \text{ m}^3/\text{s}$$
$$W_1/W_2 = (N_1/N_2)^2(D_1/D_2)^2$$
$$W_2 = W_1(N_2/N_1)^2$$
$$= (12 \text{ m})(2100 \text{ rpm}/1760 \text{ rpm})^2 = 17.1 \text{ m}$$

Since the efficiency remains the same at the corresponding points on the performance curves,

$$P_{i1}/P_{i2} = (N_1/N_2)^3(D_1/D_2)^5(\rho_1/\rho_2)$$

P(W) H(m)   **E(%)**

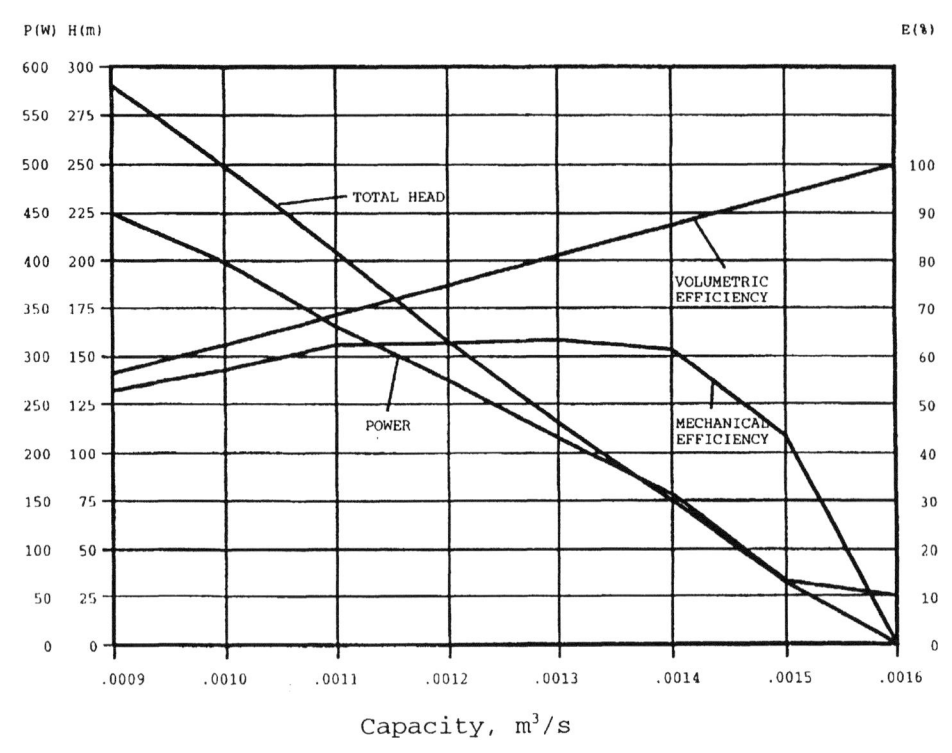

Figure 4.17. Performance of a gear pump. Fluid density is 928 kg/m³.

and

$$P_{i2} = P_{i1}(N_2/N_1)^3$$
$$P_{i2} = (1180 \text{ W})(2100 \text{ rpm}/1760 \text{ rpm})^3 = 2004 \text{ W}$$

Example 4.4. A pump with a 17.5 cm runner delivers 0.008 m³/s against a 12 m head and requires 1180 W. If the speed is maintained constant and the runner length is increased to 18.8 cm in a geometrically similar pump (i.e., all linear dimensions are increased proportionately), what are the new operating parameters?

Solution

$$Q_1/Q_2 = (N_1/N_2)(D_1/D_2)^3$$

or

$$Q_2 = Q_1(D_2/D_1)^3$$
$$Q_2 = (0.008 \text{ m}^3/\text{s})(18.8 \text{ cm}/17.5 \text{ cm})^3 = 0.00992 \text{ m}^3/\text{s}$$
$$W_1/W_2 = (D_2/D_1)^2$$

Pumps

or
$$W_2 = W_1(D_2/D_1)^2$$
$$W_2 = (12 \text{ m})(18.8 \text{ cm}/17.5 \text{ cm})^2 = 13.85 \text{ m}$$
$$P_{i1}/P_{i2} = (N_1/N_2)^3(D_1/D_2)^5(\rho_1/\rho_2).$$

or
$$P_{i2} = P_{i1}(D_2/D_1)^5$$
$$P_{i2} = (1180 \text{ W})(18.8 \text{ cm}/17.5 \text{ cm})^5 = 1688 \text{ W}$$

Example 4.5. A pump is to be selected geometrically similar to the pump of figure 4.15 to deliver 0.005 m³/s against a head of 19.8 m. It is desirable to operate it at a point on the performance curve comparable to the 0.010 m³/s point on the base curve, figure 4.15.

Solution
The base conditions are: flow rate $Q_1 = 0.010$ m³/s; total head delivered $W_1 = 9.1$ m; runner diameter $D_1 = 17.8$ cm; speed $N_1 = 1760$ rpm; power input $P_{i1} = 1380$ W; and efficiency $e_{m1} = 65\%$. Equations 4.30 and 4.31 must be solved simultaneously for the new operating speed and runner diameter. Equation 4.30 gives:

$$N_1/N_2 = (Q_1/Q_2)(D_2/D_1)^3$$

which may be substituted into equation 4.31 to obtain

$$W_1/W_2 = (Q_1/Q_2)^2(D_2/D_1)^4$$

or
$$D_2 = D_1(Q_2/Q_1)^{1/2}(W_1/W_2)^{1/4}$$

Then,
$$D_2 = (17.8 \text{ cm})((.005 \text{ m}^3/\text{s})/(.01 \text{ m}^3/\text{s}))^{1/2}(9.1 \text{ m}/19.8 \text{ m})^{1/4}$$
$$D_2 = 10.36 \text{ cm}$$
$$N_2 = N_1(Q_2/Q_1)(D_1/D_2)^3$$
$$= (1760 \text{ rpm})((.005 \text{ m}^3/\text{s})/(.01 \text{ m}^3/\text{s}))(17.8 \text{ cm}/10.36 \text{ cm})^3$$
$$= 4459 \text{ rpm}$$

The power required is:
$$P_{i2} = P_{i1}(N_2/N_1)^3(D_2/D_1)^5(\rho_2/\rho_1)$$
$$P_{i2} = (1380 \text{ W})(4463 \text{ rpm}/1760 \text{ rpm})^3(10.36 \text{ cm}/17.8 \text{ cm})^5$$
$$P_{i2} = 1502 \text{ W}$$

4.8 Specific Speed

It is possible that the shape of the performance curves for a centrifugal pump can be altered to some degree to suit the requisites of a particular application by a pump designer, but this flexibility is very limited. Essentially, the shape of the performance curves for a centrifugal pump is a function of designed volumetric capacity, head, and shaft speed. Hence, the standard production centrifugal pump has experimentally predetermined performance characteristics that are not usually subject to modification. Because of this reliance on experimental results to predict the actual performance of centrifugal pumps, it is very desirable to have the capacity to use the results of past performance tests on a given centrifugal pump as a predicate for anticipating the performance of other alternative designs. To achieve this objective a widely used characteristic number has been developed. This parameter is called the specific speed. The non-dimensional specific speed is given by the following relationship:

$$N_s = N Q^{1/2}/(gW)^{3/4} \tag{4.33}$$

In this equation the rotative speed, volumetric flow rate, and head are all at the point of maximum efficiency. Specific speed has two useful characteristics for the pump designer: (1) geometrically similar pumps irrespective of their volumetric capacity will have identical specific speeds (however, all pumps with identical specific speeds will not necessarily be geometrically similar), and (2) within measurement error, the geometry and performance of a specific pump can be predicted with respect to specific speed and volumetric flow rate. Figure 4.18 shows how pump efficiency and impeller characteristics change over the practical range of specific speeds.

Example 4.6. Determine the specific speed for the pump of figure 4.15 at the point of maximum efficiency.

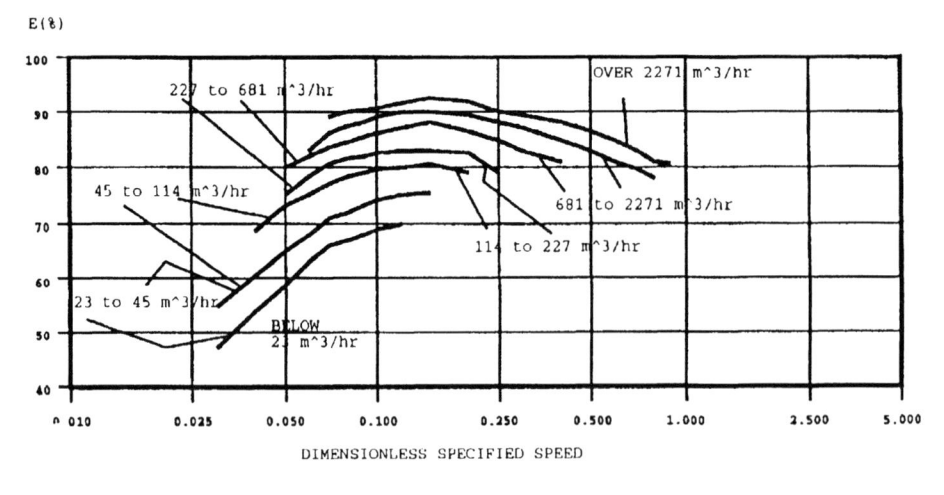

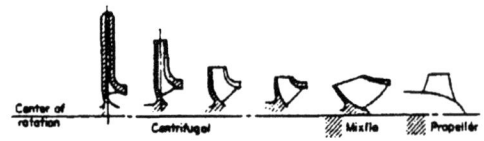

Figure 4.18. Approximate relative impeller shapes and efficiency variations with specific speed.

Solution
At maximum efficiency the pump has the following operating characteristics: $W = 8$ m and $Q = 0.0069$ m³/s.
Then,

$$N_s = NQ^{1/2}/(gW)^{3/4}$$
$$N_s = (1760 \text{ rpm})(1 \text{ min}/60 \text{ s})(0.0069 \text{ m}^3/\text{s})^{1/2}/((9.81 \text{ m/s}^2)(8 \text{ m}))^{3/4}$$
$$N_s = 0.092$$

This specific speed is typical of a primarily radial flow pump as indicated by figure 4.18.

4.9 Net Positive Suction Head

If the absolute pressure of the liquid at the suction nozzle approaches the vapor pressure of the liquid, vapor pockets will form in the impeller passages. This condition will interfere with pump performance and the collapse of the vapor pockets will be noisy and possibly destructive to the pump. This is known as cavitation, and the pump is said to be cavitating.

The amount of pressure in excess of the vapor pressure required to prevent the formation of vapor pockets is known as the *net positive suction head required* (NPSHR). The NPSHR is determined experimentally for individual pumps. It increases very rapidly at high flows as shown in the performance curves of figure 4.19.

When a pump is operating with hot liquids, it is particularly important to consider the NPSHR. The vapor pressure increases with water temperature and reduces the *net positive suction head available* (NPSHA). While the NPSHR is a characteristic of the pump, each installation has its own particular NPSHA, which is the total useful energy above the vapor pressure of the liquid available to the pump at the suction connection. The NPSHA may be calculated by the following relationship:

$$\text{NPSHA} = (p_s - p_{\text{sat}})/\gamma \qquad (4.34)$$

where

NPSHA = net positive suction pressure available (m)
p_s = pressure at suction connection of pump (Pa)
p_{sat} = saturation vapor pressure of fluid (Pa)
γ = specific weight of fluid (N/m³)

The pressure at the suction connection of the pump can be predicted for a particular system by using the Bernoulli theorem (eq. 2.13) discussed in Chapter 2.

Example 4.7. A pump is moving water at 30 C from an open reservoir. The inlet to the pump is located 3 m below the surface of the source reservoir. The galvanized type A entrance (or suction side) pipe is 10 m in length, contains two Type C elbows, and has an inside diameter of 5 cm. The galvanized outlet pipe is 20 m in length, contains two elbows, has an inside diameter of 7.5 cm, and discharges at atmospheric pressure at an

112	Principles of Process Engineering

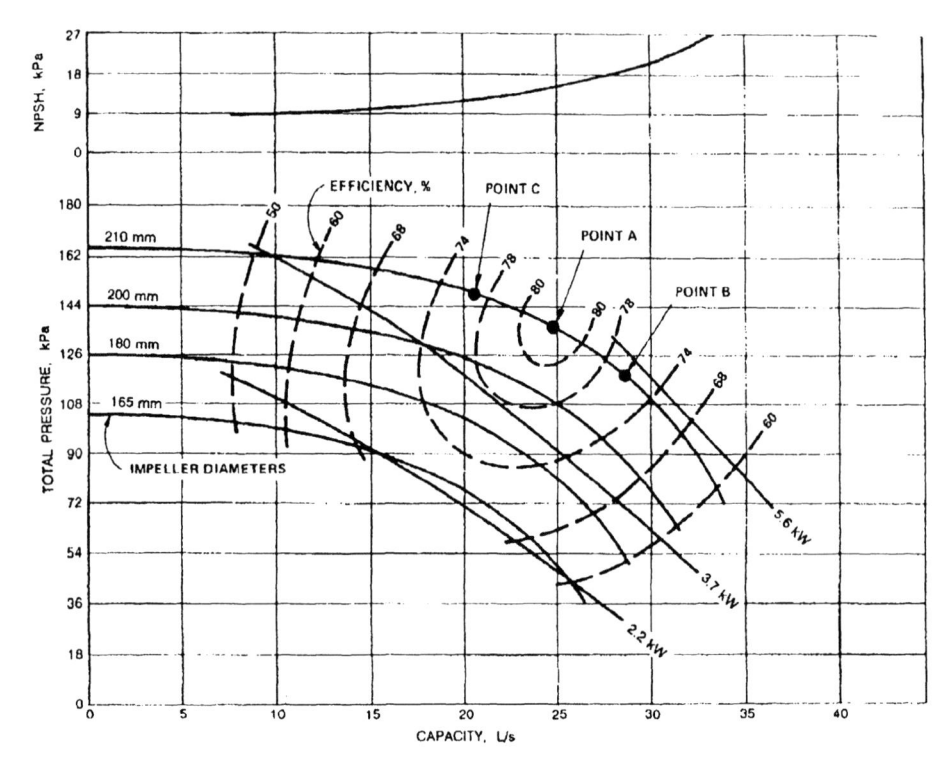

Figure 4.19. Pump performance curves.

elevation 10 m above the surface of the source reservoir. If the flow rate is 0.01 m³/s, determine the net positive suction pressure at the pump inlet.

Solution
If Bernoulli's equation is written between points at the surface of the lower reservoir and at the inlet to the pump the following is obtained:

$$h_1 - F = h_2 + p_2/\gamma + v_2^2/(2g)$$

where

h_1 = elevation of lower reservoir (m)
h_2 = elevation of pump inlet (m)
F = friction loss per unit weight of fluid between lower reservoir and pump inlet (m)
p_2 = gage pressure at pump inlet (Pa)
γ = specific weight of water (N/m³)
v_2 = velocity at pump inlet (m/s)
g = acceleration due to gravity (m/s²)

Pumps

Then,
$$p_2 = (h_1 - h_2)\gamma - F\gamma - (v_2^2\rho)/2$$

The friction term, F, is the sum of friction losses in the pipe and through the elbows:
$$F = f(l/D)(v_2^2/(2g)) + 2K(v_2^2/(2g)) + K_{ent}(v_2^2/(2g))$$

Then,
$$p_2 = (h_1 - h_2)\gamma - f(l/D)(v_2^2/(2g))\rho - Kv_2^2\rho - v_2^2\rho/2 - K_{ent}v_2^2\rho/2$$

The velocity at the inlet is:
$$v_2 = Q/A_2 = 4Q/\pi D^2 = 4(0.01 \text{ m}^3/\text{s})/\pi(.05 \text{ m})^2$$
$$v_2 = 5.093 \text{ m/s}$$
$$\text{Re} = \rho v_2 D/\mu$$
$$\text{Re} = (995.39 \text{ kg/m}^3)(5.093 \text{ m/s})(.05 \text{ m})/(0.000803 \text{ Pa s})$$
$$\text{Re} = 3.157 \times 10^5$$
$$\varepsilon/D = (0.00015 \text{ m})/(.05 \text{ m}) = 0.003$$

From figure 2.6, $f = 0.0266$
From Table 2-3, $K = 0.9$ and $K_{ent} = 0.5$.
Then,
$$p_2 = \rho((h_1 - h_2)g - f(l/D)(v_2^2/2) - Kv_2^2 - v_2^2/2 - K_{ent}v_2^2/2)$$
$$p_2 = (995.39 \text{ kg/m}^3)((3 \text{ m})(9.81 \text{ m/s}^2)$$
$$- 0.0266((10 \text{ m})/(0.05 \text{ m}))((5.093 \text{ m/s})^2/2)$$
$$- 0.9(5.093 \text{ m/s})^2 - (5.093 \text{ m/s})^2/2 - 0.5(5.093 \text{ m/s})^2/2)$$
$$p_2 = (995.39 \text{ kg/m}^3)(29.43 - 69.00 - 23.34 - 12.97 - 6.48) \text{ m}^2/\text{s}^2$$
$$p_2 = -81983 \text{ Pa} = -81.983 \text{ kPa}$$

The absolute pressure at the pump inlet is:
$$p_s = p_2 + p_{atm} = -81.983 \text{ kPa} + 101.325 \text{ kPa}$$
$$p_s = 19.342 \text{ kPa}$$

The saturation vapor pressure for 30 C water is 4.246 kPa (table 9-1). Therefore, the net positive suction head available is:

NPSHA $= (p_s - p_{sat})/\gamma$
NPSHA $= (19.342 - 4.246) \text{ kPa}(1000 \text{ Pa}/1 \text{ kPa})/(995.39 \text{ kg/m}^3)(9.81 \text{ m/s}^2)$
NPSHA $= 1.55 \text{ m}$

4.10 Pump Performance on a System

The system to which a pump is attached may be made up of lengths of pipe, valves, various joints, orifices, transitions, etc. The sum of the resistances of the various elements,

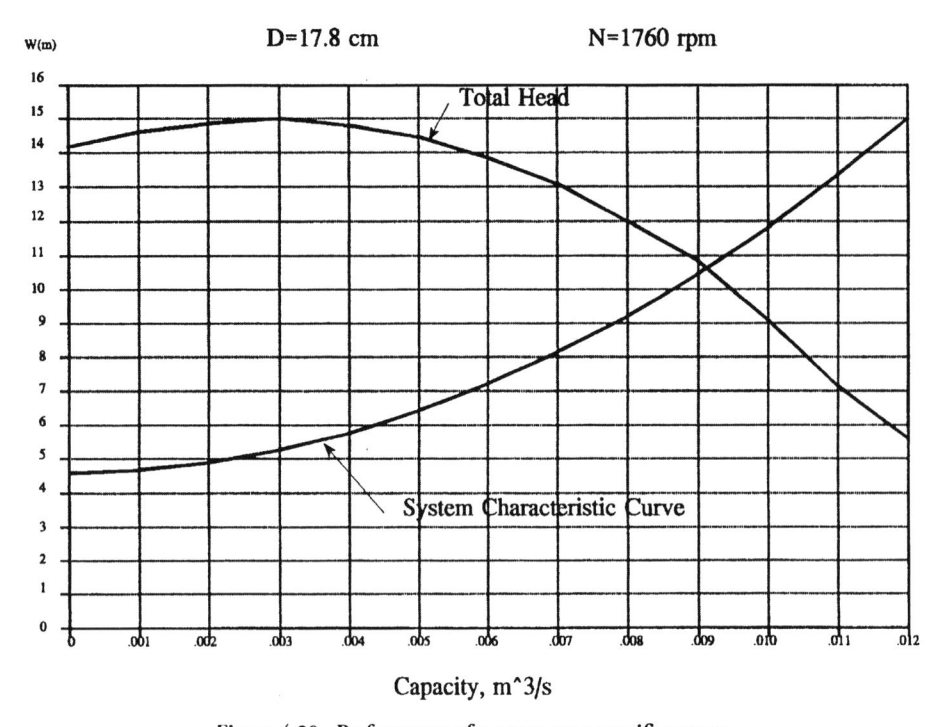

[Figure: Performance curves showing Total Head and System Characteristic Curve vs Capacity (m^3/s), with D=17.8 cm and N=1760 rpm]

Figure 4.20. Performance of a pump on a specific system.

the elevation or fluid lift, and the velocity head is the total head, which can be calculated by the Bernoulli equation (eq. 2.13). A graph of total head plotted against the capacity or rate of flow through the system is called a system characteristic curve (fig. 4.20). The total head of this system is made up of an elevation difference (lift) of 4.6 m, velocity head, and pipe and elbow friction losses. The graph shows the rate of fluid flow which will result for various total heads across the system.

When a pump is attached to a system, the head developed by the pump must equal the head across the system. The point of operation of the pump on the system is determined graphically by superimposing the system characteristic curve upon the pump performance plot. The intersection of the system characteristic curve with the total head curve of the pump defines the point of pump operation. Figure 4.20 shows such a plot for the pump of figure 4.15. Thus, the fluid flow rate through the system and the pump operating at 1760 rpm will be 0.0091 m^3/s. Reference to figure 4.15 indicates that the power requirement for the pump operating on the system of figure 4.20 will be 1310 W and the efficiency will be 72%.

4.11 Viscosity

Fluids which are pumped in processing work are frequently more viscous than water; for example, milk, cream, oils, sugar solutions, molasses. The relationship between viscosity and pump performance is not well defined, but certain important observations will help to solve pumping problems involving viscous fluids.

Pumps

The efficiency of a pump decreases as the viscosity increases. The increased fluid friction between the pump parts and the passing fluid dissipates more mechanical energy as heat energy, and less of the input energy is available to do useful work. It is not possible to provide a general correction procedure for the effect of viscosity upon efficiency. The working head increases as the viscosity increases. Reynolds number varies inversely as the viscosity. Since a large Reynolds number is desired for most satisfactory performance, fluids of high viscosity must be moved in large-diameter pipes in order to minimize friction head losses.

Nomenclature

A	cross-sectional area (m^2)
D	diameter (m)
e_j	jet efficiency (decimal)
e_m	mechanical efficiency (decimal)
e_v	volumetric efficiency (decimal)
g	acceleration due to gravity (m/s^2)
h_f	lift required (m)
h_s	submergence (m)
h_v	velocity head (m)
H	total fluid head (m)
I	moment of inertia with respect to axis of rotation (kg m^2)
J	angular momentum (kg m^2/s)
$\dot{m}$	mass flow rate (kg/s)
N	rotational speed of impeller (rev/s)
p_{atm}	atmospheric pressure (Pa)
p_s	pressure at suction connection of pump (Pa)
p_{sat}	saturation vapor pressure of fluid (Pa)
P_f	power loss to friction in jet pump (W)
P_i	power input to pump (W)
P_o	power output of pump to fluid (W)
Q	volumetric flow rate of fluid (m^3/s)
Q_d	displacement volume per unit time for pump (m^3/s)
T	torque exerted by impeller (N m)
v	peripheral velocity of vanes (m/s)
V	velocity of fluid (m/s)
w	width of impeller vane (m)
W	work done on fluid per unit weight of fluid or total head delivered by pump (J/N) or (m)
y	radial component of fluid velocity (m/s)
α	angle between fluid velocity vector and vane velocity vector (radians)
β	vane pitch angle of impeller (radians)
ε	roughness factor (m)
γ	specific weight of fluid (N/m^3)
μ	dynamic viscosity (Pa s)

ρ fluid density (kg/m^3)
ω angular velocity (s^{-1})

References

Am. Soc. Mech. Engrs. 1965. Performance Test Code 8.2. *Centrifugal Pumps*. Am. Soc. Mech. Engrs. New York.
Am. Soc. Heating, Refrigeration, and Air Conditioning Engrs. 1983. *Equipment Handbook*. Chapter 31. Am. Soc. Heat., Refrig., and Air Cond. Engrs., Inc. New York.
Babbitt, H. E., J. J. Doland, and J. L. Cleasby. 1967. *Water Supply Engineering*. New York: McGraw-Hill Book Co.
Baumeister, T., E. A. Avallone, and T. Baumeister III. 1978. *Mark's Standard Handbook for Mechanical Engineers*. New York: McGraw-Hill Book Co.
Daugherty, R. L. 1915. *Centrifugal Pumps*. New York: McGraw-Hill Book Co., Inc.
Hydraulic Institute. 1965. Standards. 11th Ed. Hydraulic Institute. New York.
Kovats, A. 1964. *Design and Performance of Centrifugal and Axial Flow Pumps and Compressors*. New York: The MacMillan Co.
Lambeck, R. P. 1983. *Hydraulic Pumps and Motors: Selection and Application for Hydraulic Power Control Systems*. New York: Marcel Dekker, Inc.
McCabe, R. E., P. G. Lanckton, and W. V. Dwyer. 1984. *Metering Pump Handbook*. New York: Industrial Press Inc.
Murdock, J. W. 1976. *Fluid Mechanics and Its Applications*. Boston, Mass.: Houghton Mifflin Co., Inc.
Shepherd, D. G. 1956. *Principles of Turbomachinery*. New York: The MacMillan Co.
Stepanoff, A. J. 1957. 2nd Ed. *Centrifugal and Axial Flow Pumps: Theory, Design, and Application*. New York: John Wiley & Sons, Inc.
Stepanoff, A. J. 1978. *Pumps and Blowers, Two-Phase Flow*. Huntington, N.Y.: Robert E. Kriefer Publishing Co.
Sullivan, J. A. 1982. *Fluid Power, Theory and Applications*. Reston, Va.: Reston Publishing Co. Inc.

Problems

1. A 1750-rpm centrifugal pump with a 12-cm impeller delivers 0.009 m^3/s against a 6-m water head and uses 750 W. What is the pump efficiency?

2. The pump of problem 1 is to operate against a 7.6 m head without changing efficiency. Specify the speed, discharge rate, and power required.

3. The pump of figure 4.15 is connected to 12.2 m of 5-cm-inside-diameter galvanized-iron pipe. The lift is 7.6 m. The system contains two Type C elbows, pumps from a tank, and discharges from a pipe to atmospheric pressure. What is the water pumping rate?

4. Specify a pump geometrically similar to that of figure 4.15 to operate at maximum efficiency at a head of 12.2 m and capacity of 0.012 m^3/s. Impeller diameter, speed, and power requirement should be estimated.

5. A whole milk homogenizer operating at a pressure of 17.2 MPa delivers 0.023 m^3/min. If the efficiency is 82 %, what size motor is required? What size motor is required if the pressure is 10.3 MPa? Pipe friction may be neglected.

6. A pump is moving soybean oil through a 2.5-cm-diameter pipe. The 110-V motor is using 285 watts. Assuming an overall pump and motor efficiency of 65%, what is the pumping rate in cubic meters per minute if the suction and discharge pressures are, respectively, -35 and 160 kPa?

7. If the pump of example 4.7 requires a net positive suction head of 1 m, determine the maximum elevation at which the pump could be located relative to the surface of the supply reservoir. Assume that the suction side pipes and the flow rate are the same as in the example problem.

8. Determine the input power requirement for the pump and system of example 4.7 if the overall pump efficiency is 55%.

9. A pump delivering Q_1 on the system shown below must be speeded up from N_1 to N_2 to deliver Q_2. Why must $N_2/N_1 = Q_2/Q_3$?

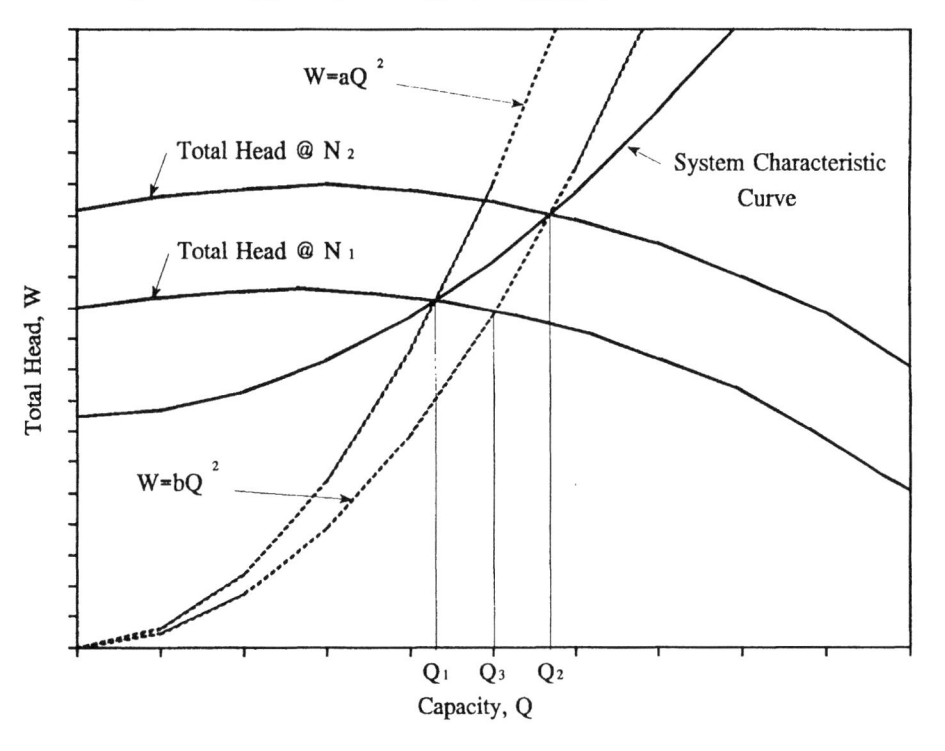

5 Fans

Fans are used in agricultural processing in connection with drying, ventilating, heating, cooling, refrigerating, aspirating, elevating, and conveying. It is important that the process engineer be able to select the best fan for a given installation with economic factors taken into consideration. This chapter will seek to describe some of the fan operating characteristics which should be considered. Although the design of fans is outside the scope of this text, some design features are discussed so that they can be better appreciated in the selection, application, and operation of fans.

Fans are similar to pumps and compressors in that all three are turbomachines that transfer energy to a flowing fluid. The distinction between pumps and fans is easy—pumps handle liquids (incompressible) while fans handle gases (compressible). The distinction between fans and compressors is not so simple. In the broad sense, the primary purpose of a fan is to move the air or gas while the primary purpose of a compressor is to increase the pressure or compress the air or gas. However, fans and compressors always both move and compress the gas to some extent. At one time a compression ratio of 1.1 was used to distinguish between fans and compressors. However, the choice of name (whether fan, compressor, or something else) is not presently regulated or standardized. Very-low-pressure-rise machines will generally be called fans while very-high-pressure-rise machines will generally be called compressors. Intermediate-pressure machines may be classified as either.

The material in this chapter will deal primarily with low-pressure-rise machines (fans) which operate in a pressure range where accuracy is not greatly reduced by assuming the gas to be incompressible. Compressors are usually applied to an agricultural processing job as packaged units such as air or refrigeration compressors. Consequently, a detailed treatment of compressors will not be included here.

5.1 Aerodynamic Classification

Fans fall into the general classification of turbomachines designated as dynamic pumps in Chapter 4. More specifically, they fit within the class specified as centrifugal pumps. As with pumps, fans may be further classified according to the manner in which the fluid enters and exits from the impeller. Figure 5.1 illustrates the four aerodynamic classes into which fans as well as centrifugal pumps may be divided. In the case

RADIAL FLOW

AXIAL FLOW

MIXED FLOW

CROSS FLOW

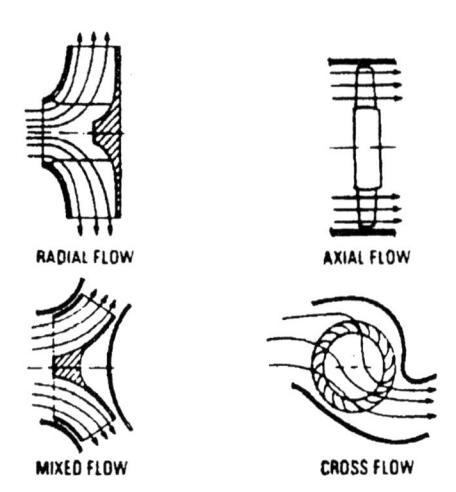

Figure 5.1. Aerodynamic classification of fans.

of fans, the term "centrifugal fan" is used specifically for those fans having radial flow characteristics with other flow types designated as axial, crossflow, or mixed-flow.

Axial-Flow Fans

As the name implies, the direction of the flow through an axial-flow fan is predominantly in the direction parallel to the axis of rotation. This fan type may be further divided into propeller, tube-axial, and vane-axial fans.

5.2 Propeller Fans

Figure 5.2 illustrates a simple propeller fan. The propeller fan may have two or more blades which may be of sheet steel or airfoil shape. The blades may be narrow or wide. They may have uniform or varied pitch. This type of fan has been developed and used to handle large volumes of air against free delivery or low pressure heads. Ventilation requirements in agriculture usually are met with propeller-type fans.

The propeller fan is essentially an air screw. The twist, angularity, or "pitch" of the blades along with the number of blades determines the distance the air would be moved due to a single rotation of the fan. For a fan blade with an angularity constant with radius, the air near the tip of the fan is moved at a faster rate than the air nearer the axis. Therefore, when the fan is operated against a significant pressure head, air is forced back through the fan near the hub and recirculation or turbulence occurs as shown in figure 5.3 and lowers the efficiency. Recirculation can be reduced or eliminated by warping the blades so that the axial air velocity is independent of radius.

5.3 Tube-Axial Fans

A tube-axial fan (fig. 5.4) consists of an axial-flow wheel or impeller within a cylinder. The blade area covers an appreciable portion of the cylinder area and the hub is of

Figure 5.2. A propeller fan. (Courtesy of The New York Blower Co.)

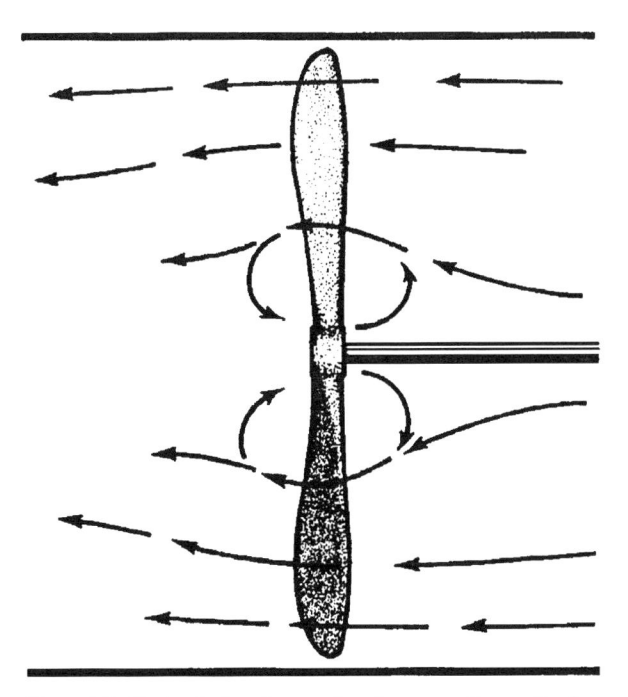

Figure 5.3. Recirculation that results when propeller or disc fan operated against too great a static head.

Figure 5.4. A tube-axial fan. (Courtesy of Cincinnati Fan.)

appreciable size. The blade is warped for efficiency, and the blades have a close radial clearance with the cylindrical housing. As a result of the design refinements, tube-axial fans will operate at higher pressures and at higher mechanical efficiencies than will propeller fans.

5.4 Vane-Axial Fans

A vane-axial fan (fig. 5.5) is similar to a tube–axial fan except that it includes a set of guide vanes either before or after the impeller. Discharge vanes transform the kinetic

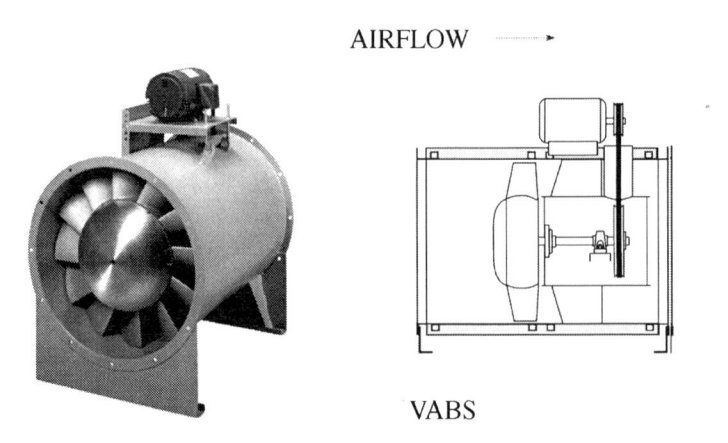

AIRFLOW ⟶

VABS

Figure 5.5. A vane-axial fan. Note the guide vanes behind the fan. (Courtesy of GreenHeck.)

energy produced by the impeller in the form of whirl into more useful pressure energy. To make this transformation efficiently, the air is guided through a gradual turn until the tangential velocity component is eliminated. Thus, the vanes enable fan operation at higher pressures and efficiencies. Pressures on the order of 15 MPa have been developed by a single axial-flow fan, with a total efficiency of more than 85%.

Centrifugal Fans

Centrifugal fans consist of a wheel or rotor within a scroll spiral-type housing (fig. 5.6). The air enters parallel to the shaft, makes a 90 degree turn in the fan wheel, and is discharged from the wheel (and housing) in a radial manner. They may be further divided according to the blade design as: straight or radial tip, backward-curved tip, and forward-curved tip fans as illustrated in figure 5.7.

5.5 Forward-Curved-Tip Fans

This type of fan has a rotor similar to a squirrel cage and a large number of blades, i.e., up to 60, narrow in the radial dimension but wide parallel to the shaft and facing forward in the direction of rotation like a scoop. It is a low-speed fan, capable of operating at moderate pressures. It is limited to handling clean air (i.e., should not be used for conveying materials suspended in air).

Figure 5.6. Centrifugal fan. (Courtesy of the New York Blower Co.)

Forward curve   Straight   Backward curve

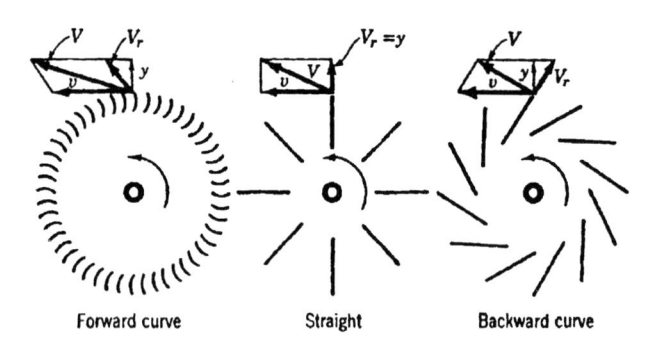

Figure 5.7. The three types of centrifugal fan rotors, with velocity diagrams.

5.6 Straight or Radial-Tip Fans

This type of fan has a smaller number of blades, from 6 to 20, and the blades are normally about two to three times as long radially as they are wide. This type of fan usually has a larger housing than other types and is more expensive; however, its price is justified for applications requiring the handling of dirty air in which materials are conveyed through the fan or where pressures must be developed which are beyond the range of lighter weight fans.

5.7 Backward-Curved-Tip Fans

This type fan has about 12 blades, essentially flat and tilted backward from the direction of wheel rotation. It is inherently a high-speed type of fan with a self-limiting power characteristic (section 5.14). It is the most efficient of the various types of centrifugal fans. With the self-limiting power feature, it is the best selection for reasonably clean air. It cannot be recommended for dirty air.

Cross-Flow Fans

5.8 Cross-flow or Tangential Fans

The cross-flow fan (fig. 5.8) has two admirable features, uniform discharge flow throughout its axial length and high air rate when referred to wheel periphery speed. Research, mainly by Eck (1973), has improved the fan's efficiency and reduced its noise level so that it is admirably suited for many air moving tasks. Examples are bakery ovens, unit heaters, baseboard heaters, air curtains, drying equipment, kitchen hoods, etc.

Mixed-Flow Fans

5.9 In-line or Tubular Centrifugal Fans

Though generally called a centrifugal fan, this type actually combines the features of centrifugal and vane-axial fans and is best described as a mixed-flow fan. Figure 5.9 shows a tubular centrifugal fan providing the straight-through air flow of vane-axials

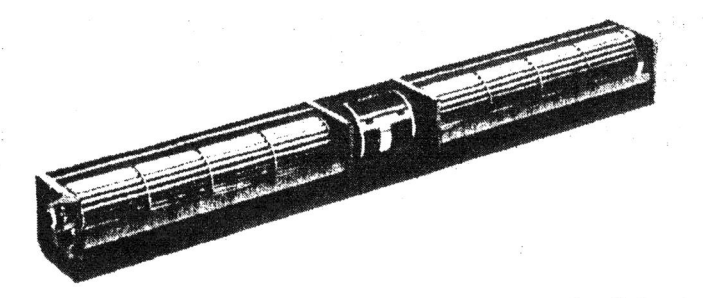

(Courtesy Lau Industries)

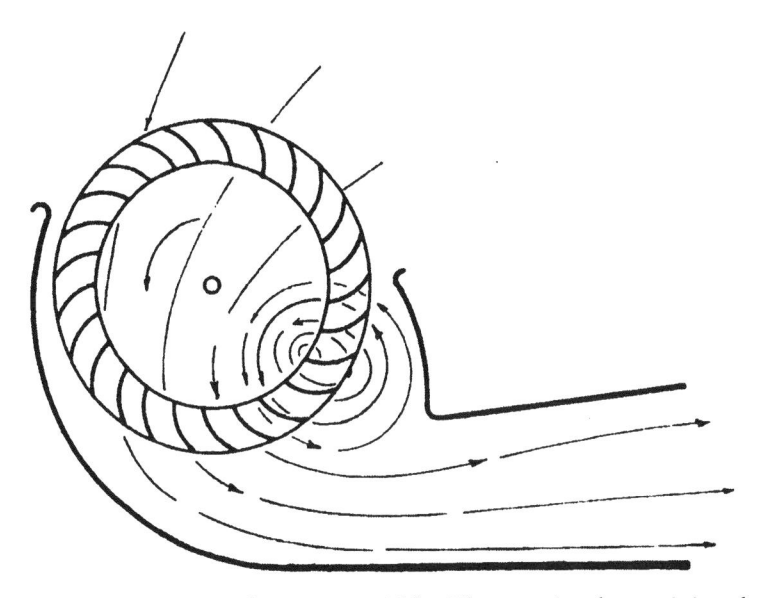

Figure 5.8. The corss-flow or tangential fan. The vortex is a characteristic and essential feature of this fan.

with the performance, service characteristics, and quietness of centrifugal fans. It can be installed directly in a straight duct with the same size and shape inlet and outlet.

Performance

5.10 Fan Testing

Fans are tested using procedures similar to those for pumps. Standards for the testing procedures are given by ASHRAE Standard 51-75 and AMCA Standard 210-74. Figure 5.10 illustrates a procedure for developing flow characteristics for a fan in which pitot tube traverses of the test duct are performed with the fan operating at a constant speed. The fan is tested from shutoff conditions to nearly free delivery conditions. Shutoff refers to the condition where the duct is completely blanked off so that there is no flow while

Figure 5.9. In-line centrifugal fan. (Courtesy of The New York Blower Co.)

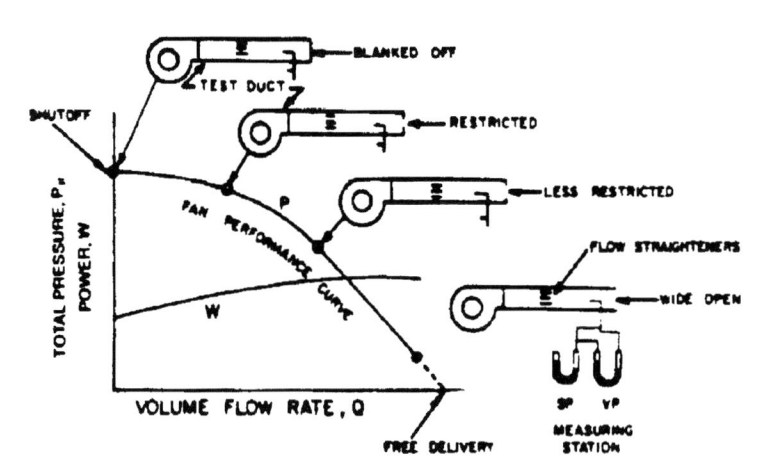

Figure 5.10. A method of obtaining fan performance curves.

free delivery refers to the condition where the outlet resistance is zero. Between these two conditions, various flow restrictions are used to simulate various conditions on the fan. Sufficient points are obtained to define the performance curves over the range from shutoff to free delivery.

A set of fan performance curves consists of plots of total pressure, static pressure, power, total efficiency, and static efficiency versus the volumetric flow rate of the fan. If either the static pressure or total pressure curve is given or experimentally determined for a fan of known outlet area, then the other can be determined from the following

Fans

relationship:
$$p_t = p_s + (v^2\rho)/2 \tag{5.1}$$

where

p_t = total pressure (Pa)
p_s = static pressure (Pa)
v = air velocity at fan outlet (m/s)
ρ = air density (kg/m³)

The velocity of air at the fan outlet is determined from the volumetric flow rate and outlet area by:
$$v = Q/A \tag{5.2}$$

where

Q = volumetric flow rate (m³/s)
A = outlet area of fan (m²)

If any one of the power, static efficiency, or total efficiency curves is given or determined, the others can then be calculated from the following relationships:
$$e_t = p_t Q/P \tag{5.3}$$

and
$$e_s = p_s Q/P \tag{5.4}$$

where

e_t = total efficiency (decimal)
p_t = total pressure (Pa)
Q = volumetric flow rate (m³/s)
e_s = static efficiency (decimal)
p_s = static pressure (Pa)
P = power input to fan (W)

Thus, a minimum of two curves may be used to calculate a set of five performance curves.

Example 5.1. A fan having an outlet area of 0.6 m² produces a volumetric air flow rate ($\rho = 1.2$ kg/m³) of 6.0 m³/s at a static pressure of 500 Pa and requires an input power of 4200 W. Determine the total pressure, total efficiency, and static efficiency.

Solution

$$v = Q/A = (6.0 \text{ m}^3/\text{s})/(0.6 \text{ m}^2) = 10 \text{ m/s}$$
$$p_t = p_s + v^2\rho/2$$
$$= 500 \text{ Pa} + (10 \text{ m/s})^2(1.2 \text{ kg/m}^3)/2$$
$$= 500 \text{ Pa} + 60 \text{ Pa} = 560 \text{ Pa}$$
$$e_t = p_t Q/P = (560 \text{ Pa})(6.0 \text{ m}^3/\text{s})/(4200 \text{ W})$$
$$= 0.8 = 80.0\%$$

$$e_s = p_s Q/P = (500 \text{ Pa})(6.0 \text{ m}^3/\text{s})/(4200 \text{ W})$$
$$= 0.714 = 71.4\%$$

5.11 Sound Power Level Ratings

The sound emitted by a fan is an inevitable byproduct of the energy-transfer process. However, careful selection of fans so that they operate at quieter points on their sound power level curves, will minimize noise.

Sound power level is expressed in decibels based on the following relationship:

$$L = 10 \log(W/W_o) \tag{5.5}$$

where

L = sound power level (dB)
W = power (W)
W_o = reference power = 1 pW = 1×10^{-12} W

Table 5-1 gives typical sound powers and sound power levels for some familiar sources. Table 5-2 lists permissible noise exposures as specified by the Occupational Safety and Health Act (OSHA). Note that the permissible sound levels are expressed in terms of equivalent A-weighted sound levels (dBA). The procedure for determining the equivalent A-weighted level is discussed in Chapter 4 of the reference edited by Jorgensen (1983). It is in effect a procedure which adjusts the sound level at a frequency other than 1000 Hz to an equivalent level at 1000 Hz. This is done because humans react less in terms of loudness to low frequency and high frequency sounds than they do to sounds of medium frequency content, when all are at the same level.

Table 5-1. Typical sound power and sound power levels

Power (watts)	Power Level (dB re 1 pW)	Source (long time avg.)
100 000	170	
10 000	160	Jet airplane
1 000	150	
100	140	
10	130	Large orchestra
1	120	
0.1	110	Blaring radio
0.01	100	
0.001	90	Shouting
0.000 1	80	
0.000 01	70	Conversational Speech
0.000 001	60	
0.000 000 1	50	Small electric clock
0.000 000 01	40	
0.000 000 001	30	Soft whisper

Table 5-2. Permissible noise exposures

Duration per day, hours	Sound Level, dBA
8	90
6	92
4	95
3	97
2	100
1.5	102
1	105
0.5	110
0.25 or less	115

If the sound power of two or more sources is combined, the overall sound power level is given by:

$$Lc = 10 \log(10^{(L1/10)} + 10^{(L2/10)} + \cdots + 10^{(Ln/10)}) \tag{5.6}$$

where

Lc = overall sound power level (dB)
$L1, L2, \ldots, Ln$ = sound power levels of individual sources (dB)

Example 5.2. A fan has a sound power of 0.002 W. Determine the sound power level in decibels for a single fan and for three identical fans operating at the same distance from the point at which sound power is determined.

Solution
For a single fan:

$$L = 10 \log(W/W_o) = 10 \log((0.002 W)/(1 \times 10^{-12} W))$$
$$= 10 \log(2 \times 10^9) = 10(9.30) = 93.0 \text{ dB}$$

For three identical fans:

$$Lc = 10 \log(10^{(L1/10)} + 10^{(L2/10)} + 10^{(L3/10)})$$
$$= 10 \log((3.0)(10)**(93/10).) = 10(9.78) = 97.8 \text{ dB}$$

5.12 Specific Speed

The same non-dimensional specific speed used for pumps is useful in characterizing fans. If the total head in equation 4.33 is replaced by its equivalent, (p_t/γ), then the specific speed equation becomes:

$$N_s = N Q^{1/2} \rho^{3/4} / p_t^{3/4} \tag{5.7}$$

where

N_s = specific speed (dimensionless)
N = fan speed (rev/s)

Q = volumetric flow rate (m³/s)
ρ = air density (kg/m³)
p_t = total pressure (Pa)

The specific speed obtained at the point of maximum total efficiency is indicative of the design features of the fan with the same approximate relationship between impeller shape and specific speed holding as is given in figure 4.18. Within the general category of axial flow fans, the specific speed at maximum efficiency decreases with an increasing hub ratio (ratio of hub radius to blade tip radius) or with an increasing solidity (ratio of blade chord length to blade spacing). Thus, propeller fans with small hubs and long narrow blades would have high specific speeds at maximum efficiency while tube-axial or vane-axial fans with enlarged hubs and wider blades would have lower specific speeds at maximum efficiency. Within the general category of centrifugal fans, the specific speed at maximum efficiency decreases with increasing blade length and with decreasing blade width. Thus, a forward-curved-tip centrifugal fan with blades narrow in the radial dimension but wide in the direction parallel to the shaft would have a higher specific speed at maximum efficiency than a radial-tip centrifugal fan having blades longer radially but shorter parallel to the shaft.

Example 5.3. Determine the specific speed of the fan of example 5.1 if the operating speed is 600 rpm.

Solution

$$N_s = NQ^{1/2}\rho^{3/4}/p_t^{3/4}$$
$$= (10 \text{ rev/s})(6.0 \text{ m}^3/\text{s})^{1/2}(1.2 \text{ kg/m}^3)^{3/4}/(560 \text{ Pa})^{3/4}$$
$$= ((10)(6.0)^{1/2}(1.2)^{3/4}/(560)^{3/4}) \text{ kg}^{3/4}\text{m}^{3/4}/\text{s}^{3/2}\text{N}^{3/4}$$
$$= 0.244$$

5.13 Axial-Flow Fan Performance

Characteristic performance curves for an axial-flow fan are given in figure 5.11. The reversal of the power and pressure curves is characteristic, although it is more pronounced in some particular designs. The point of reversal indicates the limit of stable operation of the fan. As a general rule, the power curve is relatively flat, especially within the practical operating range above approximately 40% of wide-open capacity. When this rule does not hold, the power generally decreases toward wide-open capacity and increases near the point of shutoff. Under such conditions, the power requirement increases when the fan is throttled, owing to greater resistance, and unless extra power is available for this contingency, the power unit may be overloaded and difficulties may result.

It is generally advisable to assume that the power increases as the capacity decreases, although the rate of increase may be small and a reversal in the power curve usually exists.

Fans

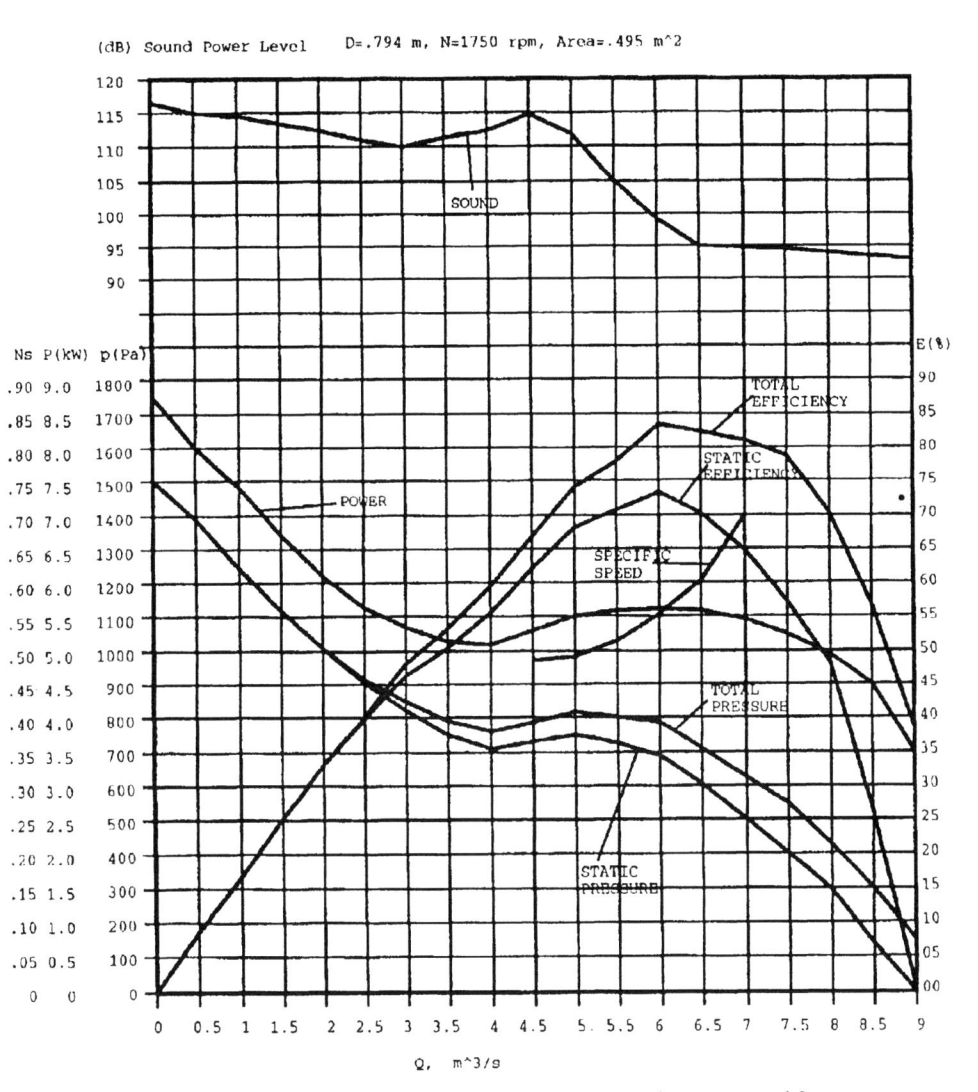

Figure 5.11. Typical constant speed performance curves for a vane-axial fan.

Figure 5.11 also gives a plot of sound power level versus volumetric flow rate of the fan. Note that the sound power level is high when the fan is operating at flows less than that corresponding to maximum efficiency but drops as maximum efficiency is approached and then remains low at higher flow rates.

A plot of specific speed is also included in figure 5.11. Note that the specific speed at the point of maximum total efficiency is approximately 0.62. As shown in figure 4.18, this value is typical of turbomachines having a primarily axial flow.

5.14 Backward-Curved-Tip Centrifugal Fan Performance

Characteristic performance curves for a backward-curved-tip centrifugal fan are given in figure 5.12. Note that (1) the maximum efficiency occurs at about 60% of wide-open capacity; (2) the pressure curves increase fairly consistently from wide-open capacity to nearly complete shutoff; and (3) the power curve is a maximum at a point nearly coincident with the maximum total efficiency.

The self-limiting power feature of the backward-curved-tip centrifugal fan is a very important feature. It is possible to choose a fan with peak efficiency and be unable to overload the motor by either increasing or decreasing the pressure or capacity as long as the speed is constant.

The backward-curved-tip centrifugal fan operates at about 1.75 to 2.0 times the speed of a forward-curved-tip fan of comparable size and capacity.

The sound power level curve for the backward-curved-tip fan of figure 5.12 shows that operation is quietest in the same flow range as highest efficiencies.

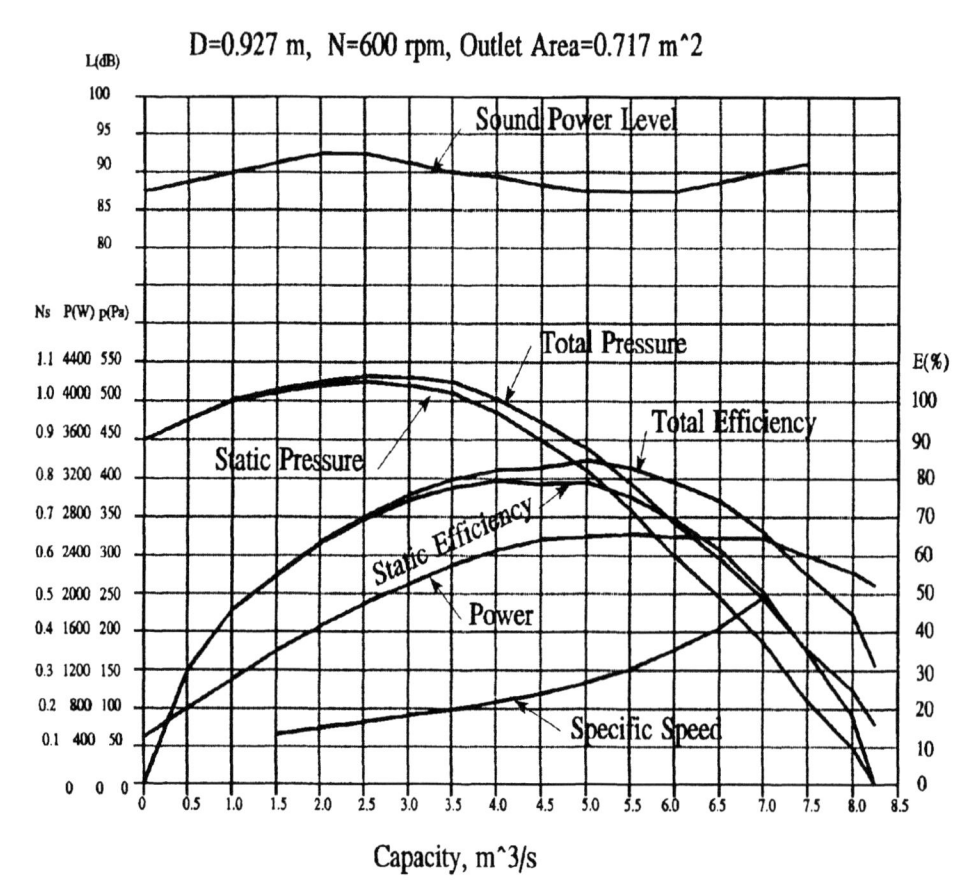

Figure 5.12. Typical constant speed performance curves for a backward-curved-tip centrifugal fan.

Fans 133

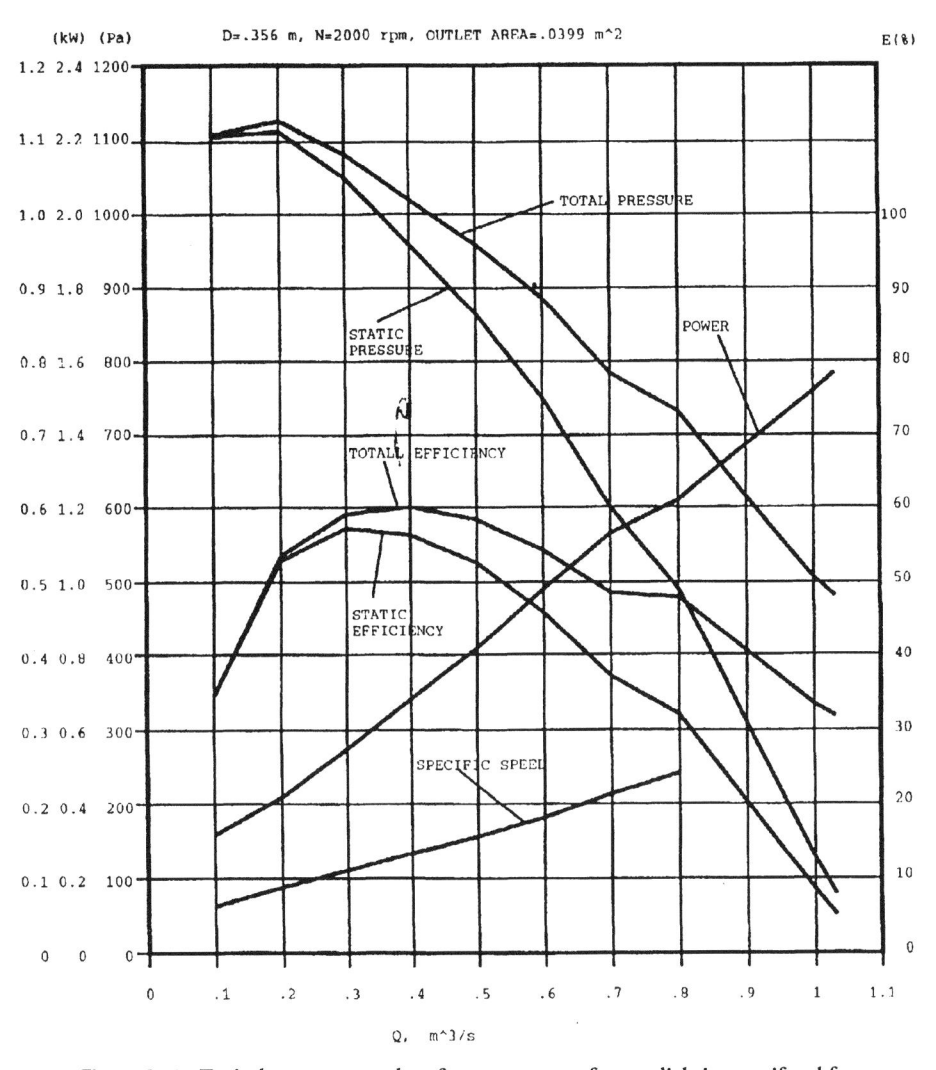

Figure 5.13. Typical constant speed performance curves for a radial-tip centrifugal fan.

The specific speed plot in figure 5.12 indicates that at the point of maximum total efficiency the fan has a specific speed of approximately 0.27. This is a value typical of turbomachines with flows in the mixed or radial mode.

5.15 Radial-Tip Centrifugal Fan Performance

Typical performance curves for a radial-tip fan are shown in figure 5.13. Note that (1) the maximum efficiency occurs at about 40% of wide-open capacity and (2) the power requirement increases as the capacity increases.

The increase in power that is required as the capacity increases may be disadvantageous, since the motor might be overloaded when decreasing the system resistance.

At the point of maximum total efficiency, the specific speed for the radial-tip fan of figure 5.13 is approximately 0.13. This is typical of turbomachines having primarily radial flow as indicated by figure 4.18.

5.16 Forward-Curved-Tip Centrifugal Fan Performance

The performance curves for a forward-curved-tip centrifugal fan are similar to those shown for the radial-tip fan of figure 5.13. This type fan does often have a pronounced dip in the pressure curves at low flow levels. This allows the possibility of three flow rates at the same pressure. If the fan is operating in this region, hunting may result between these three points.

5.17 Fan Laws

As was the case for centrifugal pumps in Chapter 4, the performance of geometrically similar fans operating at speeds other than those for which tests were conducted may be estimated by use of the affinity laws. Equations 4.30, 4.31, and 4.32 may be stated for fans as follows:

$$Q_1/Q_2 = (N_1/N_2)(D_1/D_2)^3(K_2/K_1) \tag{5.8}$$

$$p_{t1}/p_{t2} = (N_1/N_2)^2(D_1/D_2)^2(\rho_1/\rho_2)(K_2/K_1) \tag{5.9}$$

$$P_1/P_2 = (N_1/N_2)^3(D_1/D_2)^5(\rho_1/\rho_2)(K_2/K_1) \tag{5.10}$$

where

Q = volumetric flow rate (m³/s)
D = fan diameter (m)
N = fan speed (rev/s)
K = compressibility factor (dimensionless)
ρ = air density (kg/m³)
p_t = total pressure (Pa)
P = fan input power (W)

A fourth relationship may be given to express the change in sound power level:

$$L_1 = L_2 + 70 \log(D_1/D_2) + 50 \log(N_1/N_2) + 20 \log(\rho_1/\rho_2) \tag{5.11}$$

where L = sound power level (dB).

Example 5.4. A fan geometrically similar to the vane-axial fan of figure 5.11 has a diameter of 1.588 m and operates at 875 rpm. Determine its volumetric flow rate, total pressure, power requirement, and sound power level at the point where its total efficiency is maximum. Assume the compressibility factor to be 1.0.

Fans

Solution
The fan of figure 5.11 has the following characteristics at its maximum total efficiency:

$$Q_1 = 6.0 \text{ m}^3/\text{s} \qquad e_t = 83\%$$
$$p_{t1} = 780 \text{ Pa} \qquad D_1 = 0.794 \text{ m}$$
$$P_1 = 5.6 \text{ kW} \qquad N_1 = 1750 \text{ rpm}$$
$$L_1 = 99 \text{ dB}$$

Then:

$$Q_2 = Q_1(N_2/N_1)(D_2/D_1)^3(K_1/K_2)$$
$$= (6.0 \text{ m}^3/\text{s})(875 \text{ rpm}/1750 \text{ rpm})(1.588 \text{ m}/0.794 \text{ m})^3$$
$$= (6.0 \text{ m}^3/\text{s})(4) = 24.0 \text{ m}^3/\text{s}$$

$$p_{t2} = p_{t1}(N_2/N_1)^2(D_2/D_1)^2(\rho_2/\rho_1)(K_1/K_2)$$
$$= (780 \text{ Pa})(875 \text{ rpm}/1750 \text{ rpm})^2(1.588 \text{ m}/0.794 \text{ m})^2$$
$$= 780 \text{ Pa}$$

$$P_2 = P_1(N_2/N_1)^3(D_2/D_1)^5(\rho_2/\rho_1)(K_1/K_2)$$
$$= (5.6 \text{ kW})(875 \text{ rpm}/1750 \text{ rpm})^3(1.588 \text{ m}/0.794 \text{ m})^5$$
$$= (5.6 \text{ kW})(4) = 22.4 \text{ kW}$$

$$L_2 = L_1 + 70 \log(D_2/D_1) + 50 \log(N_2/N_1) + 20 \log(\rho_2/\rho_1)$$
$$= 99 \text{ dB} + 70 \log(2) + 50 \log(0.5)$$
$$= 99 \text{ dB} + 21.1 \text{ dB} - 15.1 \text{ dB} = 105 \text{ dB}$$

Example 5.5. Determine the diameter and operating speed of a vane-axial fan geometrically similar to that of figure 5.11 which will provide a volumetric flow rate of 10 m^3/s at a total pressure of 1000 Pa and a total efficiency of 75% (to right of maximum). Assume the compressibility factors to be unity and the density to be constant.

Solution
For the fan of figure 5.11 at 75% total efficiency, the characteristics are:

$$Q_1 = 7.7 \text{ m}^3/\text{s} \qquad D_1 = 0.794 \text{ m}$$
$$p_{t1} = 500 \text{ Pa} \qquad N_1 = 1750 \text{ rpm}$$
$$P_1 = 5.1 \text{ kW} \qquad L_1 = 94 \text{ dB}$$

The known parameters for the fan to be selected are:

$$Q_2 = 10.0 \text{ m}^3/\text{s and } p_{t2} = 1000 \text{ Pa}$$

Equations 5.8 and 5.9 must be solved simultaneously for D_2 and N_2. From equation (5.8) and $K_1 = K_2 = 1.0$:

$$N_1/N_2 = (Q_1/Q_2)(D_2/D_1)^3$$

Substitution into equation 5.9 yields:

$$(p_{t1}/p_{t2}) = (Q_1/Q_2)^2(D_2/D_1)^6(D_1/D_2)^2$$
$$= (Q_1/Q_2)^2(D_2/D_1)^4$$

Then,

$$D_2 = D_1(p_{t1}/p_{t2})^{1/4}(Q_2/Q_1)^{1/2}$$
$$= (0.794 \text{ m})(500 \text{ Pa}/1\,000 \text{ Pa})^{1/4}((10.0 \text{ m}^3/\text{s})/(7.7 \text{ m}^3/\text{s}))^{1/2}$$
$$= (0.794 \text{ m})(0.841)(1.14) = 0.761 \text{ m}$$

Then,

$$N_2 = N_1(Q_2/Q_1)(D_1/D_2)^3$$
$$= (1750 \text{ rpm})((10.0 \text{ m}^3/\text{s})/(7.7 \text{ m}^3/\text{s}))(0.794 \text{ m}/0.761 \text{ m})^3$$
$$= (1750 \text{ rpm})(1.299)(1.136) = 2583 \text{ rpm}$$

5.18 Compressibility

The effects of compressibility are accounted for in the fan laws by the inclusion of a compressibility factor K. This coefficient may be approximated by the following relationship:

$$K = c((p_{t2}/p_{t1})^{1/c} - 1)/(p_{t2}/p_{t1} - 1) \qquad (5.12)$$

where

K = compressibility factor (dimensionless)
c = $ne_t/(n-1)$ (dimensionless)
n = isentropic exponent = 1.4
e_t = total efficiency of fan (decimal)
p_{t1} = absolute total pressure at fan entrance (Pa)
p_{t2} = absolute total pressure at fan exit (Pa)

Figure 5.14 is a graphical representation of equation 5.12. Note that the compressibility factor has a value of 1.0 at an absolute pressure ratio of 1.0 and that at an absolute pressure ratio of 1.1 the compressibility factor ranges from 0.965 to 1.02 for total efficiencies ranging from 1.0 to 0.2, respectively.

Example 5.6. Determine the compressibility factor for the fan of figure 5.11 at its maximum total efficiency of 83%. Assume the atmospheric pressure to be 101.325 kPa.

Solution

$$p_{t1} = p_{atm} = 101.325 \text{ kPa}$$
$$p_{t2} = p_{atm} + 720 \text{ Pa} = 102.045 \text{ kPa}$$
$$e_t = 0.83 \qquad n = 1.4$$
$$p_{t2}/p_{t1} = (102.045 \text{ kPa})/(101.325 \text{ kPa}) = 1.007106$$

Fans

(Kp)Compressibility Factor

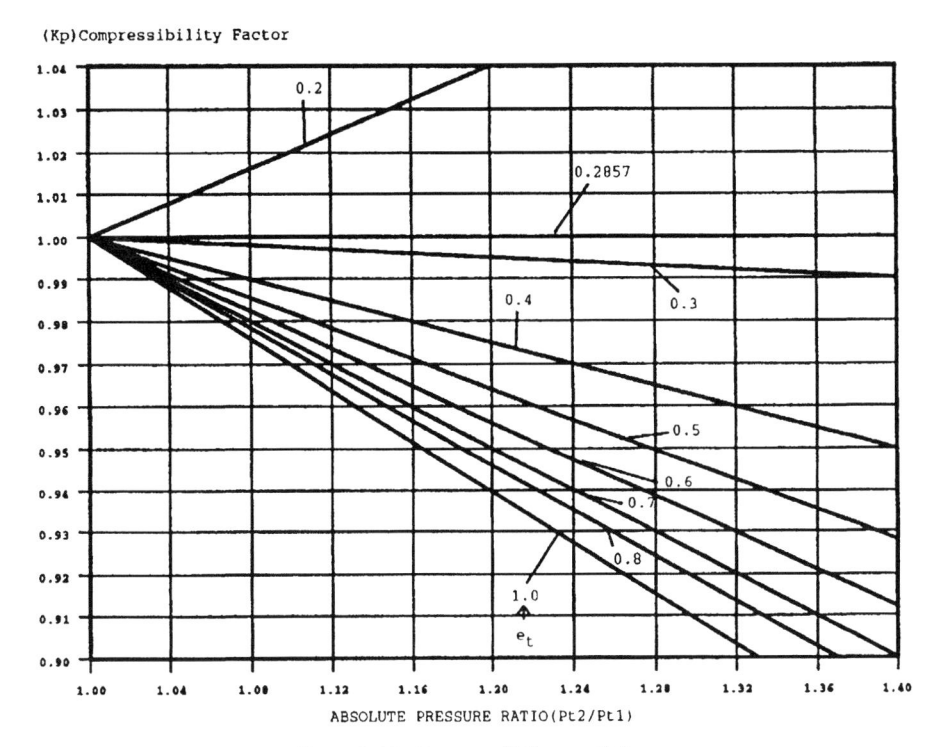

Figure 5.14. Compressibility coefficient.

$$c = ne_t/(n-1) = 1.4(.83)/(1.4-1) = 2.905$$
$$K = c((p_{t2}/p_{t1})^{1/c} - 1)/(p_{t2}/p_{t1} - 1)$$
$$= 2.905(1.007106^{1/2.905} - 1)/(1.007106 - 1)$$
$$= 0.9977$$

Fan Operation on a System

5.19 System Characteristic Curves

The overall resistance to air flow through a system will change with the volumetric flow rate. This variation can be shown graphically by plotting the total pressure required versus the volumetric flow rate through the system in the same manner that fan characteristics are presented. The total pressure required to produce a given flow can be calculated using the basic principles of fluid flow discussed in Chapter 2. Figure 5.15 illustrates two kinds of system characteristic curves which may be encountered. In figure 5.15A the total pressure varies as the square of the volumetric flow rate. This is what is usually assumed for a fan system in which the flow through each element is believed completely turbulent. This is generally a good approximation for systems involving air flow through

Figure 5.15A
$P_t = C_1 \dot{Q}^2$
(Completely Turbulent)

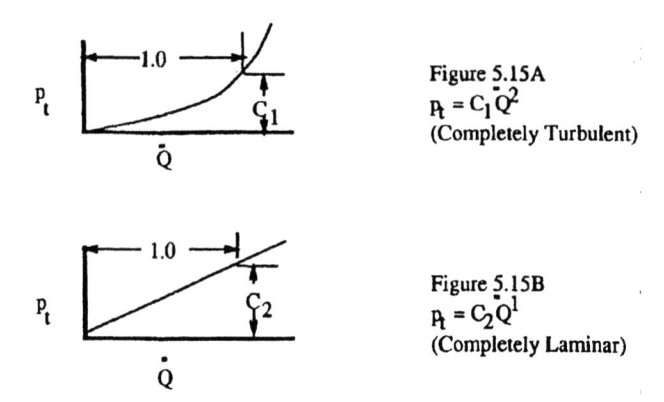

Figure 5.15B
$P_t = C_2 \dot{Q}^1$
(Completely Laminar)

Figure 5.15. System characteristic curves.

ducts, elbows, and enlargements. However, it may not be appropriate for the situation in which air is forced through a bed of material such as grain.

Figure 5.15B illustrates a system characteristic curve in which there is laminar flow throughout. Air flow through beds of materials may approach this situation but are usually somewhere between the laminar and turbulent situations. Actual system characteristic curves should be calculated using the relationships of Chapter 2.

It should be emphasized that total pressure values are generally used for the system characteristic curves. The point of operation of a fan on the system is then the intersection of the fan total pressure curve and the system characteristic curve. Fan total pressure is a true indication of the energy imparted to the airstream by the fan. By using total pressure for both fan selection and air distribution system design, the engineer is assured that correct fundamentals are being followed. Static pressure curves have often been used in low velocity ventilating systems. Use of static pressure curves is valid when the fan outlet area is equal to the fan outlet duct area but is not appropriate when fans are connected in series. In some cases, static pressure curves are more appropriate when fans are connected in parallel. This is discussed further in section 5.21.

Figure 5.16 shows system characteristic curves for three hypothetical systems along with the total pressure and total efficiency curves for the vane-axial fan of figure 5.11. The system characteristic curves are based on the relationship:

$$p_t = CQ^2 \qquad (5.13)$$

where

p_t = total pressure (Pa)
Q = volumetric flow rate (m³/s)
C = proportionality constant = 15, 10, and 5 Pa s²/m⁶ for systems A, B, and C, respectively

The operating conditions for the fan on System A are approximately: 6.7 m³/s flow rate, 680 Pa total pressure, and a total efficiency of 82.5%. Operating points for System B and

Fans

[Figure: Graph showing E(%) and p(Pa) vs Q, m^3/s with curves for TOTAL PRESSURE, TOTAL EFFICIENCY, SYSTEM A, SYSTEM B, SYSTEM C]

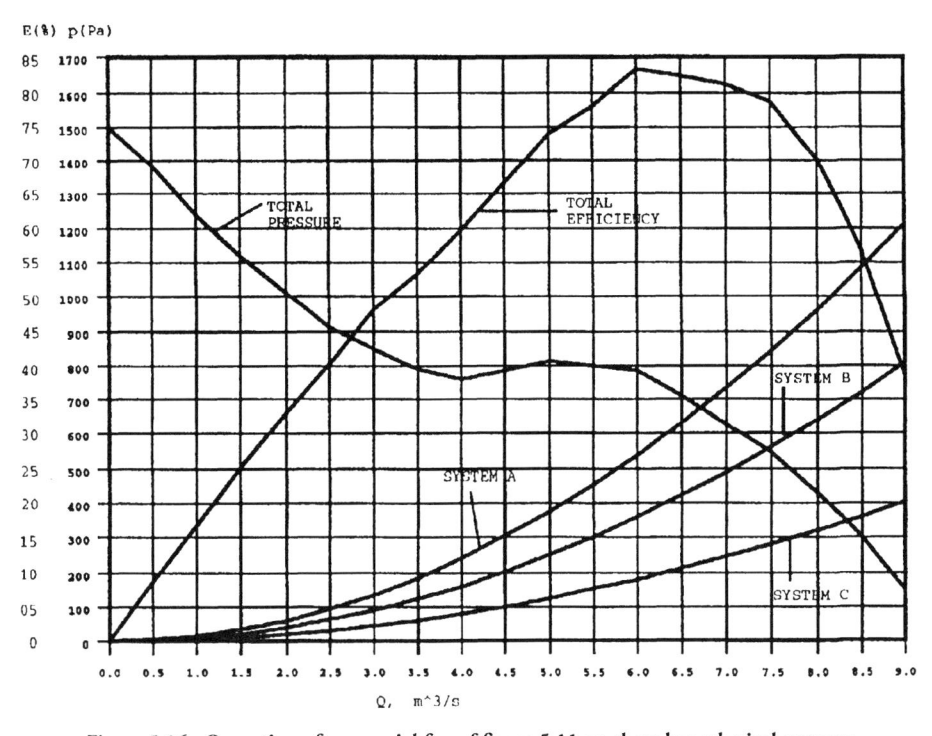

Figure 5.16. Operation of vane-axial fan of figure 5.11 on three hypothetical systems.

C are, respectively: 7.4 m^3/s at 560 Pa total pressure and 79% efficiency and 8.3 m^3/s at 350 Pa total pressure and 62.5% efficiency. Referring to figure 5.11, it may be observed that the fan operating on System A and providing 6.7 m^3/s will produce a static pressure of 580 Pa, have a power requirement of 5.6 kW, have a static efficiency of 70%, have a specific speed of 0.63, and have a sound power level of 95 dB. Note that the fan of figures 5.11 and 5.16 operates at a point near its maximum total efficiency when operating on System A. The total efficiency of 82.5% at a flow rate of 6.7 m^3/s is just to the right of the peak in the total efficiency curve. It is desirable that fans be selected with operating conditions to the right of the peak so that increases in the resistance of the system will increase rather than decrease efficiency. It is also important to avoid operating points which may lead to unstable operation. In the case of the fan of figure 5.16, unstable operation would occur at total pressures in the range of 760 to 810 Pa since the fan can produce two or more flow rates at these pressures. At total pressures above 810 Pa the operation would be stable but total efficiencies would be relatively low.

Figure 5.17 shows the same three hypothetical system characteristic curves along with the total pressure and total efficiency curves of the backward-curved-tip fan of figure 5.12. This fan operating on System A would produce a flow of approximately 5.3 m^3/s at 420 Pa total pressure, 84% total efficiency, 370 Pa static pressure, 77% static efficiency, 325 W power requirement, 0.28 specific speed, and 88 dB sound power level.

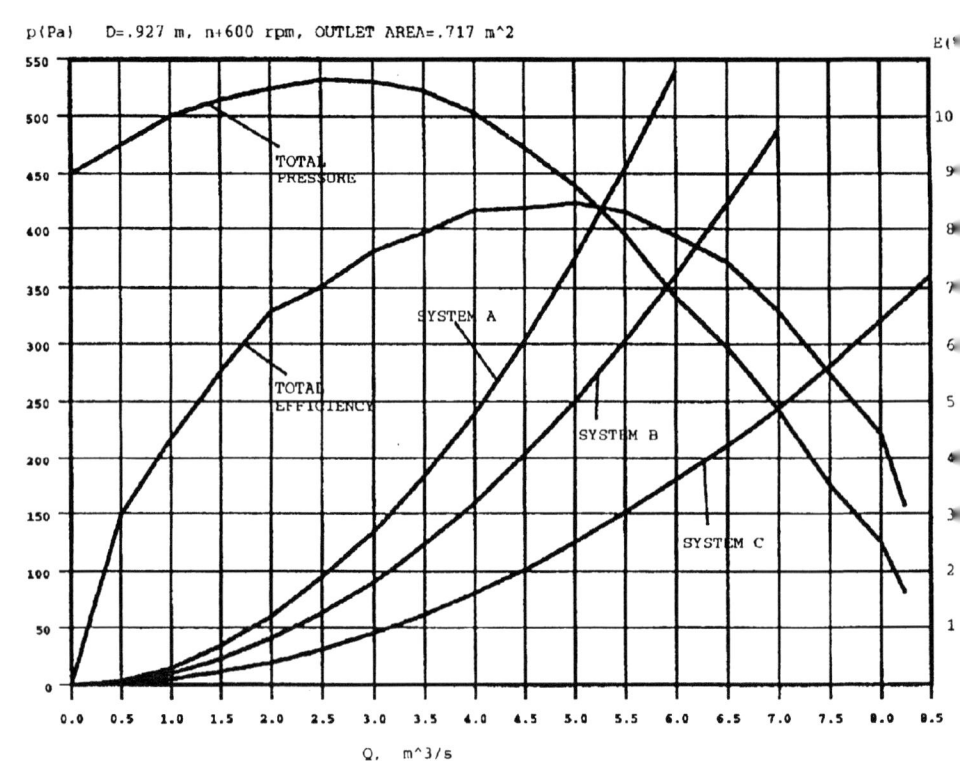

Figure 5.17. Operation of backward-curved-tip centrifugal fan of figure 5.12 on three hypothetical systems.

In the case of this fan, operation at total pressures above approximately 450 Pa can lead to problems with instability.

5.20 Fans in Series

When two or more fans are connected in series the same flow must pass through each fan and each fan will add the same total pressure to the air stream as if the fan were operating independently at the same flow rate. This is shown graphically in figure 5.18 where the total pressure curve for the backward-curved-tip centrifugal fan of figure 5.12 is reproduced along with the system characteristic curve for System A and the effective total pressure curves for two and three identical fans operating in series. Note that at any given flow rate the total pressure developed by two identical fans in series is twice the total pressure developed by a single fan. The operating points for one, two, and three fans are, respectively: 5.3 m^3/s at 420 Pa, 6.4 m^3/s at 610 Pa, and 7.0 m^3/s at 730 Pa. If the fans are not identical, then the total pressures at given flow rates must be added to obtain the effective total pressure curves for the fans operating in series.

Fans   141

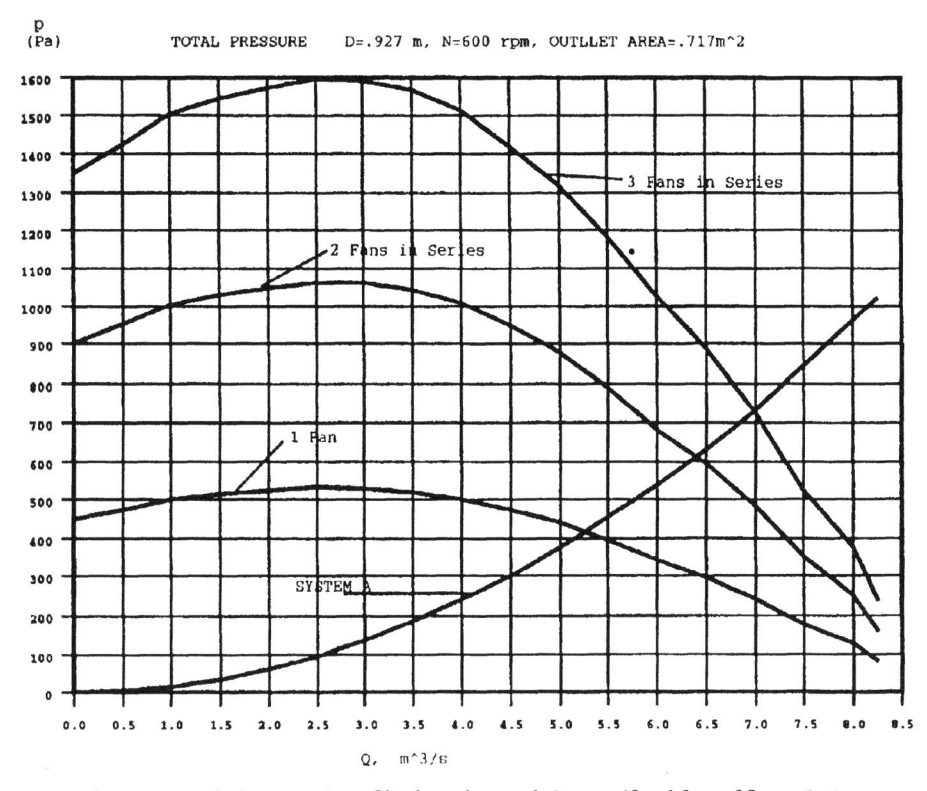

Figure 5.18. Series operation of backward-curved-tip centrifugal fans of figure 5.12.

5.21 Fans in Parallel

When two or more fans are connected in parallel, they must each operate at the same total pressure and the total flow rate will be the sum of the flows which would be provided by the fans operating individually at that total pressure. This is shown graphically in figure 5.19 where the total pressure curve for the backward-curved-tip centrifugal fan of figure 5.12 is reproduced along with the system characteristic curve for System A and the effective total pressure curves for two and three identical fans operating in parallel. Note that at any given total pressure the total flow rate developed by two identical fans in parallel is twice the flow rate produced by a single fan. These effective curves assume that all the identical fans produce equal flow rates. This is not always true since at higher total pressures different flow rates may be produced due to the peak in the total pressure curve. Thus the effective total pressure curves of figure 5.19 may be considered an idealized situation and considerable instability may be encountered at total pressures above approximately 450 Pa.

The operating points for the fan and system combinations of figure 5.19 are: 5.3 m^3/s at 420 Pa for one fan, 5.95 m^3/s at 530 Pa for two fans in parallel, and 5.9 m^3/s at 520 Pa for three fans in parallel. Note that the flow rate and total pressure for three fans is less

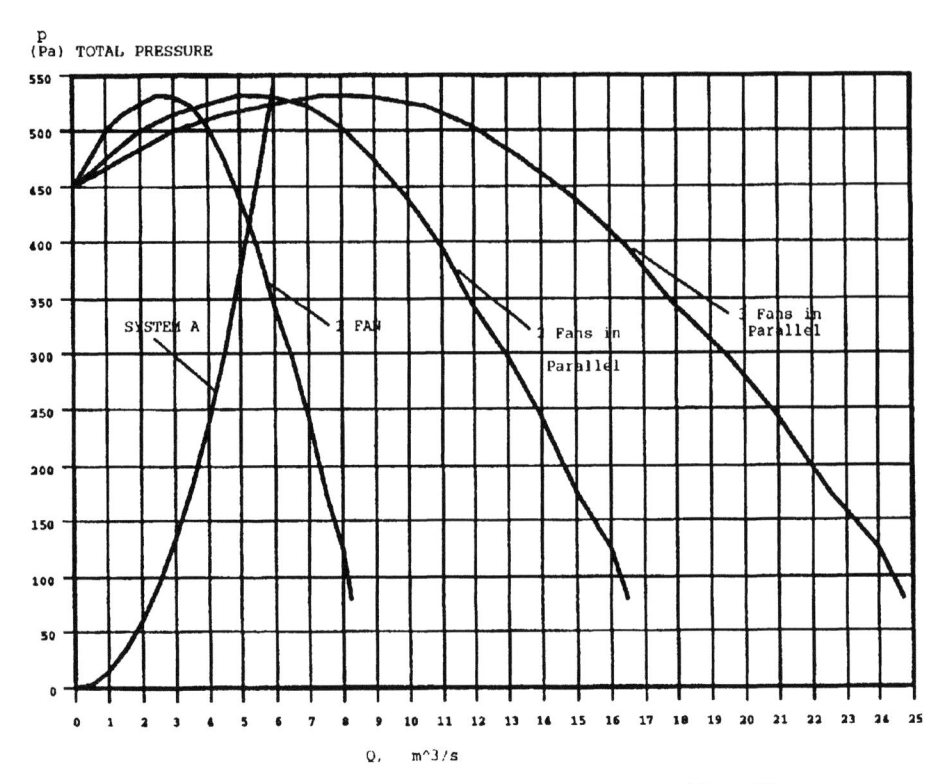

Figure 5.19. Parallel operation of backward-curved-tip centrifugal fans of figure 5.12.

than that for two fans in this case due to rising total pressure curves in the region of intersection with the system characteristic curve. Neither two nor three fans operating in parallel result in desirable operating conditions on System A due to the fact that operation will be in an unstable region.

In the cases of fans operating in series or in parallel on System A in figures 5.18 and 5.19, the operating flow rate and total pressure increased more for fans operated in series than when operated in parallel. This is not a general rule, however. Figure 5.20 shows that two fans in series produce a higher flow rate and total pressure when operating on System A while two fans in parallel produce a higher flow rate and total pressure when operating on System C. The point of intersection of the particular system curve with the single fan curve determines whether best results will be obtained by connecting multiple fans in series or in parallel. If the single fan operating point is near the maximum total pressure which can be produced by the single fan, then connecting fans in series is preferred since higher total pressures can be generated. If the single fan operating point is near the maximum flow rate which can be produced by the single fan, then connecting fans in parallel is preferred since higher flow rates can be produced at similar total pressures. For operating points between these two extremes, both series and parallel connection should be considered to determine which is most advantageous.

Fans   143

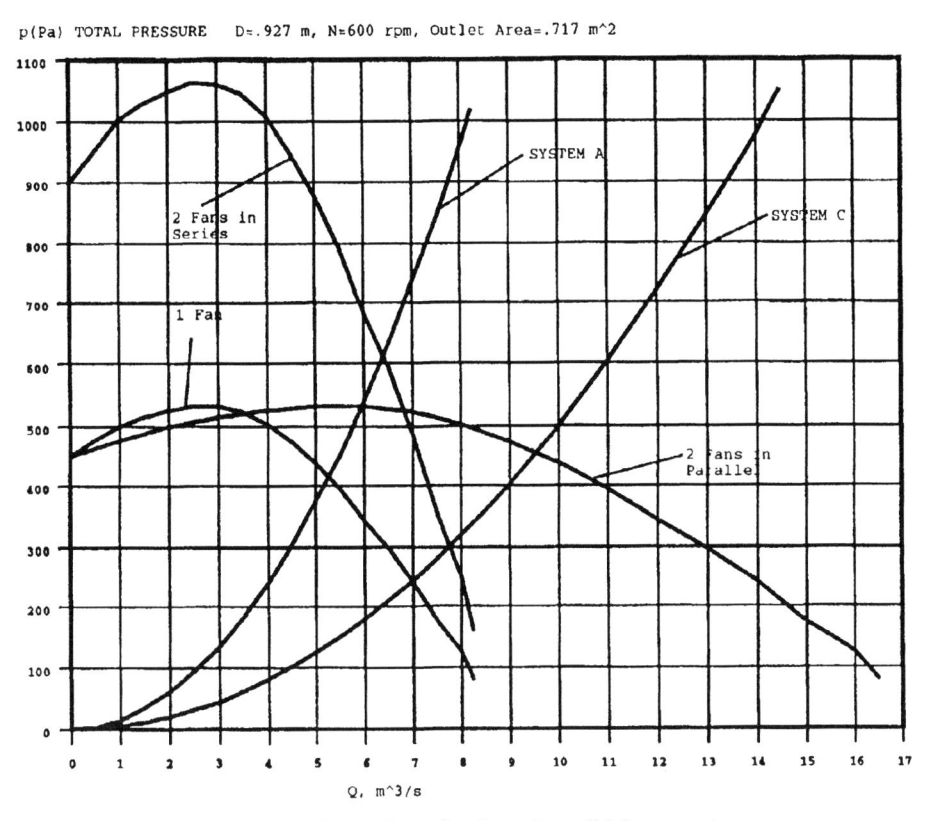

p(Pa) TOTAL PRESSURE   D=.927 m, N=600 rpm, Outlet Area=.717 m^2

Figure 5.20. Comparison of series and parallel fan operation.

In the case of parallel connection of fans to a system in which all the velocity head is immediately lost (such as is the case in which fans provide air to a drying system having a large plenum for pressure equalization prior to entering the material to be dried), the operating point becomes the intersection of the static pressure curves of the fans and the system. If the fans are of different sizes, then determination of the operating point based on static pressure curves is more accurate than if total pressure curves are used. If the velocity pressure is not totally lost, the system-fan combination becomes more complex and care should be taken in predicting the performance. In any case, the total energy provided by the fans must equal the total energy loss through the system.

5.22 Systems in Series and Parallel

When systems are connected in series or parallel, their individual system characteristic curves may be combined to form an effective combined system curve in the same manner as total pressure curves are combined for fans. This does, however, neglect any additional resistance added by whatever components are used to connect the systems. Figure 5.21 shows the total pressure curve for the backward-curved-tip centrifugal fan of figure 5.12 along with the system characteristic curve for System B, the effective system curve for

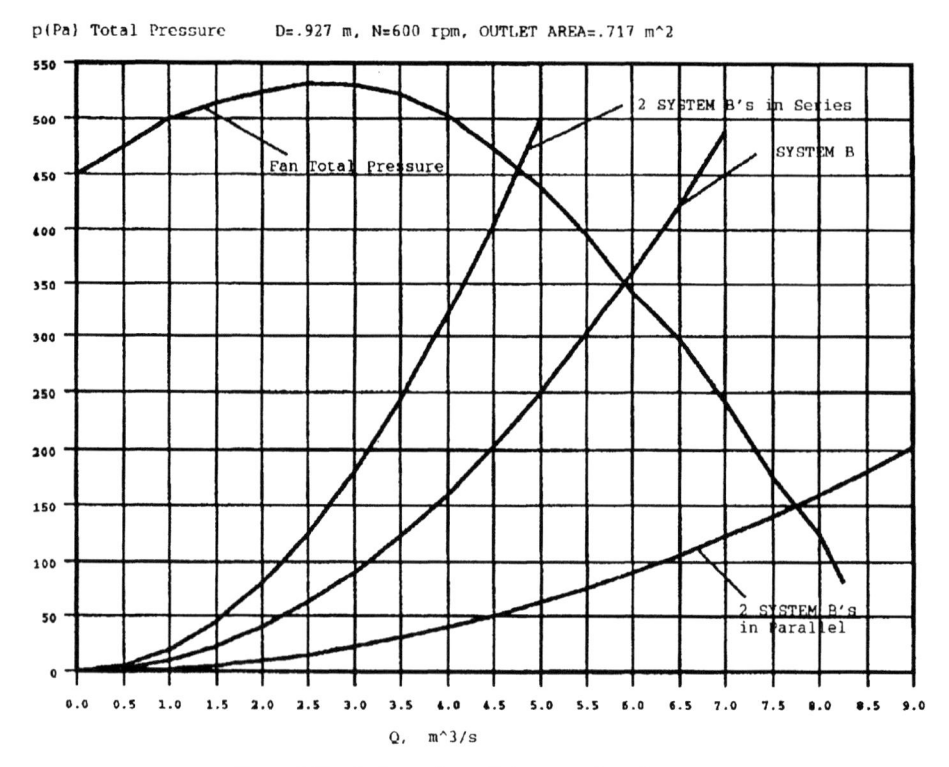

Figure 5.21. Series and parallel connection of systems.

two System B's in series, and the effective system curve for two System B's in parallel. Note that for a given air flow rate, the total pressure for two systems in series is twice that for a single system. Also, for a given total pressure, the total volumetric flow rate for two systems in parallel is twice that for a single system. The operating points in figure 5.21 are: 5.95 m^3/s at 355 Pa for one system, 4.75 m^3/s at 455 Pa for two systems in series, and 7.7 m^3/s at 150 Pa for two systems in parallel.

5.23 Effect of Varying Fan Operating Speed

The flow rate and total pressure produced by a given fan on a system can be modified by changing the operating speed of the fan. The fan laws (eqs. 5.8–5.11) can be used to predict the fan performance at new operating speeds. Figure 5.22 shows the total pressure curves for the backward-curved-tip centrifugal fan of figure 5.12 at operating speeds of 600, 500, 400, and 300 rpm. The curve given at 600 rpm was used to calculate the curves at the other speeds. Figure 5.22 also shows the hypothetical system characteristic curves given by equation 5.13. It should be noted that for a system characteristic curve in which the total pressure is proportional to the square of the volumetric flow rate, the system characteristic curves also represent constant total fan efficiency curves over a range of operating speeds. Thus, for a given fan and a given system with totally turbulent flow

```
p(Pa) Total Pressure    D=.927 m,   OUT AREA=.717 m^2
```

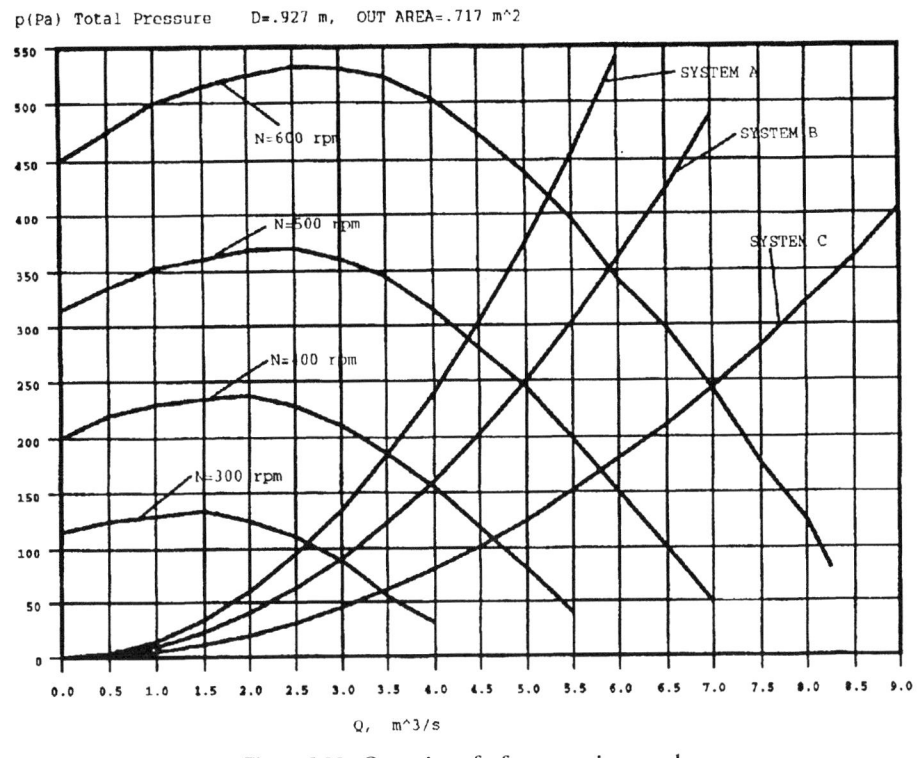

Figure 5.22. Operation of a fan at varying speeds.

characteristics, any desired flow rate may be obtained by varying the fan speed, but the fan total efficiency will not change. This will be left to the student to prove from the fan laws and equation 5.13 for a fully turbulent flow.

Example 5.7. Determine the required operating speed for the fan of figure 5.22 to provide 8.0 m³/s on System C. What will the total pressure be?

Solution
Base conditions for the fan are:

$$N_1 = 600 \text{ rpm} \qquad D_1 = 0.927 \text{ m}$$
$$Q_1 = 6.94 \text{ m}^3/\text{s} \qquad p_{t1} = 241 \text{ Pa}$$

Then,

$$Q_1/Q_2 = (N_1/N_2)(D_1/D_2)^3 = (N_1/N_2)$$
$$N_2 = N_1(Q_2/Q_1)$$
$$= (600 \text{ rpm})((8.0 \text{ m}^3/\text{s})/(6.94 \text{ m}^3/\text{s})) = 692 \text{ rpm}$$

$$p_{t2} = p_{t1}(N_2/N_1)^2$$
$$= (241 \text{ Pa})(692 \text{ rpm}/600 \text{ rpm})^2 = 320 \text{ Pa}$$

Nomenclature

A	outlet area of fan (m^2)
D	diameter (m)
e_s	static efficiency (decimal)
e_t	total efficiency (decimal)
L	sound power level (dB)
n	isentropic exponent = 1.4
N	fan speed (rev/s)
N_s	specific speed (dimensionless)
P	power input to fan (W)
p_s	static pressure (Pa)
p_t	total pressure (Pa)
Q	volumetric flow rate (m^3/s)
v	air velocity (m/s)
W	sound power (W)
W_o	reference power level = 1 pW = 1×10^{-12} W
γ	specific weight (N/m^3)
ρ	density of air (kg/m^3)

References

Air Moving and Conditioning Association. Standard Test Code for Air Moving Devices. AMCA Standard 210. Arlington Heights, Ill.

Am. Soc. Heating, Refrigerating, and Air Conditioning Engrs. 1975. Laboratory Methods of Testing Fans. ASHRAE Standard 51–75. Atlanta, Ga.: ASHRAE

———. 1983. Fans. *1983 Equipment Handbook,* Ch. 3. Atlanta, Ga.: ASHRAE.

Am. Soc. Mech. Engrs. 1965. *Compressors and Exhausters.* New York: ASME.

Eck, B. 1973. *Fans. Design and operation of centrifugal, axial-flow and cross-flow fans.* (Translated and edited R. S. Azad and D. R. Scott). Elmsford, N.Y.: Pergamon Press.

Jorgensen, R. 1983. *Fan Engineering.* Buffalo, N.Y.: Buffalo Forge Company.

Stepanoff, A. J. 1955. Turboblowers. *Theory, Design, and Application of centrifugal and axial flow compressors and fans.* New York: John Wiley & Sons.

Problems

1. A static pressure of 400 Pa is required to move air through a bin of grain at 0.1 m^3/s m^2. If the floor has an area of 16 m^2 and an axial flow fan having a diameter of 0.3 m is used, what power is required assuming a total efficiency of 75%?

2. Determine the operating speed, total pressure, static pressure, total efficiency, static efficiency, power requirement, and sound power level when the speed of the vane-axial fan of figure 5.16 (and 5.11) is adjusted to provide 8.0 m^3/s when operating on System A.

3. A fan geometrically similar to that of figure 5.12 must operate at 500 Pa total pressure, 3.0 m³/s, and a total efficiency of 80%. Specify the wheel diameter, speed, power required, and sound power level.

4. Determine the discharge velocity of the fan of figure 5.12 from the outlet area and from the velocity pressure when the fan provides 6.0 m³/s.

5. Assume that the fan of figure 5.11 is connected to a long flexible tube that permits a capacity of 5.0 m³/s. Now connect the discharge end of the tube to the fan inlet. The areas are the same. Discuss the effect on operating conditions—pressures, flow rates, power.

6. A room is to be ventilated by an axial-flow fan as shown below. Compute the conditions required for selecting a fan and the required power. Assume 70% total efficiency. Assume the discharge port to be an orifice with $K = 0.60$.

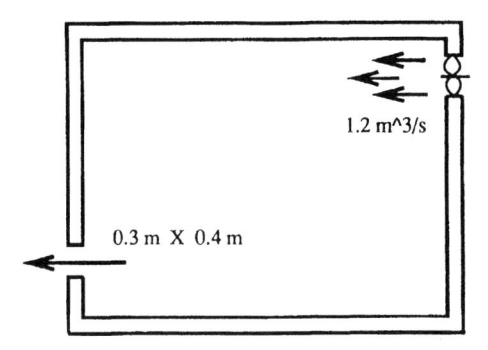

[Figure: Room diagram showing 1.2 m^3/s flow at upper right, and a 0.3 m X 0.4 m opening at lower left]

7. Prove that system characteristic curves of the form,

$$p_t = CQ^2$$

are lines of constant total efficiency for a given fan operating at different speeds.

8. Determine the flow rate through each backward-curved-tip centrifugal fan, power requirement of each fan, and total efficiency of each fan when the two fans operate in parallel as in figure 5.19.

9. Will the specific speed of a fan at maximum total efficiency vary with its operating speed? Justify your answer.

6 Heat Transfer—Conduction

Transfer of heat is the principal unit operation in the processing of many products, for example, pasteurization of milk and fruit juices, freezing of foods, cooling of fruits and vegetables for transportation and storage, and thermal sterilization of canned foods. Heat transfer is also an essential operation in providing the energy for vaporization in evaporation, distillation, and drying. Heat must be supplied to maintain desirable temperatures for bacterial growth in cottage-cheese making; on the other hand, it must be removed to avoid undesirably high temperatures in fermentation processes.

Heat is transferred by three mechanisms: conduction, convection, and radiation. In many systems all three operate simultaneously. They all have in common that (1) temperature differences must exist and (2) heat is always transferred in the direction of decreasing temperature. On the other hand, they differ entirely in the (1) physical mechanisms and (2) laws by which they are governed. Each of the heat transfer types will be briefly defined in the next three sections, the remainder of this chapter will be devoted to conduction, and Chapters 7 and 8 will further discuss convection and radiation.

6.1 Conduction

Heat conduction is the transfer of energy between adjacent molecules, not dependent on gross movement of material. The rate of heat transfer by conduction through a substance is directly proportional to the temperature gradient and to the cross-sectional area of the path, thus the Fourier-Biot Law:

$$q_x = -kA(dt/dx) \qquad (6.1)$$

where

q_x = heat flow in the x-direction (W)
A = cross-sectional area of flow path (m^2)
t = temperature (K)
x = distance through conducting medium (m)
k = thermal conductivity (W/mK)

The heat flux is the energy transported per unit area per unit time and may be expressed as:

$$q_x'' = -k(dt/dx) \qquad (6.2)$$

where q_x'' = heat flux in the x-direction (W/m^2).

The thermal conductivity, k, is a property of the conducting material. Values for different materials have been determined experimentally (table 6-1).

Equation 6.2 for heat transfer is analogous to Ohm's Law for electrical circuits. Ohm's Law may be written as:

$$i = (1/R)(dE/dx) \qquad (6.3)$$

where

i = current (amps)
R = resistance (ohms)
dE/dx = gradient of electrical potential (volts)

6.2 Convection

Heat convection is the transport of thermal energy by moving matter. The transport may be (a) natural or free convection, caused by differences in buoyancy, or (b) forced convection, accomplished by wind or mechanically with pumps, blowers, or fans.

Heat transfer by convection occurs on walls of rooms, on the outside of warm and cold pipes, and between the surfaces and fluids of all types of heat exchangers. The heat flux due to convection from the surface to the fluid is represented by:

$$q'' = h(t_s - t_f) \qquad (6.4)$$

where

h = convective heat transfer coefficient (W/m^{2}K)
t_s = surface temperature (K)
t_f = temperature of surrounding fluid (K)

6.3 Radiation

Heat radiation is the transfer of energy, without need of a conducting or convecting medium. Radiation rate depends upon the area and the nature and absolute temperature of the surface. The relationship of these factors is defined thusly:

$$q'' = \varepsilon \sigma T^4 \qquad (6.5)$$

where

q'' = heat flux emitted by a body (W/m^2)
ε = emissivity (dimensionless)
σ = Stefan-Boltzmann constant = 5.73×10^{-8} W/m^{2}K^4
T = absolute temperature (K)

Table 6-1. Coefficients of thermal conductivity of various materials

Material	Apparent Density (kg/m^3)	Temperature (C)	k (W/mK)
Air		0	0.0242
Asbestos, cement boards	1920	20	0.74
Asbestos sheets	889	51	0.166
Asbestos	580	0	0.156
	580	100	0.192
Aluminum		0	202
Aluminum foil, 7 air spaces per 6.4 cm	3.2	38	0.043
Brick, building		20	0.692
Cardboard, corrugated			0.064
Concrete 1:4 dry			0.76
Concrete, stone			0.93
Copper, pure		18	388
		100	377
Cotton wool	80	30	0.042
Cork, board	160	30	0.043
Cork, ground	150	30	0.043
Diatomaceous earth	444	204	0.114
	444	871	0.159
Fiber insulating board	237	21	0.048
Glass, boro-silicate	2230	30–75	1.09
Glass, soda			0.5–0.76
Glass, window			0.5–1.06
Ice	921	0	2.2
Iron, wrought		18	60.4
		100	59.9
Iron, cast		54	47.8
		102	46.4
Mill shavings			0.057–0.09
Mineral wool	151	30	0.0389
	316	30	0.042
Sawdust	190	21	0.05
Snow	556	0	0.47
Steel, mild		18	45.3
		100	44.8
Steel, stainless (18-8)		500	21.5
Water		0	0.571
Wood shavings	140	30	0.059
Wood, across grain, balsa	110–130	30	0.043–0.5
oak	825	15	0.21
white pine	540	15	0.151
Wool, animal	110	30	0.036

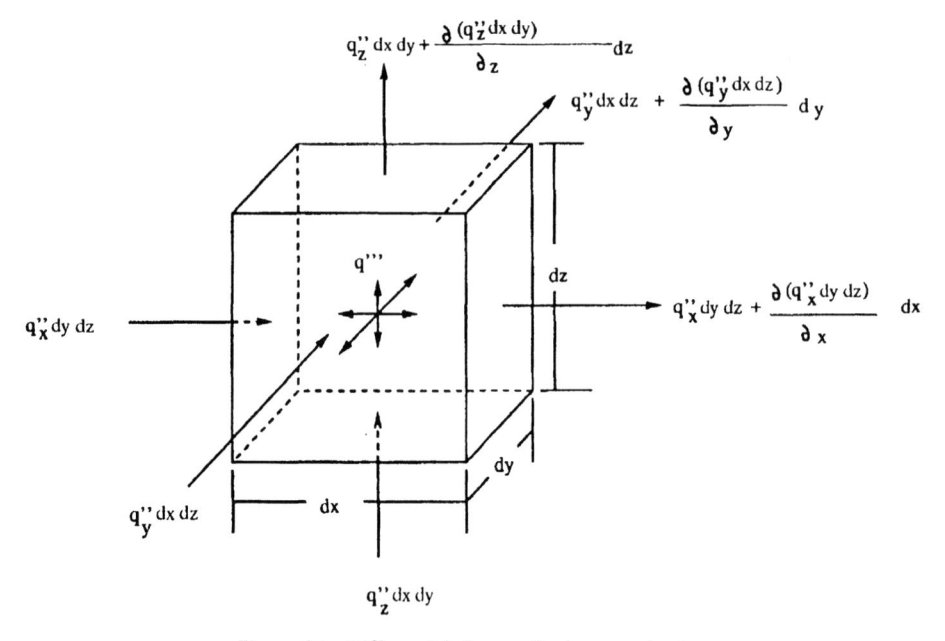

Figure 6.1. Differential element for heat conduction.

6.4 Fourier-Poisson Equation

The basic equation for describing the conservation of energy within a homogeneous, isotropic body in which heat conduction occurs may be developed by considering a differential cube as shown in figure 6.1. The heat flux vectors in the x, y, and z directions are q_x'', q_y'', and q_z'', respectively. The rate of heat generation in the body due to work done on the body is q''' (W/m^3). The first law of thermodynamics for the system may be stated as:

$$\partial Q/\partial \tau + \partial w/\partial \tau = \partial u/\partial \tau \tag{6.6}$$

where

$\partial Q/\partial \tau$ = rate at which heat is added to the body by conduction (W)
$\partial w/\partial \tau$ = rate at which work is done on the body (W)
$\partial u/\partial \tau$ = rate of change of internal energy of body (W)

The rate at which heat is added to the body is equal to the rate of heat flow in through the surfaces of the cube minus the rate of heat flow out. Then,

$$\begin{aligned}\partial Q/\partial \tau = & \; q_x'' dy\, dz + q_y'' dx\, dz + q_z'' dx\, dy \\ & - q_x'' dy\, dz - \left(\partial(q_x'' dy\, dz)/\partial x\right) dx \\ & - q_y'' dx\, dz - \left(\partial(q_y'' dx\, dz)/\partial y\right) dy \\ & - q_z'' dx\, dy - \left(\partial(q_z'' dx\, dy)/\partial z\right) dz\end{aligned}$$

Heat Transfer—Conduction

or

$$\partial Q/\partial \tau = -(\partial q_x''/\partial x + \partial q_y''/\partial y + \partial q_z''/\partial z)dx\,dy\,dz \tag{6.7}$$

If the rate at which work is done on the body per unit volume is q''', then,

$$\partial w/\partial \tau = q''' dx\,dy\,dz \tag{6.8}$$

The internal energy of the body is given by:

$$u = \rho c_p t\, dx\,dy\,dz \tag{6.9}$$

where

ρ = mass density (kg/m³)
c_p = specific heat (J/kg K)
t = temperature (K)

Then,

$$\partial u/\partial \tau = \rho c_p (\partial t/\partial \tau) dx\,dy\,dz \tag{6.10}$$

Substitution of equations 6.7, 6.8, and 6.10 into equation 6.6 yields:

$$-(\partial q_x''/\partial x + \partial q_y''/\partial y + \partial q_z''/\partial z) + q''' = \rho c_p (\partial t/\partial \tau) \tag{6.11}$$

The Fourier-Biot Law may now be used to obtain:

$$q_x'' = -k(\partial t/\partial x); \qquad q_y'' = -k(\partial t/\partial y); \qquad q_z'' = -k(\partial t/\partial z)$$

Then, for a homogeneous and isotropic material, equation 6.11 becomes:

$$k(\partial^2 t/\partial x^2 + \partial^2 t/\partial y^2 + \partial^2 t/\partial z^2) + q''' = \rho c_p (\partial t/\partial \tau) \tag{6.12}$$

or

$$\partial^2 t/\partial x^2 + \partial^2 t/\partial y^2 + \partial^2 t/\partial z^2 + q'''/k = (1/\alpha)(\partial t/\partial \tau) \tag{6.13}$$

where $\alpha = k/(\rho c_p)$ = thermal diffusivity (m²/s).

Equation 6.13 is called the Fourier-Poisson Equation and is the most general equation for heat conduction problems within a homogeneous, isotropic material. The following sections will consider special cases of the equation.

Steady-State Conduction

6.5 Steady-State Conduction in a Plane Wall with No Heat Generation

Steady-state refers to the state in which there is no change in temperature with respect to time and thus no change in heat flux with time. For this condition the right-hand side of equation 6.13 becomes zero. For a plane wall the heat transfer is only in the x-direction with partial derivatives in the y- and z-directions becoming zero. If there is no heat generation within the body, then q''' is zero. Under these conditions the Fourier-Poisson

```
x=0      x=a
t=t1     t=t2
```

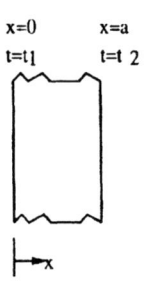

Figure 6.2.

equation reduces to:

$$(d^2t/dx^2) = 0 \tag{6.14}$$

Integration gives:

$$dt/dx = c_1$$

and

$$t = c_1 x + c_2 \tag{6.15}$$

Therefore, the temperature distribution in a plane homogeneous wall is a straight line function. The constants c_1 and c_2 must be determined from known boundary conditions. If the temperatures at the two boundaries are known as in figure 6.2, then:

$$t_1 = c_1(0) + c_2$$

or

$$c_2 = t_1 \tag{6.16}$$

and

$$t_2 = c_1 a + t_1$$

or

$$c_1 = (t_2 - t_1)/a \tag{6.17}$$

Then,

$$t = (t_2 - t_1)(x/a) + t_1 \tag{6.18}$$

If the heat flux and one temperature are known as in figure 6.3, then the constants c_1 and c_2 may be determined as follows:

$$q_x'' = -k(dt/dx) = -kc_1 \tag{6.19}$$

or

$$c_1 = -q_x''/k \tag{6.20}$$

Heat Transfer—Conduction

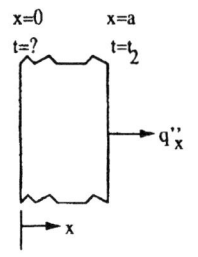

Figure 6.3.

and

$$t_2 = -(q_x''/k)a + c_2$$

or

$$c_2 = t_2 + (q_x''/k)a \tag{6.21}$$

Then,

$$t = (q_x''/k)(a - x) + t_2 \tag{6.22}$$

6.6 Steady-State Heat Conduction in a Plane Wall with Uniform Heat Generation

If there is uniform heat generation within the body, then the Fourier-Poisson equation reduces to:

$$d^2t/dx^2 + q'''/k = 0 \tag{6.23}$$

Integration yields:

$$dt/dx = -(q'''/k)x + c_1 \tag{6.24}$$

and

$$t = -(q'''/2k)x^2 + c_1 x + c_2 \tag{6.25}$$

Therefore, the temperature distribution in a plane homogeneous wall with internal heat generation is parabolic. The two constants must be evaluated from two known conditions. If the two surface temperatures are known as in figure 6.4, then

$$t_1 = c_2 \tag{6.26}$$

and

$$t_2 = -q'''a^2/2k + c_1 a + t_1$$

or

$$c_1 = (t_2 - t_1)/a + q'''a^2/2k \tag{6.27}$$

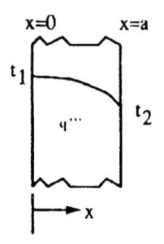

Figure 6.4.

Then,

$$t = (q'''x/2k)(a - x) + (t_2 - t_1)(x/a) + t_1 \tag{6.28a}$$

6.7 Steady-State Heat Conduction in a Composite Plane Wall with Convection at the Surfaces and No Heat Generation

If a plane wall consists of layers of two or more different materials and there is heat convection at the surfaces as in figure 6.5, then under steady-state conditions the heat flux within each material and at each surface must be constant in order to meet the requirement of conservation of energy. For the case illustrated in figure 6.5 the following equations must hold:

$$q_x'' = h_1(t_1 - t_3) \tag{6.28b}$$

$$q_x'' = -k_1(t_4 - t_3)/\Delta x_1 \tag{6.29}$$

$$q_x'' = -k_2(t_5 - t_4)/\Delta x_2 \tag{6.30}$$

$$q_x'' = -k_3(t_6 - t_5)/\Delta x_3 \tag{6.31}$$

$$q_x'' = h_2(t_6 - t_2) \tag{6.32}$$

Equations 6.28b through 6.32 may be solved for the temperature differences as follows:

$$t_1 - t_3 = q_x''/h_1 \tag{6.33}$$

$$t_3 - t_4 = q_x'' \Delta x_1 / k_1 \tag{6.34}$$

$$t_4 - t_5 = q_x'' \Delta x_2 / k_2 \tag{6.35}$$

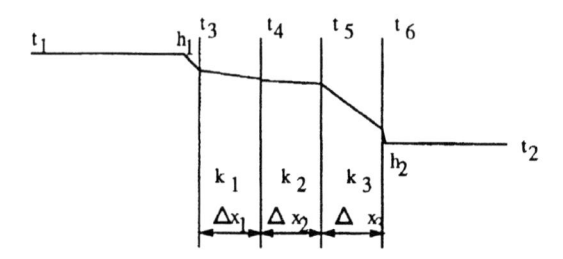

Figure 6.5. Composite plane wall with convection at surfaces

Heat Transfer—Conduction

$$t_5 - t_6 = q_x'' \Delta x_3 / k_3 \tag{6.36}$$

$$t_6 - t_2 = q_x'' / h_2 \tag{6.37}$$

Addition of equations 6.33 through 6.37 eliminates all the intermediate temperatures and leaves the following relationship:

$$t_1 - t_2 = q_x''(1/h_1 + \Delta x_1/k_1 + \Delta x_2/k_2 + \Delta x_3/k_3 + 1/h_2) \tag{6.38}$$

or

$$q_x'' = (t_1 - t_2)/(1/h_1 + \Delta x_1/k_1 + \Delta x_2/k_2 + \Delta x_3/k_3 + 1/h_2) \tag{6.39}$$

Equation 6.39 may be written as:

$$q_x'' = U(t_1 - t_2) \tag{6.40}$$

where

$$U = 1/(1/h_1 + \Delta x_1/k_1 + \Delta x_2/k_2 + \Delta x_3/k_3 + 1/h_2) \tag{6.41}$$

The U-value given by equation 6.41 is called the overall heat transfer coefficient. Note that the denominator of equation 6.41 consists of a term for each material in the wall and for each surface at which convection takes place. The units of U are W/m²K. The inverse of the overall heat transfer coefficient is the resistance value given by the following:

$$R = 1/h_1 + \Delta x_1/k_1 + \Delta x_2/k_2 + \Delta x_3/k_3 + 1/h_2 \tag{6.42}$$

Therefore, the R-values for one-dimensional heat transfer through a wall are additive and have dimensions of m²K/W. The additive nature of thermal resistances is analogous to the additive nature of electrical resistance and may be expressed as:

$$R = \sum R_i \tag{6.43}$$

where for the wall of figure 6.5:

$$R_1 = 1/h_1$$
$$R_2 = \Delta x_1/k_1$$
$$R_3 = \Delta x_2/k_2$$
$$R_4 = \Delta x_3/k_3$$
$$R_5 = 1/h_2$$

Table 6-2 lists resistance and or resistivity values for some common building materials. Resistivity is the resistance per unit thickness of the material or the inverse of thermal conductivity. The resistance values for the individual components of the wall can be added to determine the total resistance of the composite wall.

Table 6-2. Typical resistance values for building materials

	Resistivity (mK/W)	Resistance (m²K/W)
Air Spaces		
Bounded by ordinary materials		
1.9 cm or more in width		
Vertical		0.16
Horizontal, heat flow up		0.13
Horizontal, heat flow down		0.19
Bounded by aluminum foil		
Vertical		0.38
Horizontal, heat flow up		0.15
Horizontal, heat flow down		0.40
Space 3.81 cm or more divided by material reflective on both sides		
Vertical		0.76
Horizontal, heat flow up		0.65
Horizontal, heat flow down		1.96
30° slope, heat flow up		0.70
30° slope, heat flow down		1.35
Exterior finishes, frame walls		
Brick veneer, 10.2 cm thick (nominal)		0.08
Stucco	0.55	
Wooden shingles		0.14
White pine and redwood	8.93	
Southern yellow pine lap siding	8.70	
1.6 cm thick	8.70	0.14
Wood sheathing, paper and siding		
Fir sheathing—2 cm yellow pine lap siding		0.35
Asbestos shingles		0.03
Plywood siding, three-ply, 1 cm		0.07
Stone veneer	0.58	
Insulating materials		
Fill and blankets, mineral or vegetable fiber or animal hair	25.64	
Corkboard	23.26	
Insulating board	20.83	
Sawdust and shavings	16.95	
Macerated paper	25.64	
Shredded wood and cement, slab	15.15	
Shredded redwood bark (80 kg/m³)	27.03	
Paper and asbestos fiber	25.00	
Mineral wool made from rock, slag or glass	25.64	
Vermiculite, expanded mica	21.74	
Vermiculite, expanded	14.49	
Cotton, batt or blanket	28.57	
Balsa wood, 144 kg/m³	18.18	
Kapok, 16 kg/m³	28.57	

Table 6-2. (Continued)

	Resistivity (mK/W)	Resistance (m²K/W)
Interior finishes		
Composition wallboard	13.89	
Gypsum plaster, 1.3 cm thick	2.10	0.03
Gypsum wallboard, 1 cm thick		0.05
Gypsum lath and plaster, 1.3 cm plaster		0.07
Insulating fiberboard (1.3 cm)		0.27
Insulating board lath (1.3 cm) and plaster (1.3 cm)		0.29
Insulating board lath (2.5 cm) and plaster (1.3 cm)		0.57
Metal lath and plaster (1.9 cm)		0.04
Plywood, three-ply (0.6 cm)	6.94	0.04
Wood lath and plaster		0.07
Cement plaster	0.87	
Masonry materials		
Brick, common (10.2 cm)	1.39	0.14
Brick, face (10.2 cm)	0.76	0.08
Cement mortar	0.58	
10.2 cm hollow clay tile		0.18
15.2 cm hollow clay tile		0.28
20.3 cm hollow clay tile		0.29
25.4 cm hollow clay tile		0.30
30.5 cm hollow clay tile		0.44
Concrete, lightweight aggregate	2.77	
Concrete, cinder aggregate	1.39	
Concrete, ordinary	0.87	
Concrete, sand and gravel aggregate	0.58	
10.2 cm concrete blocks, hollow cinder aggregate		0.18
20.3 cm concrete blocks, hollow		
lightweight aggregate		0.35
cinder aggregate		0.29
sand and gravel aggregate		0.18
20.3 cm painted concrete blocks		0.34
20.3 cm insulated concrete blocks cores filled with rockwool		0.65
30.5 cm concrete blocks, hollow		
lightweight aggregate		0.37
cinder aggregate		0.33
sand and gravel aggregate		0.22
cinder aggregate		0.33
sand and gravel aggregate		0.22
7.6 cm gypsum tile, hollow		0.29
10.2 cm gypsum tile, hollow		0.38
Tile and terrazzo	0.58	
Stone	0.58	
2 coats p.c. paint		0.03
Window glass	1.26	

Table 6-2. (Continued)

	Resistivity (mK/W)	Resistance (m²K/W)
Roofing materials		
Asbestos shingles		0.03
Asphalt shingles		0.03
Built-up roofing (1 cm)		0.05
Heavy roll roofing		0.03
Slate (1.3 cm)	0.69	0.01
Wood shingles		0.14
Sheathing		
Asbestos-cement board	2.57	
Gypsum, 1.3 cm		0.06
Insulating fiberboard (2 cm)		0.42
Plywood, 3-ply, 0.8 cm thick		0.05
Wood (2 cm)	8.70	0.17
Wood (2.5 cm), soft	10.64	
hard	5.68	
Wood (2.5 cm) plus building paper		0.20
Surfaces		
Inside surfaces, Ordinary materials		
still air, vertical		0.11
horizontal, heat flow up		0.09
horizontal, heat flow down		0.15
Outside surfaces, Ordinary materials		
24.1 km/h wind		0.03
48.3 km/h wind		0.02
Reflective materials		
24.1 km/h wind		0.03

Example 6.1. Determine the heat flux through the composite wall of figure 6.6 if the inside temperature is 25 C and the outside temperature is 0 C.

Solution

Material	Resistivity mK/W	Resistance m²K/W
Inner Surface		0.11
Wood lath & plaster		0.07
Air Space, over 1.91 cm		0.16
Wood Sheathing, paper		0.20
Pine siding, 1.91 cm	8.93	0.17
Outer surface		0.03
		$R = \overline{0.74}$

$U = 1/(0.74 \text{ m}^2\text{K/W}) = 1.35 \text{ W/m}^2\text{K}$

$q'' = U(t_i - t_0) = (1.35 \text{ W/m}^2\text{K})(25 \text{ K}) = 33.75 \text{ W/m}^2$

Heat Transfer—Conduction 161

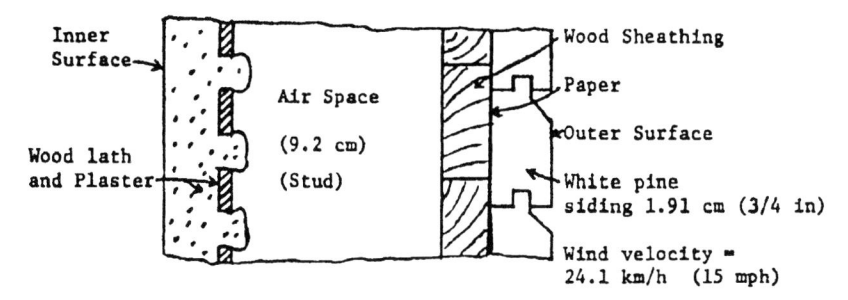

Figure 6.6. Composite wall with convection at inner and outer surfaces.

6.8 Steady-State Heat Conduction in a Cylindrical Shell

The Fourier-Poisson Equation may be written in cylindrical coordinates as:

$$\partial^2 t/\partial r^2 + (1/r)(\partial t/\partial r) + \partial^2 t/\partial z^2 + (1/r^2)(\partial^2 t/\partial \theta^2) + q'''/k$$
$$= (1/\alpha)(\partial t/\partial \tau) \tag{6.44}$$

In the case of steady-state conduction in the radial direction with no heat generation, equation 6.44 reduces to:

$$r(d^2 t/dr^2) + dt/dr = 0 \tag{6.45}$$

or

$$(d/dr)(r(dt/dr)) = 0 \tag{6.46}$$

Integration yields:

$$r(dt/dr) = c_1 \tag{6.47}$$

or

$$dt/dr = c_1/r \tag{6.48}$$

A second integration yields:

$$t = c_1 \ln r + c_2 \tag{6.49}$$

Notice that the temperature is indeterminate at $r = 0$ since $\ln r$ approaches negative infinity as r approaches zero. Physically this is true since it would be impossible to maintain steady-state heat transfer in a solid cylinder without some source of heat at the center. Thus, a hollow cylinder with inside radius r_1 and outside radius r_2 is considered as shown in figure 6.7. The constants c_1 and c_2 must be determined from two known boundary conditions. In figure 6.7 the known conditions are the temperatures at the inner and outer surfaces. Then,

$$t_1 = c_1 \ln r_1 + c_2 \tag{6.50}$$
$$t_2 = c_1 \ln r_2 + c_2 \tag{6.51}$$

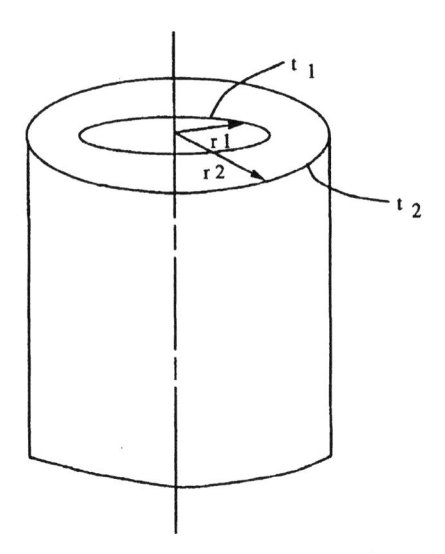

Figure 6.7. Hollow cylinder with steady-state heat transfer in radial direction.

Equations 6.50 and 6.51 may be solved for the constants to obtain:

$$c_1 = -(t_1 - t_2)/\ln(r_1/r_2) \tag{6.52}$$

$$c_2 = t_1 + (t_1 - t_2)\ln r_1/\ln(r_2/r_1) \tag{6.53}$$

Then,

$$t = t_1 - (t_1 - t_2)\ln(r/r_1)/\ln(r_2/r_1) \tag{6.54}$$

Thus, the temperature distribution in a cylindrical shell under steady-state conditions is a logarithmic function of radius.

The heat flux at any radius r is:

$$q_r'' = -k(dt/dr) \tag{6.55}$$

and from equations 6.48 and 6.52:

$$dt/dr = -(t_1 - t_2)/(r \ln(r_2/r_1)) \tag{6.56}$$

Thus,

$$q_r'' = k(t_1 - t_2)/(r \ln(r_2/r_1)) \tag{6.57}$$

The total heat transfer through a cylindrical shell is the heat flux at a given radius multiplied by the surface area at that radius. Then,

$$q_r = q_r''(2\pi r l) \tag{6.58}$$

Heat Transfer—Conduction

where

q_r = total heat transfer through cylindrical shell (W)
l = length of cylindrical shell (m)

Substitution of equation 6.57 into 6.58 gives:

$$q_r = 2\pi l k(t_1 - t_2)/\ln(r_2/r_1) \qquad (6.59)$$

The heat transfer may also be expressed per unit of length of the cylindrical shell by dividing equation 6.59 by the length to obtain:

$$q'_r = 2\pi k(t_1 - t_2)/\ln(r_2/r_1) \qquad (6.60)$$

Note that the heat transfer per unit length is independent of the radius at which it is evaluated. However, the heat flux (the heat transfer per unit area) varies with the inverse of radius.

6.9 Steady-State Heat Conduction in a Composite Cylinder with Convection at the Surfaces

Figure 6.8 illustrates a composite cylinder of three materials with convection at the inner surface at r_1 and at the outer surface at r_4. As indicated in the previous section, during steady-state conditions the heat flow per unit length is independent of radius and thus must be the same in each of the materials and at both the inner and outer surface.

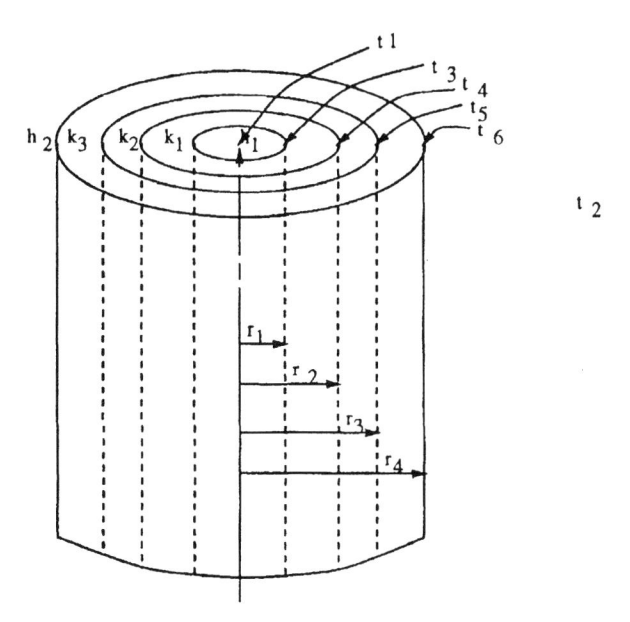

Figure 6.8. Composite cylindrical shell with steady-state heat transfer in radial direction.

Then,

$$q'_r = h_1(2\pi r_1)(t_1 - t_3) \tag{6.61}$$
$$q'_r = 2\pi k_1(t_3 - t_4)/\ln(r_2/r_1) \tag{6.62}$$
$$q'_r = 2\pi k_2(t_4 - t_5)/\ln(r_3/r_2) \tag{6.63}$$
$$q'_r = 2\pi k_3(t_5 - t_6)/\ln(r_4/r_3) \tag{6.64}$$
$$q'_r = h_2(2\pi r_4)(t_6 - t_2) \tag{6.65}$$

As in the case of the composite wall, equations 6.61 through 6.65 may be solved for the temperature differences and summed to eliminate all temperatures except the inner and outer temperatures. The resulting relationship may be solved for the heat flow per unit length:

$$q'_r = 2\pi U'(t_1 - t_2) \tag{6.66}$$

where U' = overall heat transfer coefficient per unit length (W/mK) and

$$U' = 1/R' \tag{6.67}$$

where R' = resistance per unit length (mK/W).

The total resistance per unit length is the sum of individual resistance terms for the different materials or surfaces. These are given by:

$$R'_1 = 1/r_1 h_1 \tag{6.68}$$

for the inner surface,

$$R'_2 = \ln(r_2/r_1)/k_1 \tag{6.69}$$

for the innermost material,

$$R'_3 = \ln(r_3/r_2)/k_2 \tag{6.70}$$

for the second material,

$$R'_4 = \ln(r_4/r_3)/k_3 \tag{6.71}$$

for the outer material, and

$$R'_5 = 1/r_4 h_2 \tag{6.72}$$

for the outer surface.

In general the total resistance per unit length may be written as the summation of individual resistances as:

$$R' = \sum R'_i \tag{6.73}$$

It should be noted that resistances per unit length cannot be tabulated for different materials as were total resistances in plane wall sections since the resistance varies with both thermal properties and physical dimensions of the cylindrical shells. Also, note that the resistance to convection at the surfaces decreases with the inverse of radius so that an

Heat Transfer—Conduction

increase in resistance due to increasing the thickness of the shell material may be offset by a reduction in resistance at the surface.

It is often desirable to define an overall heat transfer coefficient for a cylindrical shell based on a specified surface area. This is commonly done for heat exchangers which will be discussed in a later section. The overall heat transfer coefficient may be defined by the following relationships:

$$q_r = q'_r l = U A \Delta t \tag{6.74}$$

or

$$q_r = 2\pi U' l \Delta t = 2\pi r l U \Delta t \tag{6.75}$$

Then,

$$U = U'/r \tag{6.76}$$

where r is the radius at the surface for which the overall heat transfer coefficient is defined. This will normally be either the inner or outer surface and the corresponding area would have to be used to properly calculate total heat transfer.

6.10 Biot Number for Maximum Heat Transfer Through a Composite Cylindrical Shell

As noted in the previous section, increasing the thickness of a cylindrical shell will increase the resistance to conductance through the material, but will reduce the resistance to convection at the surface and may actually result in an increase in heat transfer from the cylinder. In this section we will consider the relationships which determine when maximum heat transfer will be obtained. Let us consider a cylindrical shell of one material with convection at both the inner and outer surface. In this analysis it will be assumed that the convection coefficient at the surface is independent of radius. Then the total resistance per unit length becomes:

$$R' = 1/(r_i h_i) + \ln(r_o/r_i)/k + 1/(r_o h_o) \tag{6.77}$$

where

r_i = radius of inner surface (m)
r_o = radius of outer surface (m)
h_i = convective heat transfer coefficient at inner surface (W/m²K)
h_o = convective heat transfer coefficient at outer surface (W/m²K)
k = thermal conductivity of material (W/mK)

When the resistance is a minimum, then the total heat transfer will be a maximum. If the outer radius is assumed to be a variable with other terms constant, then the radius at which the resistance is minimum may be obtained by differentiating with respect to r_o and setting the resulting equation equal to zero. Then,

$$dR'/dr_o = 1/(r_o k) - 1/(r_o^2 h_o) = 0 \tag{6.78}$$

and by rearrangement,

$$r_o h_o / k = 1 \tag{6.79}$$

The left-hand side of equation 6.79 is defined as the Biot number. Thus,

$$Bo = r_o h_o / k = 1 \tag{6.80}$$

at the radius at which heat transfer is either a maximum or a minimum. It will be left to the reader to take second derivatives to show that it is indeed a maximum. This means that if the Biot number is less than one, an increase in thickness of the material will increase heat transfer. If the Biot number is greater than one, an increase in thickness of the material will decrease heat transfer. Though this derivation is based on a cylindrical shell of a single material, the same relationship holds for the outer material of a composite cylinder. If the convective heat transfer coefficient at the outer surface is a function of the outer radius, then the relationship becomes much more complicated, but there will still exist a radius at which heat transfer will be a maximum.

6.11 Steady-State Heat Conduction in a Spherical Shell

The Fourier-Poisson Equation may be written in spherical coordinates as:

$$\partial^2 t/\partial r^2 + (2/r)(\partial t/\partial r) + (1/r^2 \sin^2 \phi)(\partial^2 t/\partial \theta^2) + (1/r^2)(\partial^2 t/\partial \phi^2)$$
$$+ (\cos \phi / r^2 \sin \phi)(\partial t/\partial \phi) + q'''/k = (1/\alpha)(\partial t/\partial \tau) \tag{6.81}$$

In the case of steady-state conduction in the radial direction with no heat generation, equation 6.81 reduces to:

$$r^2 (d^2 t/dr^2) + 2r(dt/dr) = 0 \tag{6.82}$$

or

$$(d/dr)(r^2 (dt/dr)) = 0 \tag{6.83}$$

Integration yields:

$$r^2 (dt/dr) = c_1 \tag{6.84}$$

or

$$dt/dr = c_1 / r^2 \tag{6.85}$$

A second integration yields:

$$t = -c_1 / r + c_2 \tag{6.86}$$

As in the case of a cylindrical body, the temperature distribution within a spherical body is indeterminate at $r = 0$ and a hollow shell must be considered. Physically this is due to the need for a heat source at the center in order to maintain steady-state conduction within the spherical body. Figure 6.9 illustrates a spherical shell with inner radius, r_1, and outer radius, r_2. The constants c_1 and c_2 must be determined from the known temperatures at

Heat Transfer—Conduction

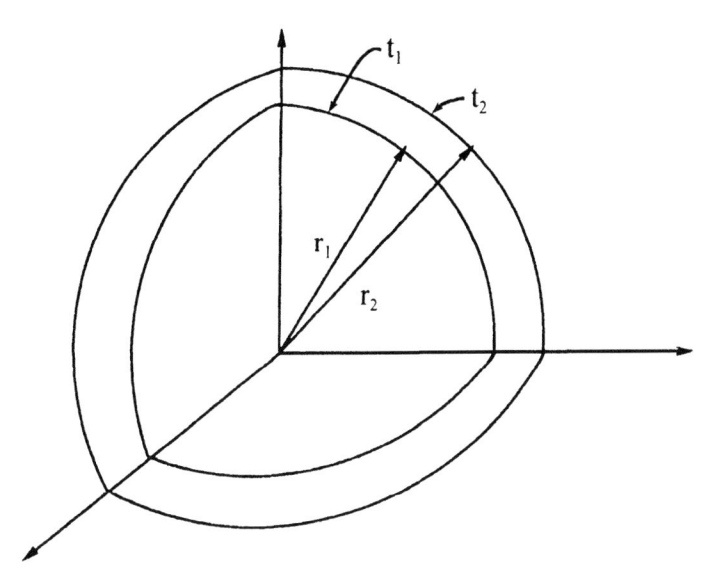

Figure 6.9. Hollow spherical shell with steady-state heat transfer in radial directrion.

the inner and outer surfaces. Then,

$$t_1 = -c_1/r_1 + c_2 \tag{6.87}$$
$$t_2 = -c_1/r_2 + c_2 \tag{6.88}$$

Equations 6.87 and 6.88 may be solved for the constants to obtain:

$$c_1 = -(t_1 - t_2)r_1r_2/(r_2 - r_1) \tag{6.89}$$
$$c_2 = t_1 - (t_1 - t_2)r_2/(r_2 - r_1) \tag{6.90}$$

Then,

$$t = t_1 + (t_1 - t_2)r_2(r_1/r - 1)/(r_2 - r_1) \tag{6.91}$$

Thus, the temperature in a spherical shell under steady-state conditions varies hyperbolically with radius.

The heat flux at any radius r is:

$$q_r'' = -k(dt/dr) \tag{6.92}$$

and from equations 6.85 and 6.89:

$$dt/dr = -(t_1 - t_2)r_1r_2/(r^2(r_2 - r_1)) \tag{6.93}$$

Thus,

$$q_r'' = k(t_1 - t_2)r_1r_2/(r^2(r_2 - r_1)) \tag{6.94}$$

The total heat transfer through the spherical shell is the heat flux at a given radius multiplied by the surface area at that radius. Then,

$$q_r = q_r''(4\pi r^2) \qquad (6.95)$$

where q_r = total heat transfer through the spherical shell (W).

Substitution of equation 6.94 into 6.95 gives:

$$q_r = 4\pi k(t_1 - t_2)r_1 r_2/(r_2 - r_1) \qquad (6.96)$$

Note that the total heat flow is independent of radius as is required for conservation of energy in a steady-state condition.

6.12 Steady-State Heat Conduction in a Composite Sphere with Convection at the Surfaces

Figure 6.10 illustrates a composite sphere of two materials with convection at the inner surface at r_1 and at the outer surface at r_3. Since the total heat flow is the same in each material and at each surface, then:

$$q_r = 4\pi h_1 r_1^2 (t_1 - t_3) \qquad (6.97)$$
$$q_r = 4\pi k_1 (t_3 - t_4) r_1 r_2/(r_2 - r_1) \qquad (6.98)$$
$$q_r = 4\pi k_2 (t_4 - t_5) r_2 r_3/(r_3 - r_2) \qquad (6.99)$$
$$q_r = 4\pi h_2 r_3^2 (t_5 - t_2) \qquad (6.100)$$

The intermediate temperatures can be eliminated from equations 6.97 through 6.100 and

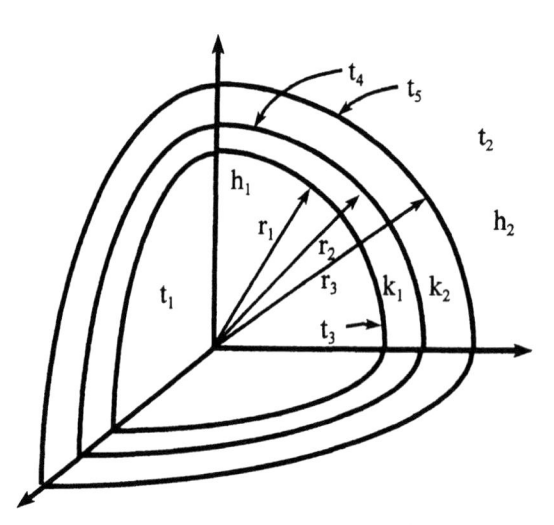

Figure 6.10. Composite spherical shell with convection at the surfaces

the following relationship results:

$$q_r = 4\pi U_s(t_1 - t_2) \tag{6.101}$$

where U_s = total heat transfer coefficient for sphere (W/K) and

$$U_s = 1/R_s \tag{6.102}$$

where R_s = total heat transfer resistance for sphere (K/W).

The total resistance for the composite sphere is the sum of individual resistances for the different materials or surfaces. These are given by:

$$R_{s1} = 1/(r_1^2 h_1) \tag{6.103}$$

for the inner surface,

$$R_{s2} = (r_2 - r_1)/(k_1 r_1 r_2) \tag{6.104}$$

for the innermost material,

$$R_{s3} = (r_3 - r_2)/(k_2 r_2 r_3) \tag{6.105}$$

for the outer material, and

$$R_{s4} = 1/(r_3^2 h_2) \tag{6.106}$$

for the outer surface.

In general the total resistance may be expressed as:

$$R_s = \sum R_{si} \tag{6.107}$$

As in the case of cylindrical shapes, the resistance terms cannot be tabulated for spherical bodies since they depend on both thermal properties and physical dimensions. It should also be noted that as the radius increases the resistance to conduction through the material increases but the resistance to convection at the outer surface decreases.

An overall heat transfer coefficient for a spherical shell based on a specific surface area may be defined from the following relationships:

$$q_r = 4\pi U_s \Delta t = q_r'' A \tag{6.108}$$

or

$$q_r = 4\pi U_s \Delta t = U \Delta t (4\pi r^2) \tag{6.109}$$

Then,

$$U = U_s/r^2 \tag{6.110}$$

where r is the radius at the surface for which the overall heat transfer coefficient is defined. This is normally the inner or outer surface and the corresponding area must be used to properly calculate total heat transfer.

6.13 Biot Number for Maximum Heat Transfer Through a Composite Spherical Shell

The same procedures used in section 6.10 for a cylindrical shell may be used to determine when heat transfer from a composite spherical shell is a maximum. Consider a spherical shell of one material with convection at both the inner and outer surface. If the convection coefficient at the surface is independent of radius, then the total resistance is:

$$R_s = 1/(r_i^2 h_i) + (r_o - r_i)/(k r_i r_o) + 1/(r_o^2 h_o) \tag{6.111}$$

where

r_i = radius of inner surface (m)
r_o = radius of outer surface (m)
h_i = convective heat transfer coefficient at inner surface (W/m^{2}K)
h_o = convective heat transfer coefficient at outer surface (W/m^{2}K)
k = thermal conductivity of material (W/mK)

Differentiating with respect to r_o and setting the result to zero gives:

$$dR_s/dr_o = 1/(k r_o^2) - 2/(r_o^3 h_o) = 0 \tag{6.112}$$

or

$$Bo = r_o h_o / k = 2 \tag{6.113}$$

It may be confirmed that this represents maximum heat transfer (minimum resistance) by evaluating the second derivative at the radius defined by equation 6.113. If the Biot number is less than two, an increase in thickness of the material will increase heat transfer. If the Biot number is greater than two, an increase in thickness of the material will decrease heat transfer.

Transient Heat Flow

6.14 Transient Heat Conduction with Convection at Surface

Transient heat transfer occurs when boundary conditions change with time so that temperature at any given point does not remain constant. Cooling of meats, fruits, and vegetables, and thermal processing of canned foods are examples. We will consider three different situations which may occur. They are: (1) negligible internal resistance to heat flow, (2) negligible surface resistance to heat flow, and (3) finite internal and surface resistance to heat flow.

Case 1

Negligible Internal Resistance (Newtonian Heating or Cooling). If the thermal conductivity of a material being heated or cooled is very high in comparison to the convection coefficient at the surface then the internal temperature gradient may be assumed negligible and the rate of heat transfer to or from the material is controlled by the convection

Heat Transfer—Conduction

ate at the surface. This is a good approximation for cases where:

$$Bo = h\delta/k < 0.2 \tag{6.114}$$

where

Bo = Biot number (dimensionless)
h = surface convective heat transfer coefficient (W/m²K)
k = thermal conductivity of material (W/mK)
δ = characteristic length (one-half thickness of a slab or the radius of a cylinder or sphere) (m)

Consider an object with a high thermal conductivity which is suddenly placed in surroundings of a different temperature t_0 after having been at a temperature t_1. Then if there is negligible internal resistance to heat flow, the mean temperature, t, of the body will at no time differ appreciably from the surface temperature. The heat energy required to raise the temperature of the object is acquired from the surroundings by convection. Thus,

$$q = hA(t_o - t) \tag{6.115}$$

where

q = heat transferred to object (W)
h = convective heat transfer coefficient (W/m²K)
A = surface area of object (m²)
t_o = temperature of surroundings (K)
t = mean temperature of object (K)

This amount of heat must also equal the rate of change of the internal energy of the object,

$$q = \rho c_p V dt/d\tau \tag{6.116}$$

where

ρ = density of object (kg/m³)
c_p = specific heat of object (J/kg K)
V = volume of object (m³)
t = mean temperature (K)
τ = time (s)

Thus,

$$\rho c_p V dt/d\tau = hA(t_0 - t) \tag{6.117}$$

or

$$dt/(t_0 - t) = hA \, d\tau/(\rho c_p V) \tag{6.118}$$

Integration yields:

$$-\ln(t_0 - t) = c_1 + hA\tau/(\rho c_p V) \tag{6.119}$$

The constant of integration, c_1, must be evaluated from the boundary condition that $t = t_1$ at $\tau = 0$. Then,

$$(t - t_0)/(t_1 - t_0) = \exp(-(hA\tau/(\rho c_p V))) \tag{6.120}$$

Thus, the mean temperature of the body varies exponentially with time after being exposed to a new temperature of the surroundings. The temperature ratio given in the left side of equation 6.120 has a value of 1.0 at $\tau = 0$ and approaches 0.0 exponentially as time, τ, approaches infinity.

Case 2

Negligible Surface Resistance. If the surface convection coefficient is high compared to the thermal conductivity of the material being heated or cooled then the surface temperature is approximately the same as the temperature of the surroundings and the temperature within the body is a function of position and time. For this case the Fourier-Poisson equation can be solved for the temperature distribution within the object. For an infinite plate with one-dimensional heat transfer equation 6.13 becomes:

$$d^2 t/dx^2 = (1/\alpha)\, dt/d\tau \tag{6.121}$$

The solution to this equation is:

$$(t - t_0)/(t_1 - t_0) = (4/\pi) \sum (1/n) \exp(-(n\pi/2)^2 \Theta) \sin(n\pi x/L);$$
$$n = 1, 3, 5, \ldots \tag{6.122}$$

where

t = temperature at a given time and position (K)
t_1 = initial temperature of slab (K)
t_0 = surface temperature after time zero (K)
Θ = Fourier Modulus = $\alpha\tau/\delta^2$ (dimensionless)
α = thermal diffusivity (m^2/s)
τ = time (s)
δ = half thickness of slab (m)
L = thickness of slab (m)
x = distance from center of slab (m)

The Fourier-Modulus is a non-dimensional ratio which may be used to generalize solutions of equation 6.121. It is a function of time along with the thermal and physical properties of the material.

As in the previous case, the temperature ratio has a value of 1.0 at time zero and approaches 0.0 as time approaches infinity. Equation 6.122 is an infinite series solution to the Fourier-Poisson equation for this case. However, the solution may usually be approximated by evaluating a relatively small number of terms.

Heat Transfer—Conduction

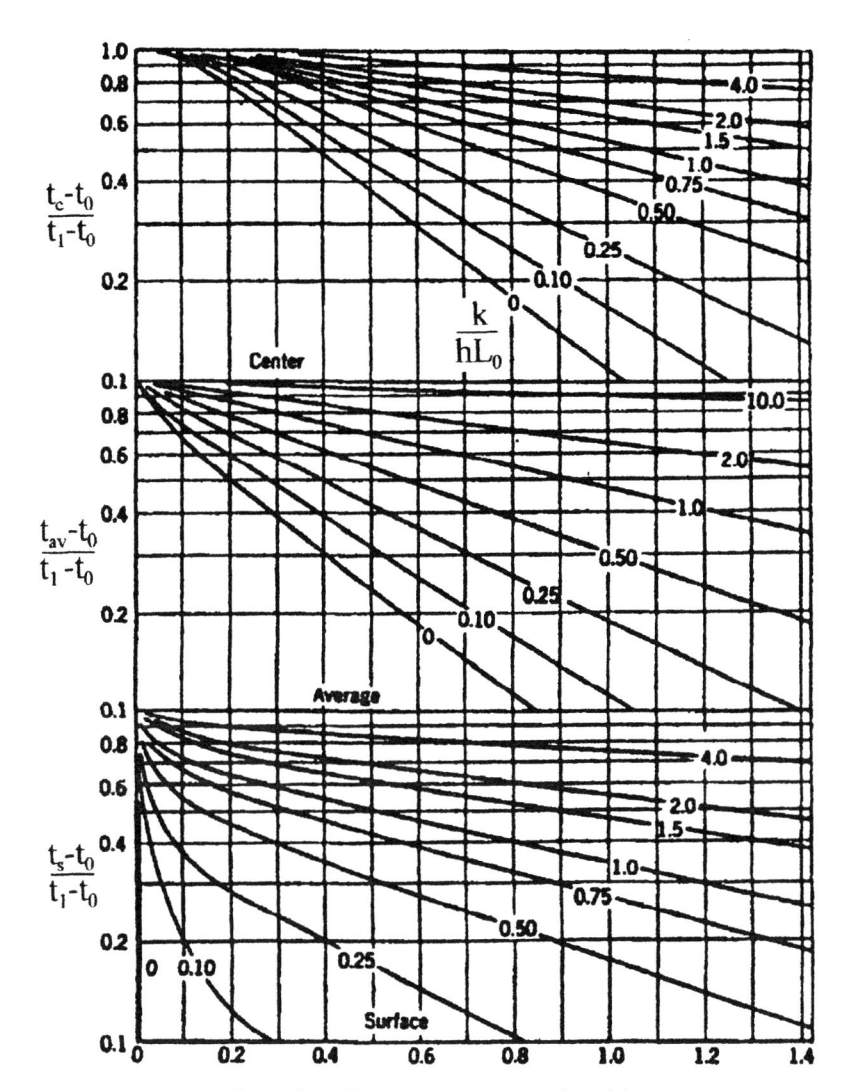

Figure 6.11. Transient temperatures in a slab.

Case 3

Finite Internal and Surface Resistance. The case where both the internal and surface resistance must be considered has been solved and charts have been prepared for infinite slabs, infinite cylinders, and spheres. They are called Heisler charts and are given in figures 6.11, 6.12, and 6.13, respectively. Each chart is a plot of temperature ratio versus Fourier-Modulus for various values of the inverse of the Biot number. Each of the figures is actually a composite of three figures with the top portion of the figure plotting the temperature ratio at the center of the body, the center portion plotting the temperature ratio based on the average temperature, and the bottom portion plotting the temperature

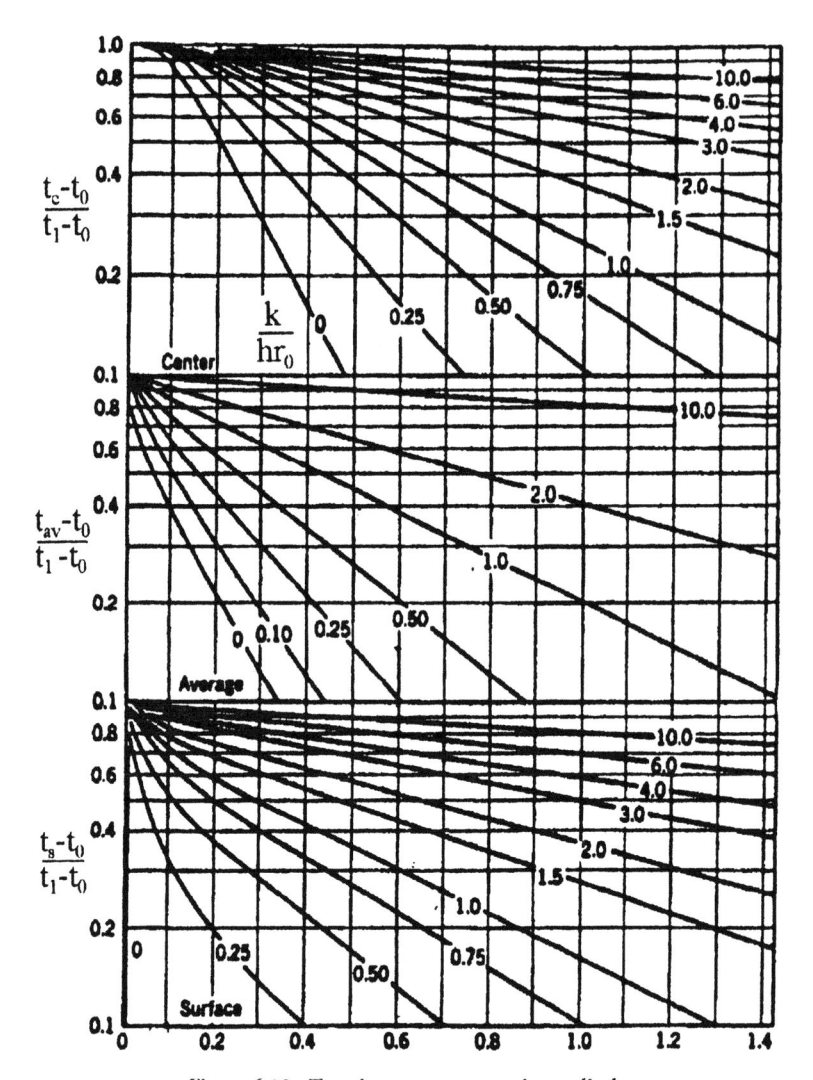

Figure 6.12. Transient temperatures in a cylinder.

ratio at the surface. It should be noted that Case 2, negligible surface resistance, may be solved using the Heisler charts by using the curves for the inverse Biot number of 0.0. Thus, numerical solution of equation 6.122 is not necessary unless the temperature at some location other than those plotted is desired.

The temperature ratio for the short cylinder has been shown to be the product of the ratio for the infinitely long cylinder of the same radius multiplied by the ratio for the infinite slab of a thickness equal to the length of the short cylinder. Similarly, the ratio for a rectangular solid with all surfaces exposed is the product of the ratios for three slabs of thicknesses equal, respectively, to the three dimensions of the solid.

Heat Transfer—Conduction

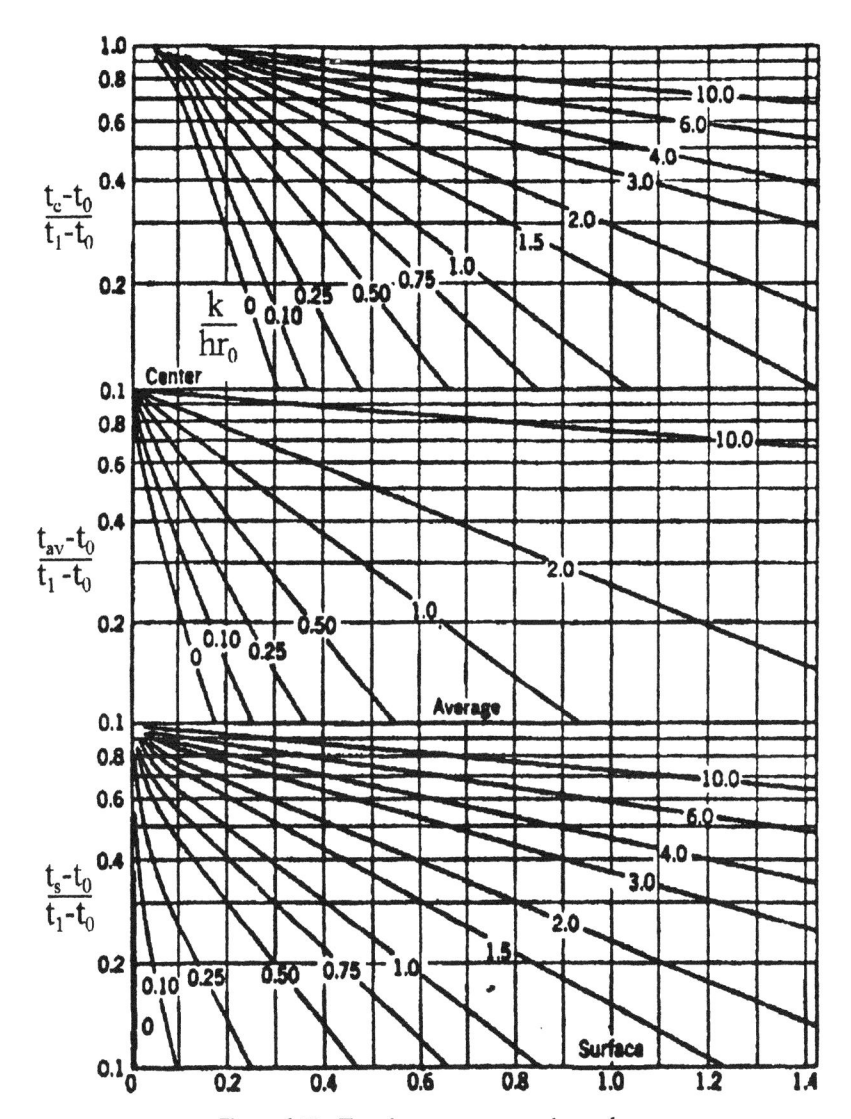

Figure 6.13. Transient temperatures in a sphere.

Example 6.2. A number 2 can, net diameter 8.41 cm, net length 10.64 cm, contains 567 g of a solid-pack food product which has a moisture content of 80% and a density of 1089 kg/m^3. The can, with its contents initially at 82 C, is placed in a retort where the surrounding temperature is promptly raised to 116 C with steam. Find the temperature at the center of the can at the end of 30 min. Surface convection coefficient, h, for condensing steam is 5 678 W/m^{2}K.

The specific heat is estimated to be 3.5 kJ/kg K; the thermal conductivity is 0.43 W/mK.

Solution

$$\alpha = k/\rho c_p$$
$$= (0.43 \text{ W/mK})/(1\,089 \text{ kg/m}^3)(3.5 \text{ kJ/kg K})(1\,000 \text{ J/kJ})$$
$$= 1.128 \times 10^{-7} \text{ m}^2/\text{s}$$

For the cylinder

$$1/Bo = k/hr_o$$
$$= (0.43 \text{ W/mK})/(5\,678 \text{ W/m}^2\text{K})(4.205 \text{ cm})(1 \text{ m}/100 \text{ cm})$$
$$= .0018$$

at 30 min

$$\Theta = k\tau/c_p\rho r_o^2 = \alpha\tau/r_o^2$$
$$= 1.128 \times 10^{-7} \text{m}^2/\text{s})(30 \text{ min})(60 \text{ s/min})/(4.205 \text{ cm})^2(1 \text{ m}/100 \text{ cm})^2$$
$$= 0.1148$$

From figure 6.12, the residual temperature ratio at the center of an infinite cylinder would be 0.8.

For the slab

$$1/Bo = k/hL_o$$
$$= (0.43 \text{ W/mK})/(5\,678 \text{ W/m}^2\text{K})(5.32 \text{ cm})(1 \text{ m}/100 \text{ cm})$$
$$= 0.0014$$

at 30 min

$$\Theta = k\tau/c_p\rho L_o^2 = \alpha\tau/L_o^2$$
$$= (1.128 \times 10^{-7} \text{m}^2/\text{s})(30 \text{ min})(60 \text{ s/min})/(5.32 \text{ cm})^2(1 \text{ m}/100 \text{ cm})^2$$
$$= 0.0717$$

From figure 6.11, the residual temperature ratio at the center of an infinite slab with a thickness equal to the length of the finite cylinder would be 0.95.

For the finite cylinder

$$(t - t_0)/(t_1 - t_0)_{\text{fin.cyl}} = (t - t_0)/(t_1 - t_0)_{\text{inf.cyl}}(t - t_0)/(t_1 - t_0)_{\text{slab}}$$
$$= (0.8)(0.95) = 0.76$$

Then

$$t = 0.76(t_1 - t_0) + t_0$$
$$= 0.76(82 - 116) \text{ C} + 116 \text{ C} = 90.2 \text{ C}$$

Heat Transfer—Conduction

Numerical Solutions

Many heat transfer problems do not yield to analytical treatment due to the complexity of either the geometry or the boundary conditions. Thus, other means of solution must be sought. Numerical solution techniques may be classified into two major types: (1) finite difference and (2) finite element methods. These may be used to solve either steady-state or transient heat transfer problems. However, the scope of this book will be limited to the use of a finite difference method for the solution of steady-state conduction problems with convection at the surface. For additional information on numerical solutions the reader may refer to one of many texts on the subject.

6.15 Finite Difference Solutions for Steady-State Heat Transfer in Two Dimensions with No Heat Generation

The Fourier-Poisson equation for steady-state heat conduction in two dimensions with no heat generation reduces to:

$$\partial^2 t/\partial x^2 + \partial^2 t/\partial y^2 = 0 \quad (6.123)$$

In order to solve this equation numerically, the material is replaced by a grid network (fig. 6.14) and approximations of partial derivatives at each node may be developed mathematically. The finite difference approximations for $\partial^2 t/\partial x^2$ and $\partial^2 t/\partial y^2$ at node (i, j) are:

$$\partial^2 t/\partial x^2 = (t(i-1, j) - 2t(i, j) + t(i+1, j))/(\Delta x)^2 \quad (6.124)$$

and

$$\partial^2 t/\partial y^2 = (t(i, j-1) - 2t(i, j) + t(i, j+1))/(\Delta y)^2 \quad (6.125)$$

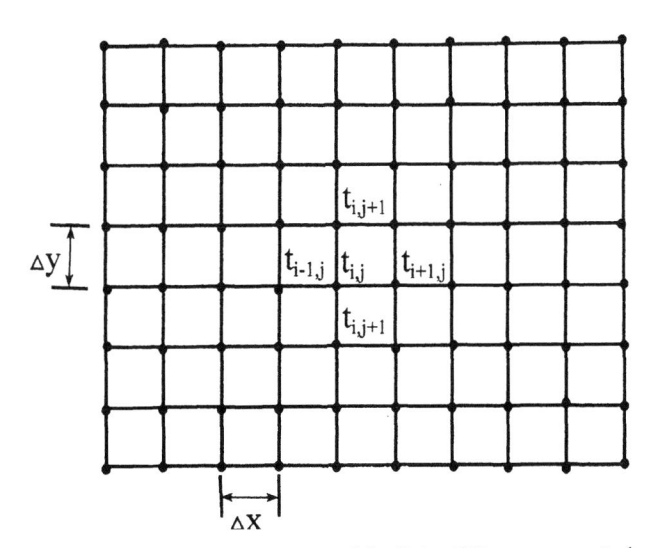

Figure 6.14. Rectangular grid used for finite difference numerical solutions of conduction heat transfer.

If these are substituted into equation 6.123 and Δx and Δy are assumed equal (i.e., a square grid), the following equation results:

$$t(i, j) = (t(i - 1, j) + t(i + 1, j) + t(i, j - 1) + t(i, j + 1))/4 \qquad (6.126)$$

Equation 6.126 may be written for each interior node within the material for which a two-dimensional heat transfer solution is desired. The temperatures at the boundaries must either be specified or defined by heat convection equations.

Although equation 6.126 was developed strictly mathematically, it is beneficial to consider a physical approach to the development of the same relationship. Let us assume that heat is conducted only along the fictitious rods which connect the nodes of the grid and develop an energy balance for the node (i, j). Then, at steady-state:

$$kZ\Delta y(t(i-1, j) - t(i, j))/\Delta x + kZ\Delta x(t(i, j-1) - t(i, j))/\Delta y$$
$$= kZ\Delta y(t(i, j) - t(i+1, j))/\Delta x + kZ\Delta x(t(i, j) - t(i, j+1))/\Delta y \qquad (6.127)$$

where

$Z =$ dimension perpendicular to the plane (m)
$k =$ thermal conductivity (W/mK)

The two terms on the left-hand side of equation 6.127 represent the heat conducted into node (i, j) from nodes $(i-1, j)$ and $(i, j-1)$, respectively. The two terms on the right-hand side of the equation represent the heat conducted away from node (i, j) to nodes $(i + 1, j)$ and $(i, j + 1)$, respectively. For a constant k and a square grid, equation 6.127 reduces to equation 6.126.

If there is heat convection at the surface, an energy balance can be written for a left surface node as follows:

$$hZ\Delta y(t_f - t(i, j)) + kZ(\Delta x/2)(t(i, j-1) - t(i, j))/\Delta y$$
$$= kZ\Delta y(t(i, j) - t(i+1, j))/\Delta x + kZ(\Delta x/2)(t(i, j) - t(i, j+1))/\Delta y$$
$$(6.128)$$

where

$t(i, j) =$ temperature at a left surface node (K)
$h =$ convective heat transfer coefficient (W/m^{2}K)
$t_f =$ temperature of surrounding fluid (K)

Note that the conduction from the node below and to the node above the node (i, j) is reduced by one-half since the area available for conduction is one-half that available at an interior node. For a square grid and constant k, equation 6.128 reduces to:

$$t(i, j) = (t(i, j - 1) + t(i, j + 1) + 2t(i + 1, j) + 2Nt_f)/(4 + 2N) \qquad (6.129)$$

where

$$N = h\Delta x/k \qquad (6.130)$$

Heat Transfer—Conduction

Equation 6.130 may be generalized for a node on any outer surface as follows:

$$t_0 = (t_1 + t_2 + 2t_3 + 2Nt_f)/(4 + 2N) \qquad (6.131)$$

where

t_0 = the surface node temperature (K)
t_1 and t_2 = two adjacent surface node temperatures (K)
t_3 = adjacent interior node temperature (K)
t_f = surrounding fluid temperature (K)

Figure 6.15 gives the finite difference equations for nodes located at a number of different positions within a grid. When an energy balance equation has been written for each node of a grid, the number of independent equations will equal the number of unknown node temperatures. There are many techniques for solving the system of linear equations.

Interior Node

t0 = (t1+t2+t3+t4)/4

Outer surface Node

t0 = (t1+t2+2t3+2Ntf)/(4+2N)

N = hΔx/k

Outer Surface Corner Node

t0 = (t1+t2+2Ntf)/(2+2N)

N = hΔx/k

Inner Surface Corner Node

t0 = (2t1+2t3+t2+t4+2Ntf)/(6+2N)

N = hΔx/k

Insulated Surface Node

t0 = (t1+t2+2t3)/4

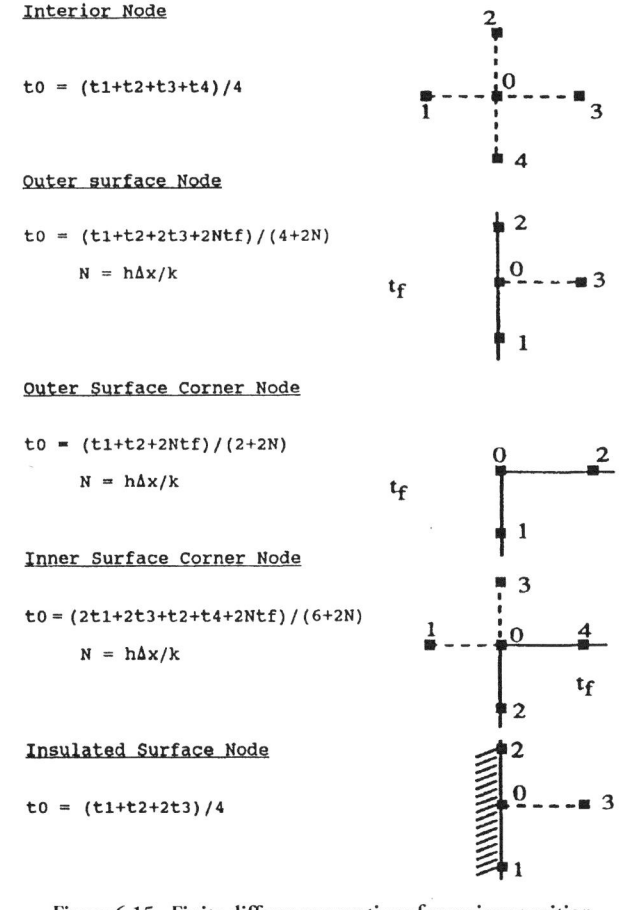

Figure 6.15. Finite difference equations for various position.

Node at Interface Between Two Materials

$$t_0 = \frac{(2Kt_1 + (1+K)t_2 + (1+K)t_3 + 2t_4)}{(4+4K)}$$

$$K = k_1/k_2$$

Node at corner Intersection of Two Materials

$$t_0 = \frac{((1+K)t_1 + (1+K)t_2 + 2t_3 + 2t_4)}{(6+2K)}$$

$$K = k_1/k_2$$

Node on Line of Symmetry (Vertical)

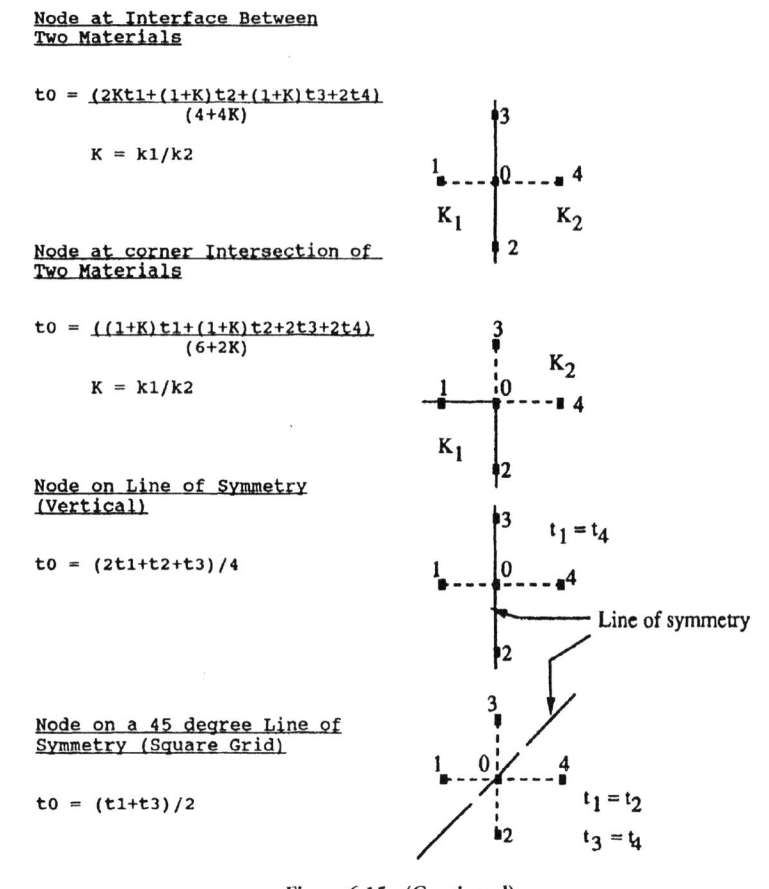

$$t_0 = (2t_1 + t_2 + t_3)/4$$

Node on a 45 degree Line of Symmetry (Square Grid)

$$t_0 = (t_1 + t_3)/2$$

Figure 6.15. (Continued)

A convenient solution technique which requires a minimum of computer programming time is the use of spreadsheet programs on a microcomputer. Spreadsheet programs such as SuperCalc, Lotus 1-2-3, Quattro, and Excel are among the best selling programs for microcomputers and may be readily used for solution of the set of equations defining the node temperatures. Harbach (1987) gives some advantages and disadvantages of this technique and describes the desirable features of the spreadsheet programs. Two features which are desirable are: (1) an iteration feature and (2) the ability to insert desired initial values.

Example 6.3. Consider the cross-section of a tall chimney as shown in figure 6.16. The gases flowing through the chimney are at 100 C while the surrounding atmosphere is 0 C. The convective heat transfer coefficients at the inner and outer surfaces are 68.0 and 17.0 W/m^{2}K, respectively. The thermal conductivity of the chimney wall is 1.7 W/mK. Determine the temperature distribution and the rate of heat flow per meter of height of the chimney.

Heat Transfer—Conduction

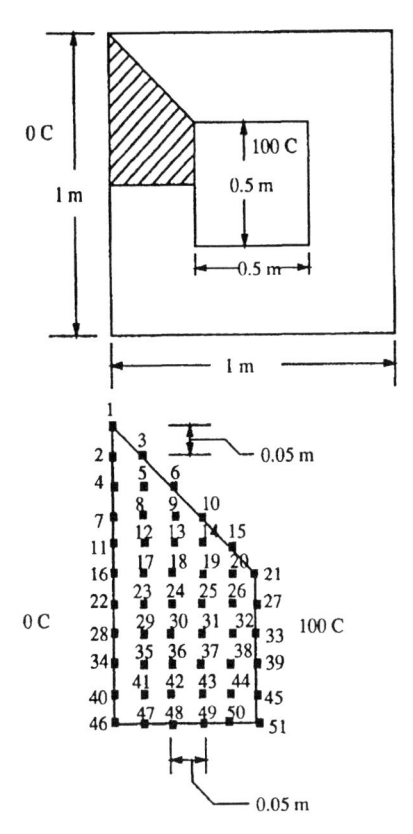

Figure 6.16. Top view of chimney and node layout for example 6.3.

Solution

Since the chimney is symmetrical it is only necessary to solve for the temperatures in a 1/8th section as indicated by the shaded area of figure 6.16. This 1/8th section may be divided into the grid with nodes 1 to 51 as shown. Examples of the different node equations obtained follow:

Interior node 5:

$$t_5 = (t_4 + t_8 + t_3 + t_6)/4$$

Outer surface node 4:

$$t_4 = (t_7 + t_2 + 2t_5 + 2N_o t_{fo})/(4 + 2N_o)$$

where $N_o = h_o \Delta x / k = (17 \text{ W/m}^2\text{K})(0.05 \text{ m})/(1.7 \text{ W/mK}) = 0.5$

Outer corner node 1:

$$t_1 = (t_2 + N_o t_{fo})/(1 + N_o)$$

(Note that node 2 and 2' are at the same temperature.)

Line of symmetry node 3:
$$t_3 = (t_2 + t_5)/2$$

Inner corner node 21:
$$t_{21} = (2t_{20} + t_{27} + N_i t_{fi})/(3 + N_i)$$

(Note: $t_{27} = t_{27'}$ and $t_{20} = t_{20'}$)
where $N_i = h_i \Delta x / k = (68 \text{W/m}^2\text{K}) (0.05 \text{ m}) / (1.7 \text{ W/mK}) = 2.0$

Inner surface node 27:
$$t_{27} = (t_{21} + t_{33} + 2t_{26} + 2N_i t_{fi})/(4 + 2N_i)$$

where $N_i = h_i \Delta x / k = (68 \text{ W/m}^2\text{K})(0.05 \text{ m})/(1.7 \text{ W/mK}) = 2.0$

Line of symmetry node 50:
$$t_{50} = (t_{49} + 2t_{44} + t_{51})/4$$

Line of symmetry node 51:
$$t_{51} = (t_{50} + t_{45} + N_i t_{fi})/(2 + N_i)$$

Line of symmetry node 46:
$$t_{46} = (t_{40} + t_{47} + N_o t_{fo})/(2 + N_o)$$

Figure 6.17 is a printout of results from a Quatro program to solve for the temperatures at the nodes of figure 6.16. In writing the spreadsheet program it is not necessary to develop separate equations for each node since relative references may be used and the equation copied for all other similarly placed nodes. Since spreadsheet programs will vary slightly, it will be left to the reader to develop an actual program which will give the results shown in figure 6.17.

tfo	tfi	ho	hi	dx	k	No	Ni
0.0	100	17	68	0.05	1.7	0.5	2
5.9							
8.8	13.2						
11.7	17.6	23.5					
14.5	21.9	29.5	37.3				
17.2	26.0	35.2	45.1	55.6			
19.6	29.6	40.3	52.2	66.1	84.5		
21.5	32.5	44.3	57.3	72.2	90.1		
23.0	34.7	47.1	60.6	75.3	91.6		
23.9	36.1	48.9	62.4	76.9	92.3		
24.5	36.9	49.8	63.4	77.7	92.5		
24.6	37.1	50.1	63.7	77.9	92.6		
qpin =		1223.6					
qpout =		1223.6					

Figure 6.17. Spreadsheet solution of example 6.3.

Heat Transfer—Conduction

The rate of heat flow per meter of height of the chimney can be approximated by summing the rates of heat flow through the rods terminating either at the inner or outer surface. At the inner surface we have:

$$q' = 8(0.5 h_i \Delta y(t_{fi} - t_{21}) + h_i \Delta y(t_{fi} - t_{27}) + h_i \Delta y(t_{fi} - t_{33}) + h_i \Delta y(t_{fi} - t_{39}) \\ + h_i \Delta y(t_{fi} - t_{45}) + 0.5 h_i \Delta y(t_{fi} - t_{51}))$$

or

$$q' = 8(68 \text{ W/m}^2\text{K})(0.05 \text{ m})(0.5(100 - 84.5) + (100 - 90.1) + (100 - 91.6) \\ + (100 - 92.3) + (100 - 92.5) + 0.5(100 - 92.6)) \text{ K}$$
$$q' = 8(68 \text{ W/m}^2\text{K})(0.05 \text{ m})(44.95 \text{ K}) = 1223.6 \text{ W/m}$$

At the outer surface we have:

$$q' = 8(0.5 h_o \Delta y(t_1 - t_{fo}) + h_o \Delta y(t_2 - t_{fo}) + h_o \Delta y(t_4 - t_{fo}) \\ + h_o \Delta y(t_7 - t_{fo}) + h_o \Delta y(t_{11} - t_{fo}) + h_o \Delta y(t_{16} - t_{fo}) \\ + h_o \Delta y(t_{22} - t_{fo}) + h_o \Delta y(t_{28} - t_{fo}) + h_o \Delta y(t_{34} - t_{fo}) \\ + h_o \Delta y(t_{40} - t_{fo}) + 0.5 h_o \Delta y(t_{46} - t_{fo}))$$

or

$$q' = 8((17 \text{ W/m}^2 \text{ K})(0.05 \text{ m})(0.5(5.9) + 8.8 + 11.7 + 14.5 + 17.2 \\ + 19.6 + 21.5 + 23.0 + 23.9 + 24.5 + 0.5(24.6)) \text{ K}$$
$$q' = 1223.6 \text{ W/m}$$

Note the agreement between the two numerical estimates of heat transfer per unit height. Any differences between these estimates would indicate either round-off errors (differences should be small) or programming errors.

If additional accuracy is required, a finer node mesh can be generated for the solution. However, the additional accuracy will come at the expense of greater computation times.

Nomenclature

- A area (m²)
- Bo Biot number (dimensionless)
- c_p specific heat (J/kg K)
- Gr Grashof number (dimensionless)
- h convective heat transfer coefficient (W/m²K)
- i current (amps)
- k thermal conductivity (W/mK)
- q heat flow (W)
- q' heat transfer per unit length (W/m)
- q'' heat flux (W/m²)
- q''' heat generation per unit volume (W/m³)
- Q heat added to body (J)

R   resistance (ohms for electrical resistance or $m^2 K/W$ for thermal resistance)
R'   resistance per unit length (mK/W)
R_s resistance to heat transfer in a sphere (K/W)
t   temperature (C)
T   absolute temperature (K)
u   internal energy (J)
U   overall heat transfer coefficient ($W/m^2 K$)
U'   overall heat transfer coefficient per unit length (W/m)
U_s total heat transfer coefficient for a sphere (W/K)
w   work done on body (J)
α thermal diffusivity (m^2/s)
δ characteristic length (m)
ε emissivity (dimensionless)
ρ mass density (kg/m^3)
σ Stefan-Boltzmann constant = 5.73×10^{-8} $W/m^2 K^4$
τ time (s)
Θ Fourier-Modulus (dimensionless)

References

Harbach, J. A. 1987. Solving Elliptic Partial Differential Equations Using a Spreadsheet Program. COED. Computers in Education Division of Am. Soc. of Engr. Ed. 7:14–16.

Jakob, M. and G. A. Hawkins. 1957. *Elements of Heat Transfer*. New York: John Wiley & Sons, Inc.

Kreith, F. 1958. *Principles of Heat Transfer*. Scranton, Pa.: International Textbook Co.

Neubauer, L. W. and H. B. Walker. 1961. *Farm Building Design*. Englewood Cliffs, N.J.: Prentice-Hall, Inc.

Problems

1. The inside surface temperature of a 45 cm furnace wall is 1 100 C. The heat flux is constant at 1 700 W/m^2 in the outward direction of the plane surface. Develop an expression for the temperature at any distance outward from the inner surface of the wall. Evaluate the temperature at distances of 11.25, 22.5, 33.75, and 45.0 cm from the inner surface. Assume the thermal conductivity of the wall to be 1.06 W/mK. If the ambient air temperature is 35 C, what is the surface convective heat transfer coefficient, h?

2. Starting with the first law of thermodynamics, derive an equation similar to equation 6.13 for the case where the thermal conductivity is not a constant but is given by the equation:

$$k = k_0 + k_1 x$$

where k_0 and k_1 are constants.

Heat Transfer—Conduction

3. Repeat problem 2 for the case where:

$$k = k_0 + k_1(t - t_0)$$

where

k_0 and k_1 = constants
t = temperature
t_0 = a constant reference temperature

4. Find the rate of heat loss per unit area from the rectangular hot air supply duct of a spray drying chamber. The duct is insulated with a 5 cm layer of glass-fiber batts. The inner surface temperature of the insulation is 150 C, the outer, 25 C. Thermal conductivity of the batts = 0.39 W/mK.

5. A furnace wall is to be built of firebrick 20 cm thick and building brick of the same thickness. The thermal conductivities are 1.6 and 0.7 W/mK, respectively. The inner surface of the firebrick is at 980 C, the outer surface of the structural brick at 30 C. Find the heat flux and the temperature of the brick interface. What fraction of the resistance is provided by the firebrick?

6. Determine the percentage reduction in heat loss through the example composite wall (fig. 6.6) if the wall is filled with mineral wool.

7. Determine the heat flux through the wall below:
Inside Temperature = 20 C

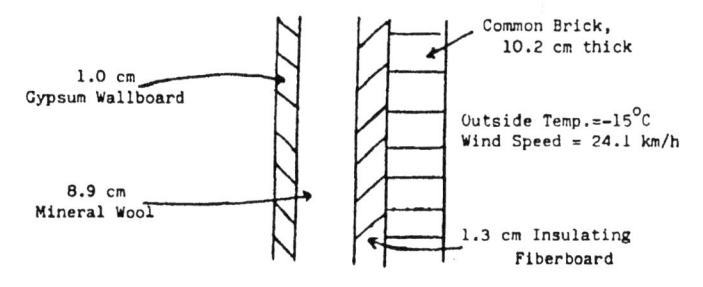

- 1.0 cm Gypsum Wallboard
- 8.9 cm Mineral Wool
- Common Brick, 10.2 cm thick
- Outside Temp. = -15°C
- Wind Speed = 24.1 km/h
- 1.3 cm Insulating Fiberboard

8. A steam main (outside diameter = 16.8 cm, inside diameter = 15.4 cm) is covered with 5 cm of high temperature insulation (k = 0.095 W/mK) and 3.8 cm of lower temperature insulation (k = 0.07 W/mK). Calculate the heat loss from 150 m of pipe, assuming that the inner and outer surface temperatures of the insulation are 450 and 30 C, respectively. Also, determine the temperature at the interface between the two layers of insulation.

9. A stainless steel tube having an outside diameter of 0.6 cm and a wall thickness of 0.05 cm is to be insulated with a material having a thermal conductivity of 0.065 W/mK. If the inner and outer convective heat transfer coefficients are 5.9 W/m²K, what will be the heat loss per meter of length for insulation thicknesses of 0, 0.25, 0.5, 0.75, 1.00, and 1.25 cm if the inside temperature is 95 C and the outside temperature is 10 C? Plot heat loss per meter versus insulation

thickness. Also, plot the surface and interface temperatures as a function of insulation thickness.

10. The annular space between two thin concentric spherical shells having radii of 7.5 and 15.0 cm is filled with bulk powdered insulating material. What wattage is required from an electric heater located in the center of the smaller sphere in order to maintain a temperature difference between the two spherical shells equal to 20 C? Assume the average thermal conductivity of the insulating material to be 0.07 W/mK. Plot the temperature versus radius if the inside surface is at 80 C.

11. Find the temperature of a steel sphere, 20 cm in diameter, initially at 20 C, 40 min. after being placed in an annealing furnace where the temperature is 315 C, if the surface convection coefficient is estimated to be 34 W/m²K. The specific heat of steel is 0.50 kJ/kgK, the density is 7210 kg/m³, and the thermal conductivity is 45 W/mK.

12. How long will be required for the sphere of problem 11 to reach a temperature of 150 C?

13. Estimate the time required to bring the center of a spherical melon, 23 cm in diameter, to a temperature of 5 C after the melon, initially at 25 C, is placed in a refrigerator at 2 C. The surface convective heat transfer coefficient will be 28.4 W/m²K, the density = 998 kg/m³, the specific heat = 3.8 J/gK, and the thermal conductivity = 0.43 W/mK. Also estimate the mean temperature of the melon when the center has reached 5 C.

14. Apples are to be cooled after harvest by immersing in water at 0 C. If the initial temperature of the apples is 30 C, determine the time required for the average temperature to reach 6 C. Assume the convection coefficient at the surface to be 42.0 W/m²K. The thermal conductivity is 0.42 W/mK, specific heat is 3.8 kJ/kgK, density is 787 kg/m³, and radius is 4 cm for the apples.

15. A long wooden rod, 2.5 cm outside diameter, is placed at 35 C into an airstream at 815 C. The surface convection coefficient between the rod and the air is 28.4 W/m²K. If the ignition temperature of the wood is 425 C, $\rho = 800$ kg/m³, $k = 0.17$ W/mK, and $c_p = 2.5$ kJ/kgK determine the time between initial exposure and ignition of the wood.

16. Develop an equation analogous to equation 6.126 for an internal node of a body in which steady-state heat transfer is three-dimensional.

17. Use an energy balance approach to derive the equation given in figure 6.15 for a node at an insulated surface.

18. Use a spreadsheet program to solve example 6.3 if the outer surface heat transfer coefficient is increased to 41 W/m²K.

19. Solve problem number 18 using a grid in which $\Delta x = \Delta y = 0.025$ m.

20. Use an energy balance approach to derive an equation for the temperature at an internal node if there is internal heat generation. Assume two-dimensional, steady-state heat flow and uniform heat generation per unit volume.

7 Heat Transfer—Convection

Heat convection is the transfer of thermal energy by fluids in ordinary or macroscopic motion which is either directly visible or can be visualized by the usual instruments for the measurement of the flow of fluids. This broad definition of convection would include the transfer of thermal energy from one location to another simply by the movement of a warm fluid to a cooler location. However, discussions of convection are usually limited to the exchange of heat energy between the fluid and surfaces of different temperatures. This energy transfer at the surface is due to movement of the fluid, however, rather than simply the molecular energy exchange which occurs in conduction.

Since in this process heat is conveyed mechanically, it is obvious that the transfer of energy depends upon the motion of the fluid and is governed by the laws of fluid dynamics, in addition to the laws of heat conduction and heat storage which must be considered at the same time. From this it will be understood that heat convection is a very complex process and that the simplicity of equation 6.4 is delusive. As a matter of fact, the film coefficient h, defined by that equation, is a function of many variables, such as shape and dimensions of the surface, roughness, direction, and velocity of the flow, temperature, density, viscosity, specific heat, and thermal conductivity of the fluid.

The differential equations which describe convective heat transfer belong to the most difficult class in theoretical physics and it has been possible to solve them for only a few very specialized cases. An empirical treatment of heat convection has been likewise unsatisfactory because of the many variables involved.

Thus, the knowledge of heat transfer by convection was extremely meager until the early 1900s when the application of the principle of similarity or the procedure of dimensional analysis to heat transfer problems began to change the situation entirely.

7.1 Dimensional Analysis

The basic principle upon which dimensional analysis is founded is dimensional homogeneity; that is, dissimilar quantities cannot be added together to form a valid physical relation. Mathematically, this means that, in any physical equation both sides of which can be written as power functions or as algebraic sums of power functions, the sum of the exponents of the basic units must be the same on the left and right sides. For instance, according to the laws of freely falling bodies:

$$v = \sqrt{2gl} \qquad (7.1)$$

where

v = velocity of body (m/s)
g = acceleration due to gravity (m/s^2)
l = distance through which body has fallen (m)

As a numerical example,

$$8.86 \text{ m/s} = \sqrt{2(9.807 \text{ m/s}^2)(4 \text{ m})}$$

Considering the basic units above,

$$m^{+1}s^{-1} = m^{1/2}s^{-1}m^{1/2}$$

Obviously, the above rule for the sum of the exponents yields:

$1 = 1/2 + 1/2$ for the exponents of m
$-1 = -1$ for the exponents of s

Now assume that equation 7.1 is not known, but that it is known that v depends only on g and l. The latter fact may be expressed by:

$$v = f(g, l) \qquad (7.2)$$

From the principle of dimensional homogeneity it is apparent that the quantities involved cannot be added or subtracted since their dimensions are different. This principle limits the equation to a combination of products of powers of the quantities involved, which may be expressed in the general form:

$$v = Cg^a l^b \qquad (7.3)$$

in which C is a dimensionless constant which may exist in the equation, but cannot be obtained by dimensional methods. Representing the fundamental units of length and time by L and T, respectively, an equation for the dimensions of equation 7.3 is:

$$LT^{-1} = C(LT^{-2})^a(L)^b \qquad (7.4)$$

and the rule of the sum of the exponents yields:

for L $1 = a + b$ (7.5)
for T $-1 = -2a$ (7.6)

Thus solving for a and b gives:

$$a = 1/2 \quad \text{and} \quad b = 1/2$$

By substitution in the right side of equation 7.3,

$$v = Cg^{1/2}l^{1/2} = C\sqrt{gl} \qquad (7.7)$$

Heat Transfer—Convection

Therefore, the law given by equation 7.1 has been found by the method of dimensional analysis, and only the constant C remains unknown and must be determined by experiments. According to equation 7.1, $C = \sqrt{2}$.

There are four basic dimensions with which we will be concerned in our study of heat convection. They are: (1) length—L, (2) mass—M, (3) time—T, and (4) temperature—Θ.

7.2 Application of Dimensional Analysis to Free Convection

Free convection concerns thermal energy transfer due to the movement of a fluid which moves only due to differences in density which are caused by temperature differences within the fluid. In 1915, W. Nusselt used the dimensional analysis procedure to develop basic relationships for free convection. Before any valid relations may be established, it is essential to know what variables are involved. Correct selection of the variables comes about primarily through experience. Let us consider the pertinent variables for the case of free convection from the outer surface of a cylindrical pipe. A list of the possible pertinent variables and their basic dimensions follows:

h = convective heat transfer coefficient $(MT^{-3}\Theta^{-1})$
D = outer diameter of pipe (L)
Δt = temperature difference between the surface and the fluid (Θ)
ρ = density of fluid (ML^{-3})
μ = dynamic viscosity of fluid $(MT^{-1}L^{-1})$
β = coefficient of thermal expansion of the fluid (Θ^{-1})
g = acceleration due to gravity (LT^{-2})
c_p = specific heat of the fluid $(L^2T^{-2}\Theta^{-1})$
k = thermal conductivity of the fluid $(MLT^{-3}\Theta^{-1})$
l = position along pipe (L)

A general relationship which includes all these variables is:

$$h = C D^a (\Delta t)^b \rho^c \mu^d \beta^e g^f c_p^i k^j l^m \tag{7.8}$$

The corresponding equation of dimensions is:

$$MT^{-3}\Theta^{-1} = C L^a \Theta^b (ML^{-3})^c (MT^{-1}L^{-1})^d (\Theta^{-1})^e (LT^{-2})^f \\ (L^2T^{-2}\Theta^{-1})^i (MLT^{-3}\Theta^{-1})^j L^m \tag{7.9}$$

The equations for the sums of the exponents become:

$$\text{for } T \quad -3 = -d - 2f - 2i - 3j \tag{7.10}$$
$$\text{for } L \quad 0 = a - 3c - d + f + 2i + j + m \tag{7.11}$$
$$\text{for } M \quad 1 = c + d + j \tag{7.12}$$
$$\text{for } \Theta \quad -1 = b - e - i - j \tag{7.13}$$

Thus there are four equations with nine unknowns. The four equations may be used to eliminate four of the unknowns by expressing them in terms of the five remaining unknowns. If we choose to express exponents a, c, d, and j in terms of b, e, f, i, and m, then we can obtain:

$$j = 1 + b - e - i \tag{7.14}$$
$$d = -2f + i - 3b + 3e \tag{7.15}$$
$$c = 2f + 2b - 2e \tag{7.16}$$
$$a = -1 + 3f + 2b - 2e - m \tag{7.17}$$

By substituting into equation 7.8 we obtain:

$$h = CD^{-1+3f+2b-2e-m}(\Delta t)^b \rho^{2f+2b-2e} \mu^{-2f+i-3b+3e} \beta^e \\ g^f c_p^i k^{1+b-e-i} l^m \tag{7.18}$$

Collecting the terms with like exponents gives:

$$hD/k = C(D^3\rho^2 g/\mu^2)^f (D^2 \Delta t \rho^2 k/\mu^3)^b (\mu^3 \beta / D^2 \rho^2 k)^e \\ (\mu c_p/k)^i (l/D)^m \tag{7.19}$$

Equation 7.19 can be further simplified if the relation between g, β, and Δt is considered. Free convection is due to a thermal buoyancy effect. If in a fluid at temperature t_1 and density ρ_1 a unit of volume is heated to temperature t_2, and, as a result, has a smaller density ρ_2, then according to Archimedes' law, the buoyancy $B = \rho_1 g - \rho_2 g$ acts upon the mentioned unit volume. However, with respect to the definition of the coefficient of thermal expansion,

$$V_2 = V_1 + \Delta V = V_1 + V_1 \beta \Delta t = V_1(1 + \beta(t_2 - t_1))$$

From the definition of density,

$$\rho_1 = m/V_1$$

or

$$\rho_1 V_1 = m = \rho_2 V_2$$

and

$$\rho_1 V_1 = \rho_2 V_1(1 + \beta(t_2 - t_1))$$

Then,

$$\rho_1 = \rho_2(1 + \beta(t_2 - t_1)) \tag{7.20}$$

By substitution of equation 7.20 into the buoyancy relationship,

$$B = g\rho_2(1 + \beta(t_2 - t_1)) - g\rho_2$$

or

$$B = \rho_2 \beta g(t_2 - t_1) = \rho_2 \beta g \Delta t \tag{7.21}$$

Since β, g, and Δt enter the equation for free convection only due to their effect on buoyancy, then they must always appear in the form of a product $\beta g \Delta t$. According to

Heat Transfer—Convection

equation 7.19 this is possible only if $f = b = e$. Therefore, it simplifies to:

$$hD/k = C(D^3\rho^2\beta g \Delta t/\mu^2)^f (\mu c_p/k)^i (l/D)^m \tag{7.22}$$

Equation 7.22 involves three important dimensionless groups. They are:

$$\text{Nusselt Number} = Nu = hD/k \tag{7.23}$$
$$\text{Grashof Number} = Gr = D^3\rho^2\beta g \Delta t/\mu^2 \tag{7.24}$$
$$\text{Prandtl Number} = Pr = \mu c_p/k = \nu/\alpha \tag{7.25}$$

Thus,

$$Nu = C(Gr)^m (Pr)^n (l/D)^p \tag{7.26}$$

represents a general expression for free convection from a cylinder. The equation also holds for a plane surface where D is then the length of the surface and a different value of C, m, n, and p would apply.

It may be noted that the Nusselt number is very similar to the Biot number for a cylinder. The difference aside from the use of the diameter in the Nusselt number and the radius in the Biot number is that in the Biot number, k is the thermal conductivity of the solid while in the Nusselt number it is the thermal conductivity of the surrounding fluid.

7.3 Application of Dimensional Analysis to Forced Convection

In forced convection, thermal energy transfer is due to the movement of a fluid which moves due to some external force such as a pump, fan, or the wind. A dimensional analysis can be made for the case of forced convection as was shown by Nusselt in 1909. If we consider the convection between the inside surface of a pipe and a fluid flowing through it, the pertinent quantities are:

h = convective heat transfer coefficient $(MT^{-3}\Theta^{-1})$
D = diameter of pipe (L)
ρ = density of fluid (ML^{-3})
μ = dynamic viscosity of the fluid $(MT^{-1}L^{-1})$
c_p = specific heat of fluid $(L^2T^{-2}\Theta^{-1})$
k = thermal conductivity of the fluid $(MLT^{-3}\Theta^{-1})$
l = position along pipe (L)
v = velocity of fluid (LT^{-1})

A general relationship among these variables is:

$$h = CD^a \rho^b \mu^c c_p^d k^e l^f v^g \tag{7.27}$$

The corresponding equation of dimensions is:

$$MT^{-3}\Theta^{-1} = CL^a (ML^{-3})^b (MT^{-1}L^{-1})^c (L^2T^{-2}\Theta^{-1})^d \\ (MLT^{-3}\Theta^{-1})^e L^f (LT^{-1})^g \tag{7.28}$$

The equations for the sums of the exponents become:

For T	$-3 = -c - 2d - 3e - g$	(7.29)
For L	$0 = a - 3b - c + 2d + e + f + g$	(7.30)
For Θ	$1 = -d - e$	(7.31)
For M	$1 = b + c + e$	(7.32)

We now have four equations with seven unknowns. Thus four of the unknowns can be expressed in terms of the other three. If a, b, c, and e are expressed in terms of d, g, and f, we obtain:

$$e = 1 - d \tag{7.33}$$
$$c = d - g \tag{7.34}$$
$$b = g \tag{7.35}$$
$$a = -1 + g - f \tag{7.36}$$

Then:
$$h = C D^{-1+g-f} \rho^g \mu^{d-g} c_p^d k^{1-d} l^f v^g \tag{7.37}$$

or
$$hD/k = C(D\rho v/\mu)^g (\mu c_p/k)^d (l/D)^f \tag{7.38}$$

or
$$Nu = C(Re)^m (Pr)^n (l/D)^p \tag{7.39}$$

where Re = Reynolds Number = $D\rho v/\mu$ (dimensionless) and C, m, n, and p are constants which must be determined experimentally.

7.4 Main Advantages of Dimensional Analysis

The time and work which may be saved by setting up an equation of dimensionless groups such as equation 7.39 can be seen by the following estimation. Assume that in order to find h for all thinkable variations, we would proceed by changing D, taking for instance, five different cylinder diameters, then conducting experiments with each of these and with fluids of five thermal conductivities k, further varying ρ five times, and so on each of the seven independent variables D, k, ρ, v, μ, c_p, and l. This would mean $5^7 = 78,125$ tests. According to equation 7.39 on the other hand it would be sufficient to make experiments with five different values of each group on the right side of the equation, that is, $5^3 = 125$ experiments, in order to get about the same result as with more than 78,000 tests with the direct method.

7.5 Main Disadvantages of Dimensional Analysis

The primary disadvantage of dimensional analysis is that a considerable amount of prior knowledge and insight into the problem must be used in the selection of the pertinent quantities. If a pertinent quantity is overlooked the final results may be erroneous.

Heat Transfer—Convection

A second disadvantage of the dimensional analysis procedure is its failure to demonstrate the mechanics of how each variable affects the desired dependent quantity. That is, it is possible to develop a working equation without really understanding what is happening.

7.6 Experimental Equations for Free Convection

Many experimental investigations have been conducted to determine the constants of equation 7.26 for many different geometric configurations.

For the average convective heat transfer coefficient for free convection about a horizontal cylinder, equation 7.26 becomes:

$$Nu = C(Gr)^m(Pr)^n = 0.56(GrPr)^{0.25} \tag{7.40}$$

in the laminar range where $(GrPr)$ is between 10^4 and 10^8 and

$$Nu = 0.13(GrPr)^{1/3} \tag{7.41}$$

in the turbulent range where $(GrPr)$ is between 10^8 and 10^{12}.

The product $(GrPr)$ may also be referred to as the Rayleigh Number:

$$\text{Rayleigh Number} = Ra = \beta \Delta t g D^3/(\alpha v) \tag{7.41a}$$

For convection from a horizontal cylinder to air at 20 C, equations 7.40 and 7.41 become approximately:

$$h = 1.32(\Delta t/D)^{0.25} \tag{7.42}$$

$$h = 1.24(\Delta t)^{1/3} \tag{7.43}$$

For shapes other than cylinders, similar equations have been developed. For plane surfaces in air at 20 C:

1. Horizontal, heated facing upward, or cooled facing downward:

$$h = 1.32(\Delta t/L)^{0.25} \quad \text{Laminar range (small plates)} \tag{7.44}$$

$$h = 1.52(\Delta t)^{1/3} \quad \text{Turbulent range (large plates)} \tag{7.45}$$

2. Horizontal, heated facing downward, or cooled facing upward, square:

$$h = 0.59(\Delta t/L)^{0.25} \quad \text{Laminar range (small plates)} \tag{7.46}$$

3. Vertical surfaces:

$$h = 1.42(\Delta t/L)^{0.25} \quad \text{Laminar range (small plates)} \tag{7.47}$$

$$h = 1.31(\Delta t)^{1/3} \quad \text{Turbulent range (large plates)} \tag{7.48}$$

In each of these relationships the laminar range is considered to be the range where the product of the Grashof and Prandtl numbers is between 10^4 and 10^8 while the turbulent range has $(GrPr)$ between 10^8 and 10^{12}.

Example 7.1. Estimate the convective heat transfer coefficient from a 2-cm-diameter copper tube to 30 C water in which it is immersed. The surface temperature of the tube is 90 C. The thermal and physical properties of water are given in table 7-1.

Table 7-1. Properties of saturated water*

Temperature t (C)	Density ρ (kg/m³)	Dynamic Viscosity μ (Pa∗s)	Kinematic Viscosity ν (m²/s)	Specific Heat c_p (kJ/kg K)	Thermal Conductivity k (W/mK)	Thermal Diffusivity α (m²/s)	Prandtl Number Pr
0	999.55	0.001790	0.0000017910	4.208	0.5598	0.0000001316	13.60
5	999.55	0.001518	0.0000015195	4.206	0.5653	0.0000001345	11.30
10	999.55	0.001306	0.0000013058	4.187	0.5746	0.0000001373	9.52
15	999.55	0.001139	0.0000011386	4.187	0.5824	0.0000001391	8.19
20	998.27	0.001005	0.0000010064	4.187	0.5916	0.0000001416	7.11
25	996.83	0.000895	0.0000008968	4.187	0.6009	0.0000001440	6.23
30	995.39	0.000803	0.0000008062	4.187	0.6092	0.0000001462	5.52
35	993.95	0.000725	0.0000007303	4.187	0.6170	0.0000001483	4.92
40	992.50	0.000658	0.0000006637	4.187	0.6241	0.0000001501	4.41
45	990.58	0.000600	0.0000006057	4.187	0.6303	0.0000001519	3.98
50	988.02	0.000552	0.0000005579	4.187	0.6369	0.0000001540	3.63
55	986.42	0.000508	0.0000005148	4.187	0.6445	0.0000001561	3.30
60	983.53	0.000471	0.0000004800	4.187	0.6508	0.0000001579	3.03
65	980.65	0.000438	0.0000004475	4.187	0.6570	0.0000001600	2.79
70	977.77	0.000408	0.0000004170	4.187	0.6618	0.0000001617	2.58
75	974.08	0.000381	0.0000003905	4.201	0.6665	0.0000001628	2.40
80	972.00	0.000357	0.0000003680	4.208	0.6701	0.0000001639	2.24
85	969.12	0.000336	0.0000003471	4.208	0.6733	0.0000001650	2.10
90	965.59	0.000317	0.0000003283	4.216	0.6764	0.0000001661	1.98
95	961.75	0.000299	0.0000003120	4.235	0.6790	0.0000001667	1.87
100	958.87	0.000283	0.0000002963	4.250	0.6805	0.0000001669	1.77
105	955.82	0.000269	0.0000002823	4.248	0.6821	0.0000001679	1.67
110	951.50	0.000255	0.0000002684	4.229	0.6836	0.0000001698	1.58
115	947.17	0.000243	0.0000002568	4.229	0.6852	0.0000001710	1.50
120	942.85	0.000232	0.0000002460	4.245	0.6854	0.0000001711	1.44
125	939.64	0.000222	0.0000002358	4.264	0.6854	0.0000001711	1.38
130	935.80	0.000212	0.0000002263	4.271	0.6854	0.0000001716	1.32
135	931.47	0.000204	0.0000002183	4.281	0.6854	0.0000001719	1.27
140	927.15	0.000196	0.0000002115	4.291	0.6854	0.0000001723	1.23
145	922.82	0.000190	0.0000002053	4.298	0.6854	0.0000001729	1.19

* Adapted from data given by M. Jakob and G. A. Hawkins. 1957. *Elements of Heat Transfer*, 3rd Ed. New York: John Wiley & Sons, Inc.

Solution
The film temperature at which property values should be evaluated is (30 C + 90 C)/2 = 60 C.
The coefficient of thermal expansion for water at 60 C can be approximated by solving equation 7.20 for β and substituting the density values at temperatures slightly above and below 60 C. Equation 7.20 yields:

$$\beta = (\rho_1 - \rho_2)/\rho_2(t_2 - t_1)$$

Heat Transfer—Convection

From table 7-1, $\rho = 980.65$ kg/m³ at 65 C and $\rho = 986.42$ kg/m³ at 55 C. Thus,

$$\beta = (980.65 \text{ kg/m}^3 - 986.42 \text{ kg/m}^3)/(986.42 \text{ kg/m}^3 * (55 \text{ C} - 65 \text{ C}))$$
$$\beta = 0.000585 \text{ K}^{-1}$$

Then,

$$Gr = D^3 \rho^2 \beta g \Delta t / \mu^2$$
$$Gr = (0.02 \text{ m})^3 (983.53 \text{ kg/m}^3)^2 (0.000585 \text{ K}^{-1})(9.81 \text{ m/s}^2)(60 \text{ K})/$$
$$(0.000471 \text{ Pa s})^2$$
$$Gr = 1.202 \times 10^7$$

From table 7-1, $Pr = 3.03$.
Thus, $GrPr = (1.202 \times 10^7)(3.03) = 3.64 \times 10^7$ and equation 7.40 is applicable.

$$Nu = 0.56(GrPr)^{0.25}$$
$$= 0.56(3.64 \times 10^7)^{0.25} = 43.5$$

Then,

$$hD/k = Nu = 43.5$$
$$h = 43.5 \, k/D = 43.5(0.6508 \text{ W/mK})/(0.02 \text{ m}) = 1415 \text{ W/m}^2\text{K}$$

Example 7.2. Estimate the convective heat transfer coefficient from the 2-cm-diameter copper tube of example 7.1 if the tube is in 30 C air rather than water. The thermal and physical properties of air are given in table 7-2.

Solution
The film temperature is still 60 C.
The coefficient of thermal expansion may be estimated for the air at 60 C by assuming it to be an ideal gas. In that case, it can be shown that $\beta = 1/T$ where T is the absolute temperature.
Then,

$$\beta = 1/(60 + 273) \text{ K} = 0.003003 \text{ K}^{-1}$$

Then,

$$Gr = D^3 \rho^2 \beta g \Delta t / \mu^2$$
$$Gr = (0.02 \text{ m})^3 (1.0604 \text{ kg/m}^3)^2 (0.003003 \text{ K}^{-1})(9.81 \text{ m/s}^2)(60 \text{ K})/$$
$$(0.00002005 \text{ Pa s})^2 = 3.955 \times 10^4$$

From table 7-2, $Pr = 0.7072$.
Thus, $GrPr(3.955 \times 10^4)(0.7072) = 2.797 \times 10^4$ and equation 7.40 is applicable.

$$Nu = 0.56(GrPr)^{0.25} = 0.56(2.797 \times 10^4)^{0.25} = 7.24$$
$$hD/k = Nu = 7.24$$
$$h = 7.24 k/D = 7.24(0.02856 \text{ W/mK})/(0.02 \text{ m}) = 10.34 \text{ W/m}^2\text{K}$$

Table 7-2. Properties of dry air at atmospheric pressure (101.35 kPa)*

Temperature t (C)	Density ρ (kg/m³)	Dynamic Viscosity μ (Pa∗s)	Kinematic Viscosity ν (m²/s)	Specific Heat c_p (kJ/kg K)	Thermal Conductivity k (w/mK)	Thermal Diffusivity α (m²/s)	Prandtl Number Pr
−70	1.7390	0.00001357	0.000007812	1.003	0.01878	0.00001054	0.7245
−60	1.6573	0.00001409	0.000008501	1.003	0.01956	0.00001115	0.7224
−50	1.5331	0.00001461	0.000009244	1.003	0.02034	0.00001282	0.7206
−40	1.5153	0.00001513	0.000009987	1.003	0.02111	0.00001388	0.7188
−30	1.4534	0.00001565	0.000010777	1.004	0.02189	0.00001502	0.7174
−20	1.3965	0.00001617	0.000011587	1.004	0.02267	0.00001618	0.7158
−10	1.3436	0.00001669	0.000012436	1.004	0.02345	0.00001740	0.7145
0	1.2946	0.00001719	0.000013295	1.004	0.02413	0.00001857	0.7155
10	1.2486	0.00001767	0.000014168	1.005	0.02404	0.00001981	0.7148
20	1.2059	0.00001816	0.000015061	1.005	0.02561	0.00002114	0.7125
30	1.1663	0.00001864	0.000015990	1.006	0.02634	0.00002246	0.7117
40	1.1293	0.00001912	0.000016965	1.006	0.02700	0.00002378	0.7110
50	1.0936	0.00001960	0.000017941	1.007	0.02778	0.00002523	0.7106
60	1.0604	0.00002005	0.000018916	1.007	0.02856	0.00002674	0.7072
70	1.0345	0.00002050	0.000019892	1.008	0.02918	0.00002811	0.7081
80	1.0021	0.00002091	0.000020888	1.009	0.02980	0.00002952	0.7080
90	0.9734	0.00002135	0.000021941	1.010	0.03055	0.00003108	0.7058
100	0.9470	0.00002179	0.000023024	1.011	0.03122	0.00003262	0.7057
110	0.9227	0.00002222	0.000024090	1.012	0.03185	0.00003412	0.7062
120	0.8996	0.00002263	0.000025166	1.013	0.03254	0.00003572	0.7047
130	0.8770	0.00002304	0.000026281	1.014	0.03326	0.00003741	0.7025
140	0.8557	0.00002345	0.000027401	1.016	0.03389	0.00003901	0.7027
150	0.8355	0.00002386	0.000028552	1.017	0.03451	0.00004063	0.7032
160	0.8153	0.00002431	0.000029806	1.018	0.03513	0.00004232	0.7044
170	0.7980	0.00002472	0.000030968	1.020	0.03576	0.00004393	0.7049
180	0.7807	0.00002513	0.000032191	1.022	0.03638	0.00004561	0.7055
190	0.7646	0.00002551	0.000033368	1.023	0.03700	0.00004731	0.7055
200	0.7487	0.00002590	0.000034606	1.025	0.03763	0.00004905	0.7057
210	0.7328	0.00002629	0.000035884	1.027	0.03825	0.00005084	0.7058
220	0.7176	0.00002666	0.000037161	1.029	0.03887	0.00005266	0.7056
230	0.7032	0.00002703	0.000038454	1.031	0.03950	0.00005451	0.7055

* Adapted from data given by M. Jakob and G. A. Hawkins. 1957. *Elements of Heat Transfer*, 3rd Ed. New York: John Wiley & Sons, Inc.

Example 7.3. Estimate the convective heat transfer coefficient from a large vertical surface to air at 20 C. The temperature of the surface is 30 C.

Solution
Assume that $GrPr > 10^8$ and equation 7.48 is applicable.

Heat Transfer—Convection

Then,

$$h = 1.31(\Delta t)^{1/3}$$
$$= 1.31(10 \text{ K})^{1/3} = 2.82 \text{ W/m}^2\text{K}$$

7.7 Experimental Equations for Forced Convection

Following are a few of many equations which have been developed experimentally relating the dimensionless ratios of equation 7.39 for forced convection in different geometries. For the average heat transfer coefficient in long pipes to a fluid flowing within the pipe the equation is:

$$Nu = 0.023(Re)^{0.8}(Pr)^{0.4} \qquad (7.49)$$

for Reynolds numbers above 2300 to insure turbulent flow.

For forced convection to gases flowing normal to single cylinders in the range of Re from 0.1 to 1 000:

$$Nu(1/Pr)^{0.3} = 0.35 + 0.47(Re)^{0.52} \qquad (7.50)$$

while from $Re = 1\,000$ to $Re = 50\,000$:

$$Nu = 0.26(Re)^{0.6}(Pr)^{0.3} \qquad (7.51)$$

For liquids flowing normal to single cylinders with Re between 0.1 and 200:

$$Nu = 0.86(Re)^{0.43}(Pr)^{0.3} \qquad (7.52)$$

while for Re above 200, equation 7.50 is recommended.

For spheres, in the range of $Re = 20$ to $Re = 150\,000$ the equation is:

$$Nu = 0.36(Re)^{0.6}(Pr)^{0.3} \qquad (7.53)$$

In each of these equations, the fluid properties are to be evaluated at the film temperature, the arithmetic average of surface and bulk mean fluid temperatures. Additional experimental equations are given by ASHRAE (1997) and various other heat transfer references.

Example 7.4. Estimate the convective heat transfer coefficient between the surface of a tube having an inside diameter of 3 cm and water flowing in the tube at a velocity of 3 m/s. Assume the water to be at 50 C and the surface to be at 40 C.

Solution

The Reynolds number for the water at the film temperature of 45 C is:

$$Re = \rho v D/\mu = (990.58 \text{ kg/m}^3)(3 \text{ m/s})(0.03 \text{ m})/(0.000600 \text{ Pa s})$$
$$= 148587$$

Thus, equation 7.49 for turbulent flow is applicable.
From table 7-1, $Pr = 3.98$.

Then,
$$Nu = 0.023(Re)^{0.8}(Pr)^{0.4}$$
$$= 0.023(148587)^{0.8}(3.98)^{0.4} = 548.6$$
$$hD/k = Nu = 548.6$$
$$h = 548.6k/D = 548.6(0.6303 \text{ W/mK})/(0.03 \text{ m}) = 11\,526 \text{ W/m}^2\text{K}$$
$$= 11.5 \text{ kW/m}^2\text{K}$$

Example 7.5. Estimate the convective heat transfer coefficient for a sphere having a 15 cm diameter to air flowing over the surface at a velocity of 5 m/s. The sphere has a surface temperature of 50 C and the air is at 30 C.

Solution
The film temperature is 40 C.

$$Re = \rho v D/\mu = (1.1293 \text{ kg/m}^3)(5 \text{ m/s})(0.15 \text{ m})/(0.00001912 \text{ Pa s})$$
$$= 44\,300$$

Thus, equation 7.53 is applicable.
From table 7-2, $Pr = 0.7127$.
$$Nu = 0.36(Re)^{0.6}(Pr)^{0.3}$$
$$= 0.36(44{,}300)^{0.6}(0.7127)^{0.3} = 199.5$$
$$hD/k = Nu = 199.5$$
$$h = 199.5\,k/D = 199.5(0.027 \text{ W/mK})/(0.15 \text{ m}) = 35.9 \text{ W/m}^2\text{K}$$

Heat Exchangers

A heat exchanger is a device for transferring heat from a hot stream of fluid to a cold stream. The fluids are prevented from mixing with each other by a heat-conducting partition such as a pipe wall. Examples include refrigeration evaporators and condensers, automotive radiators (in reality convectors), and continuous milk pasteurizers and coolers.

7.1 Types of Heat Exchangers

Heat exchangers can be divided into the four classes shown in figure 7.1. In figure 7.1, subscript a denotes the hot fluid, whereas subscript b is for the cold fluid. Subscript 1 denotes the position where the hot fluid enters the exchanger and subscript 2 the position where the hot fluid leaves. The mass flow rate of the fluids are given by w_a and w_b, respectively. The exchanger where one fluid is constant in temperature is shown in figure 7.1a. The constant temperature fluid may be maintained at a constant temperature due to a change of state, either gaining or losing energy due to evaporation or condensation. This is also a special case where the constant temperature fluid has such a high mass flow rate that it experiences practically no change in temperature.

In the counterflow exchanger, figure 7.1b, fluids a and b move in opposite directions. Fluid a can thus be brought nearly to the temperature at which fluid b enters if enough surface area is provided. This possibly is the most favorable kind of heat exchanger.

In the parallel-flow heat exchanger, figure 7.1c, both fluids move in the same direction. Many devices such as water heaters and oil heaters and coolers fall in this group.

Heat Transfer—Convection

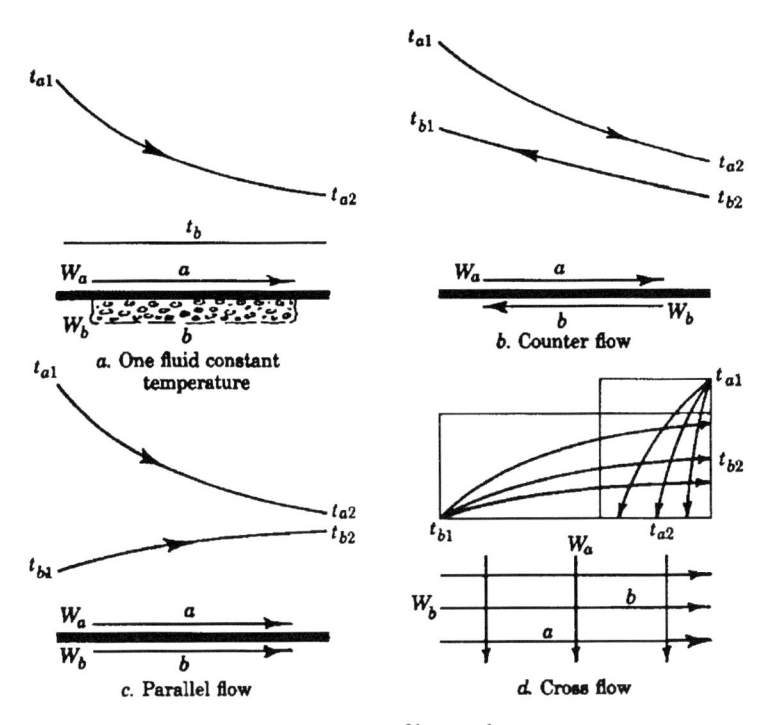

Figure 7.1. Types of heat exchangers.

a. One fluid constant temperature
b. Counter flow
c. Parallel flow
d. Cross flow

The mean temperature difference between the fluids is smaller in the parallel-flow arrangement than in counterflow, so that a greater surface area is required for the same heat transfer.

In the cross-flow exchanger, figure 7.1d, two fluids move in a number of separate parallel channels arranged so that the streams of fluid a cross those of fluid b, as in the automobile radiator. The several streams of fluid a do not mix with each other until after leaving the heat-exchange surface, as is also true for fluid b. This arrangement is less effective than counterflow, but better than parallel flow. It is used because of convenience in providing for the supply and removal of the fluid streams at relatively large rates of flow and short traverses of surface.

7.2 Log Mean Temperature Difference

As the two fluids pass through a heat exchanger the temperature difference between the fluids varies. In order to describe the heat transfer rate between the fluids it is necessary to determine some mean temperature difference, the effective surface area of the exchanger, and an overall heat transfer coefficient between the two fluids. The rate of heat transfer can then be described by:

$$q_a = U A (\Delta t)_{\ln} \tag{7.54}$$

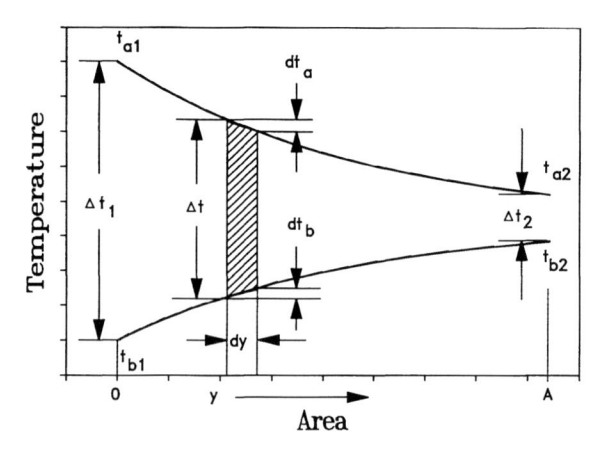

Figure 7.2. Temperature profile in a parallel-flow heat exchanger.

where

q_a = total rate of heat transfer between fluids (W)
U = overall heat transfer coefficient (W/m²K)
A = effective surface area of exchanger (m²)
$(\Delta t)_{\ln}$ = mean temperature difference (K)

In order to determine an effective average temperature difference it is necessary to consider heat transfer through differential areas of the exchanger as shown in figure 7.2. The rate of heat transferred through area dy of the exchanger in figure 7.2 is:

$$dq = U\,dy(t_a - t_b) \qquad (7.55)$$

where

dq = element of heat transferred from fluid a to fluid b (W)
U = overall heat exchange coefficient (W/m²K)
dy = element of area (m²)
t_a = temperature of fluid a (K)
t_b = temperature of fluid b (K)

The rate of heat loss by the hot fluid, a, in passing over the element of area dy may also be expressed as:

$$dq_a = -w_a c_a dt_a \qquad (7.56)$$

where

dq_a = heat loss by hot fluid a (W)
w_a = mass flow rate of a (kg/s)

Heat Transfer—Convection

c_a = specific heat of a (J/kgK)
dt_a = temperature change of a (K)

Likewise, the rate of heat gain by the cold fluid, b, in passing over the element of area dy may be expressed by:

$$dq_b = w_b c_b dt_b \qquad (7.57)$$

where

dq_b = heat gained by cold fluid b (W)
w_b = mass flow rate of fluid b (kg/s)
c_b = specific heat of fluid b (J/kgK)
dt_b = temperature change of b (K)

Energy conservation requires that the heat lost by a be equal to the heat gained by b. Thus, equations 7.56 and 7.57 may be combined to give:

$$dt_a = -dq/(w_a c_a) \qquad (7.58)$$
$$dt_b = dq/(w_b c_b) \qquad (7.59)$$
$$dt_a - dt_b = -dq(1/(w_a c_a) + 1/(w_b c_b)) \qquad (7.60)$$

Substituting $d(t_a - t_b)$ for $(dt_a - dt_b)$ and defining a new term,

$$N = 1/(w_a c_a) + 1/(w_b c_b) \qquad (7.61)$$

the following is obtained:

$$d(t_a - t_b) = -dq N \qquad (7.62)$$

Substitution of equation 7.55 for dq yields:

$$d(t_a - t_b) = -NU dy (t_a - t_b) \qquad (7.63)$$

or

$$d(t_a - t_b)/(t_a - t_b) = -\dot{N}U dy \qquad (7.64)$$

Integration yields:

$$\ln(t_a - t_b) = -NU y + c_1 \qquad (7.65)$$

The integration constant c_1 can be evaluated from the boundary condition that at $y = 0$, $(t_a - t_b) = (t_{a1} - t_{b1})$.
Thus,

$$c_1 = \ln(t_{a1} - t_{b1}) \qquad (7.66)$$

Then,

$$\ln((t_a - t_b)/(t_{a1} - t_{b1})) = -NU y \qquad (7.67)$$

A second boundary condition is that at $y = A$, $(t_a - t_b) = (t_{a2} - t_{b2})$. Substitution of this boundary condition in equation (7.67) yields:

$$(1/N)\ln((t_{a2} - t_{b2})/(t_{a1} - t_{b1})) = -UA \tag{7.68}$$

Combining equations 7.54 and 7.68 gives:

$$q_a/(\Delta t)_{\ln} = UA = -(1/N)\ln((t_{a2} - t_{b2})/(t_{a1} - t_{b1})) \tag{7.69}$$

or

$$(\Delta t)_{\ln} = -q_a N/\ln((t_{a2} - t_{b2})/(t_{a1} - t_{b1})) \tag{7.70}$$

If the definition of N from equation 7.61 is substituted in equation 7.70 then:

$$(\Delta t)_{\ln} = -(q_a/(w_a c_a) + q_a/(w_b c_b))/\ln((t_{a2} - t_{b2})/(t_{a1} - t_{b1})) \tag{7.71}$$

But,

$$dq_a = -w_a c_a dt_a = w_b c_b dt_b \tag{7.72}$$

and by integration

$$q_a = -w_a c_a (t_{a2} - t_{a1}) = w_b c_b (t_{b2} - t_{b1}) \tag{7.73}$$

Substitution into equation 7.71 and reorganization yields:

$$(\Delta t)_{\ln} = ((t_{a2} - t_{b2}) - (t_{a1} - t_{b1}))/\ln((t_{a2} - t_{b2})/(t_{a1} - t_{b1})) \tag{7.74}$$

The mean temperature difference given by equation 7.74 is known as the "log mean temperature difference". It is often written in the following form:

$$(\Delta t)_{\ln} = ((\Delta t)_{\max} - (\Delta t)_{\min})/\ln((\Delta t)_{\max}/(\Delta t)_{\min}) \tag{7.75}$$

where

$(\Delta t)_{\max}$ = maximum temperature difference between the fluids at a specific location in the exchanger (K)
$(\Delta t)_{\min}$ = minimum temperature difference between the fluids at a specific location in the exchanger (K)

This simple relation for the log mean temperature difference may be used for any of the first three types of heat exchangers and with appropriate correction factors for more complex cross-flow exchangers.

The log mean temperature difference is convenient for working heat exchanger problems in which the entrance and exit temperatures are known and the required area or the overall heat transfer coefficient are to be determined.

Example 7.6. Hot water enters a counter-flow heat exchanger at a temperature of 60 C and a flow rate of 15 000 kg/h and is cooled to 40 C by a cold stream of water which enters at 20 C with a flow rate of 20 000 kg/h. The overall heat transfer coefficient is 2 100 W/m²K. Determine: (a) the exit temperature of the cold water stream and (b) the surface area required for the exchanger.

Heat Transfer—Convection

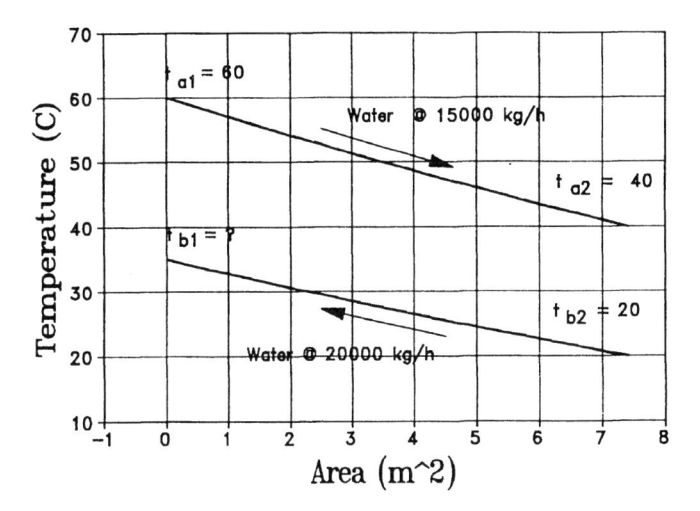

Figure 7.3. Temperature profile for counter-flow heat exchanger of example 7.6.

Solution
The first step in a heat exchanger problem should be to sketch the process such as is done for this example in figure 7.3.

Then, by conservation of energy,

$$w_a c_a \Delta t_a = w_b c_b \Delta t_b$$
$$\Delta t_b = (w_a c_a / w_b c_b) \Delta t_a$$
$$= (15\,000 \text{ kg/h}/20\,000 \text{ kg/h})(60-40) \text{ K}$$
$$= 15 \text{ K}$$
$$t_{b(\text{out})} = t_{b1} = 20 \text{ C} + 15 \text{ C} = 35 \text{ C}$$
$$q = w_a c_a \Delta t_a$$
$$= (15\,000 \text{ kg/h})(4.18 \text{ kJ/kgK})(20 \text{ K})(1 \text{ h}/3\,600 \text{ s})$$
$$= 348.3 \text{ kW}$$
$$(\Delta t)_{\ln} = ((\Delta t)_{\max} - (\Delta t)_{\min}) / \ln((\Delta t)_{\max}/(\Delta t)_{\min})$$
$$= ((60-35) - (40-20)) / \ln((60-35)/(40-20))$$
$$= (25-20)/\ln(25/20) = 22.4 \text{ K}$$
$$q = U A (\Delta t)_{\ln} = 348.3 \text{ kW}$$
$$A = q/(U(\Delta t)_{\ln})$$
$$= (348.3 \text{ kW})(1000 \text{ W}/1 \text{ kW})/((2100 \text{ W/m}^2 \text{ K})(22.4 \text{ K}))$$
$$= 7.404 \text{ m}^2.$$

7.3 Heat Exchanger Effectiveness Ratio

The log mean temperature difference can be used in the analysis of heat exchangers when all of the terminal fluid temperatures are known. However, if it is desired to predict the performance of a given heat exchanger for other conditions, the use of the log mean temperature difference is inconvenient, requiring an iterative solution technique. For this situation it is advantageous to use the heat exchanger effectiveness ratio which will be defined as:

$$E = (t_{a(\text{in})} - t_{a(\text{out})})/(t_{a(\text{in})} - t_{b(\text{in})}) \tag{7.76}$$

where

E = heat exchanger effectiveness ratio for cooling of the hot fluid (decimal)
$t_{a(\text{in})}$ = entering temperature of hot fluid (K)
$t_{a(\text{out})}$ = exit temperature of hot fluid (K)
$t_{b(\text{in})}$ = entering temperature of cold fluid (K)

This ratio describes the effectiveness of the exchanger for cooling the hot fluid to the entering temperature of the cold fluid.

7.3.1 Parallel-Flow Heat Exchanger Effectiveness Ratio

Equations 7.56 and 7.57 may be equated for a parallel-flow exchanger to obtain:

$$-w_a c_a dt_a = w_b c_b dt_b = dq \tag{7.77}$$

or

$$dt_a = -w_b c_b dt_b / (w_a c_a) \tag{7.78}$$

The general temperature difference expression $(t_a - t_b)$ will be replaced by the symbol Θ. Differentiating gives:

$$d\Theta = dt_a - dt_b \tag{7.79}$$

Substituting the relationship for $d\Theta$ into equation 7.78 gives:

$$dt_a = -(w_b c_b / w_a c_a)(dt_a - d\Theta) \tag{7.80}$$

or

$$dt_a = R d\Theta / (1 + R) \tag{7.81}$$

where

$$R = w_b c_b / (w_a c_a) \tag{7.82}$$

and a indicates the hot fluid and b indicates the cold fluid.

Substituting equations 7.81 and 7.77 into equation 7.55 gives:

$$-w_a c_a dt_a = U \Theta dy \tag{7.83}$$

and

$$-w_a c_a R d\Theta / (1 + R) = U \Theta dy \tag{7.84}$$

Heat Transfer—Convection

Separation of variables gives:

$$d\Theta/\Theta = -(1+R)U\,dy/(w_a c_a R) \tag{7.85}$$

Integration yields:

$$\ln \Theta = -(Uy/w_a c_a)(1+R)/R + c_1 \tag{7.86}$$

at $y = 0$, $\Theta = t_{a1} - t_{b1}$
Therefore, $\ln(t_{a1} - t_{b1}) = c1$
at $y = A$, $\Theta = t_{a2} - t_{b2}$
Therefore,

$$\ln(t_{a2} - t_{b2}) = -(UA/w_a c_a)(1 + 1/R) + \ln(t_{a1} - t_{b1}) \tag{7.87}$$

or

$$\ln((t_{a2} - t_{b2})/(t_{a1} - t_{b1})) = -(UA/(w_a c_a))(1 + 1/R) \tag{7.88}$$

and

$$(t_{a2} - t_{b2})/(t_{a1} - t_{b1}) = \exp(-(UA/(w_a c_a))(1 + 1/R)) \tag{7.89}$$

The dimensionless ratio, $UA/w_a c_a$, is often termed the number of transfer units (NTU) and is used for rating heat exchangers.

Equation 7.78 can be integrated as follows:

$$q = -w_a c_a (t_{a2} - t_{a1}) = w_b c_b (t_{b2} - t_{b1}) \tag{7.90}$$

This may be solved for t_{b2}:

$$t_{b2} = t_{b1} + (t_{a1} - t_{a2})/R \tag{7.91}$$

Substituting in equation 7.89 gives:

$$(t_{a2} - t_{b1})/(t_{a1} - t_{b1}) - (t_{a1} - t_{a2})/(R(t_{a1} - t_{b1}))$$
$$= \exp(-(UA/(w_a c_a))(1 + 1/R)) \tag{7.92}$$

Subtracting 1 from both sides gives:

$$-(t_{a1} - t_{a2})/t_{a1} - t_{b1}) - (t_{a1} - t_{a2})/(R(t_{a1} - t_{b1}))$$
$$= \exp(-(UA/(w_a c_a))(1 + 1/R)) - 1 \tag{7.93}$$

But,

$$E_p = (t_{a1} - t_{a2})/(t_{a1} - t_{b1}) \tag{7.94}$$

where E_p = effectiveness ratio for a parallel flow heat exchanger.
Thus,

$$-E_p - E_p/R = \exp(-(UA/(w_a c_a))(1 + 1/R)) - 1 \tag{7.95}$$

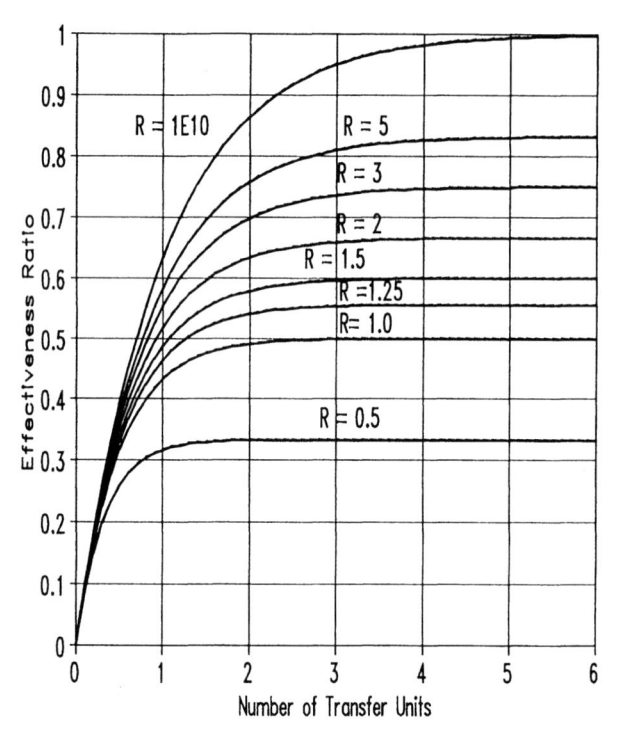

Figure 7.4. Effectiveness ratio vs. number of transfer units for a parallel-flow heat exchanger.

or

$$E_p = (1 - \exp(-(UA/(w_a c_a))(1 + 1/R)))/(1 + 1/R) \qquad (7.96)$$

Figure 7.4 plots the effectiveness ratio for a parallel-flow heat exchanger as a function of the NTU for different values of the heat capacity ratio, R.

Example 7.7. The direction of flow of the cold water of example 7.6 is reversed so that the exchanger becomes a parallel-flow exchanger. Assuming that the overall heat transfer coefficient is unchanged, determine the exit temperatures of the two water streams.

Solution
Figure 7.5 is a sketch of the process.

$$R = w_b c_b / w_a c_a = (20\,000 \text{ kg/h})/(15\,000 \text{ kg/h}) = 1.333$$

$$\text{NTU} = UA/w_a c_a$$
$$= (2\,100 \text{ W/m}^2\text{K})(7.404 \text{ m}^2)/(15\,000 \text{ kg/h})(4.18 \text{ kJ/kgK})$$
$$(1 \text{ h}/3\,600 \text{ s})(1\,000 \text{ J/kJ})$$
$$= 0.8927$$

Heat Transfer—Convection

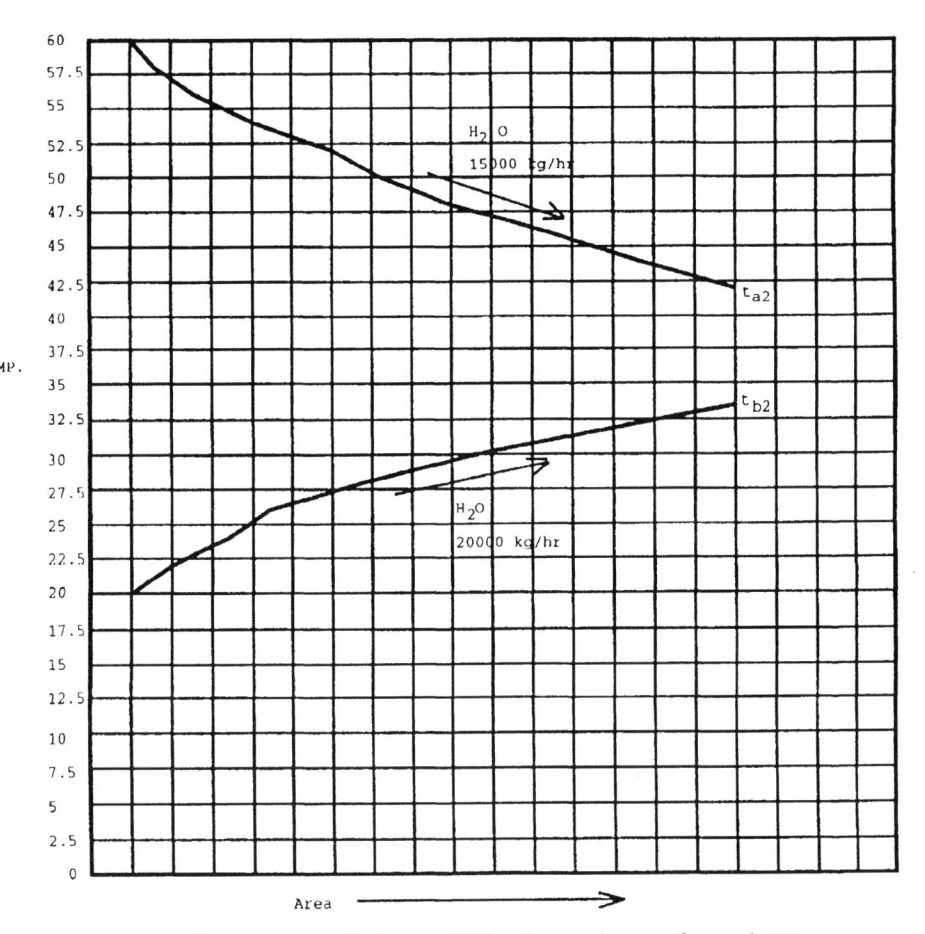

Figure 7.5. Temperature profile for parallel-flow heat exchanger of example 7.7.

$$E_p = (1 - \exp(-\text{NTU}(1 + 1/R)))/(1 + 1/R)$$
$$= (1 - \exp(-0.8927(1 + 1/1.333))/(1 + 1/1.333)$$
$$= (1 - \exp(-0.8927(1.75))/(1.75) = 0.4516$$
$$E_p = (t_{a1} - t_{a2})/(t_{a1} - t_{b1}) = 0.4516$$
$$t_{a1} - t_{a2} = 0.4516(t_{a1} - t_{b1})$$
$$t_{a2} = t_{a1} - 0.4516(t_{a1} - t_{b1})$$
$$= 60\ \text{C} - 0.4516(60 - 20)\ \text{C}$$
$$= 60\ \text{C} - 18.1\ \text{C} = 41.9\ \text{C}$$
$$w_a c_a \Delta t_a = w_b c_b \Delta t_b$$

$$\Delta t_b = (w_a c_a/(w_b c_b))\Delta t_a = \Delta t_a/R$$
$$= (60 - 41.9)\ \text{K}/(1.333)$$
$$= 13.5\ \text{K}$$

$$t_{b2} = t_{b1} + 13.5\ \text{K} = 20\ \text{C} + 13.5\ \text{C} = 33.5\ \text{C}$$

7.3.2 Counter-Flow Heat Exchanger Effectiveness Ratio

For a counter-flow heat exchanger the equation analogous to equation 7.77 is:

$$-w_a c_a dt_a = -w_b c_b dt_b = dq \tag{7.97}$$

Then,

$$dt_a = (w_b c_b/(w_a c_a))dt_b \tag{7.98}$$

Using these relationships the following temperature relationship may be obtained for a counter-flow exchanger:

$$(t_{a2} - t_{b2})/(t_{a1} - t_{b1}) = \exp(-(UA/(w_a c_a))(1 - 1/R)) \tag{7.99}$$

From this relationship it can be shown that the effectiveness ratio for a counter-flow exchanger is:

$$E_c = (1 - \exp(-\text{NTU}(1 - 1/R)))/(1 - (1/R)\exp(-\text{NTU}(1 - 1/R))) \tag{7.100}$$

where

E_c = effectiveness ratio of counter-flow heat exchanger
NTU = number of transfer units = $UA/(w_a c_a)$

There are two special cases of equation 7.100 which need to be considered. The first special case is when the heat capacity ratio, $R = w_b c_b/(w_a c_a)$, is infinite. This is one method for obtaining one constant temperature. In this case equation 7.100 becomes:

$$E_c = 1 - \exp(-\text{NTU}) \tag{7.101}$$

The second special case is when $R = 1$, where equation 7.100 becomes indeterminate. Therefore, l'Hospital's rule must be used to determine the effectiveness ratio. Applying l'Hospital's rule results in:

$$E_c = \text{NTU}/(\text{NTU} + 1) \tag{7.102}$$

or

$$E_c = (UA/(w_a c_a))/(1 + (UA/(w_a c_a))) \tag{7.103}$$

Figure 7.6 plots the counter-flow-heat exchanger effectiveness ratio versus the number of transfer units for several different values of the heat capacity ratio.

Example 7.8. The temperature of the cold water stream in example 7.6 is reduced to 10 C. Assuming the overall heat transfer coefficient to remain unchanged, determine the new exit temperatures of the two water streams.

Heat Transfer—Convection

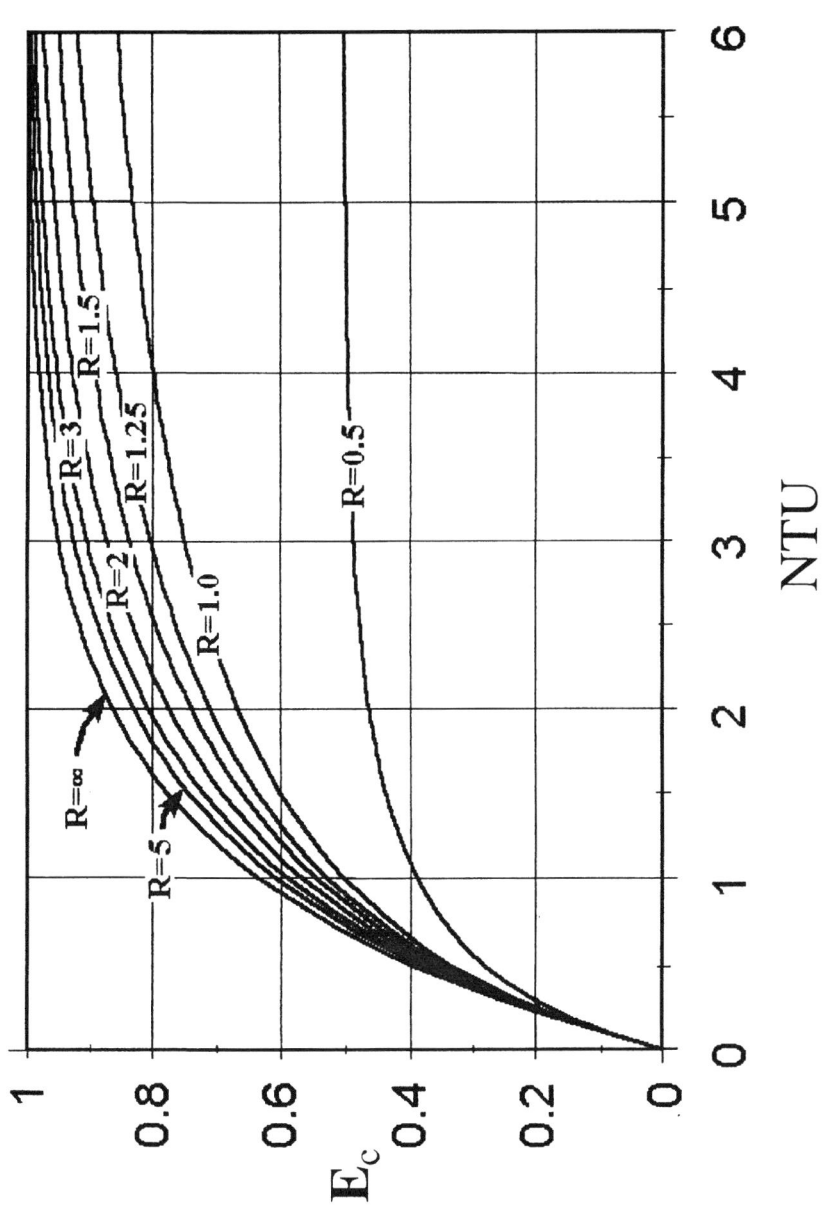

Figure 7.6. Effectiveness curves for counter-flow heat exchangers.

Solution
Figure 7.7 is a sketch of the process.

$$R = 1.333 \text{ from example 7.7}$$
$$NTU = 0.8927 \text{ from example 7.7}$$
$$E_c = (1 - \exp(-NTU(1 - 1/R)))/(1 - (1/R)\exp(-NTU(1 - 1/R)))$$
$$= (1 - \exp(-0.8927(1 - 0.75)))/(1 - 0.75\exp(-0.8927(1 - 0.75)))$$
$$= 0.500$$
$$E_c = (t_{a1} - t_{a2})/(t_{a1} - t_{b2}) = 0.500$$
$$t_{a1} - t_{a2} = 0.500(t_{a1} - t_{b2})$$
$$t_{a2} = t_{a1} - 0.500(t_{a1} - t_{b2})$$
$$= 60 \text{ C} - 0.500(60 - 10) \text{ C}$$
$$= 60 \text{ C} - 25.0 \text{ C} = 35 \text{ C}$$

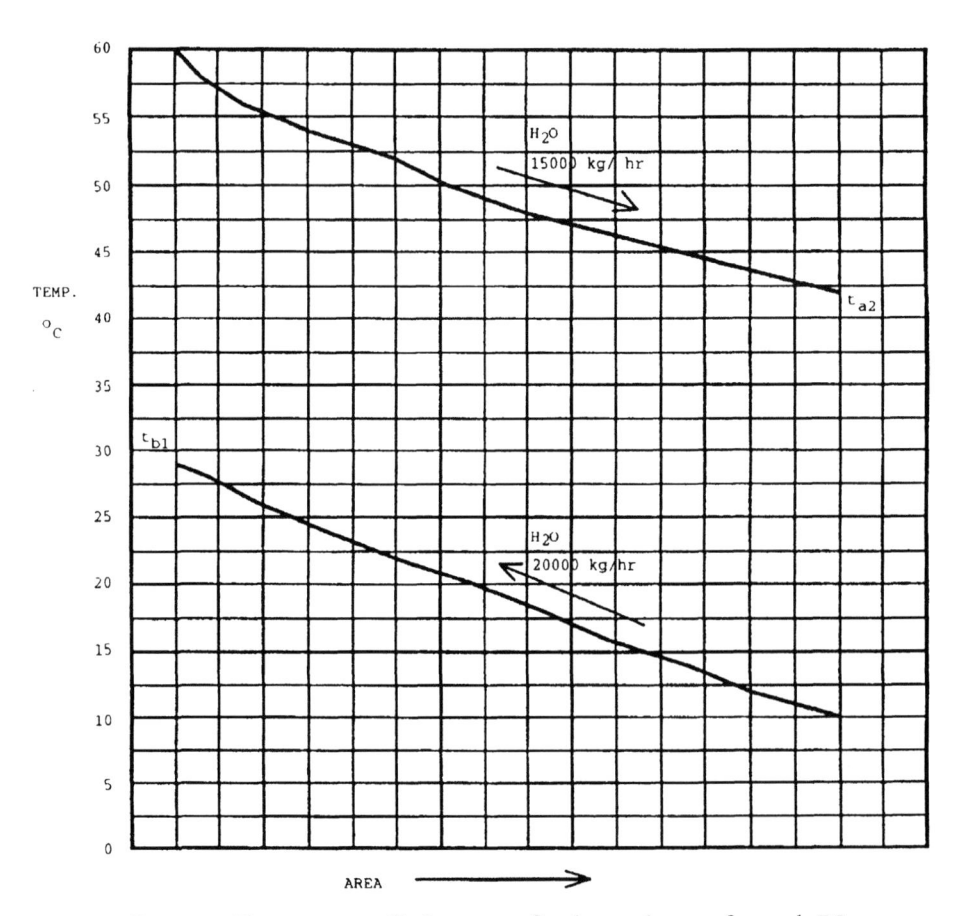

Figure 7.7. Temperature profile for counter-flow heat exchanger of example 7.8.

Heat Transfer—Convection

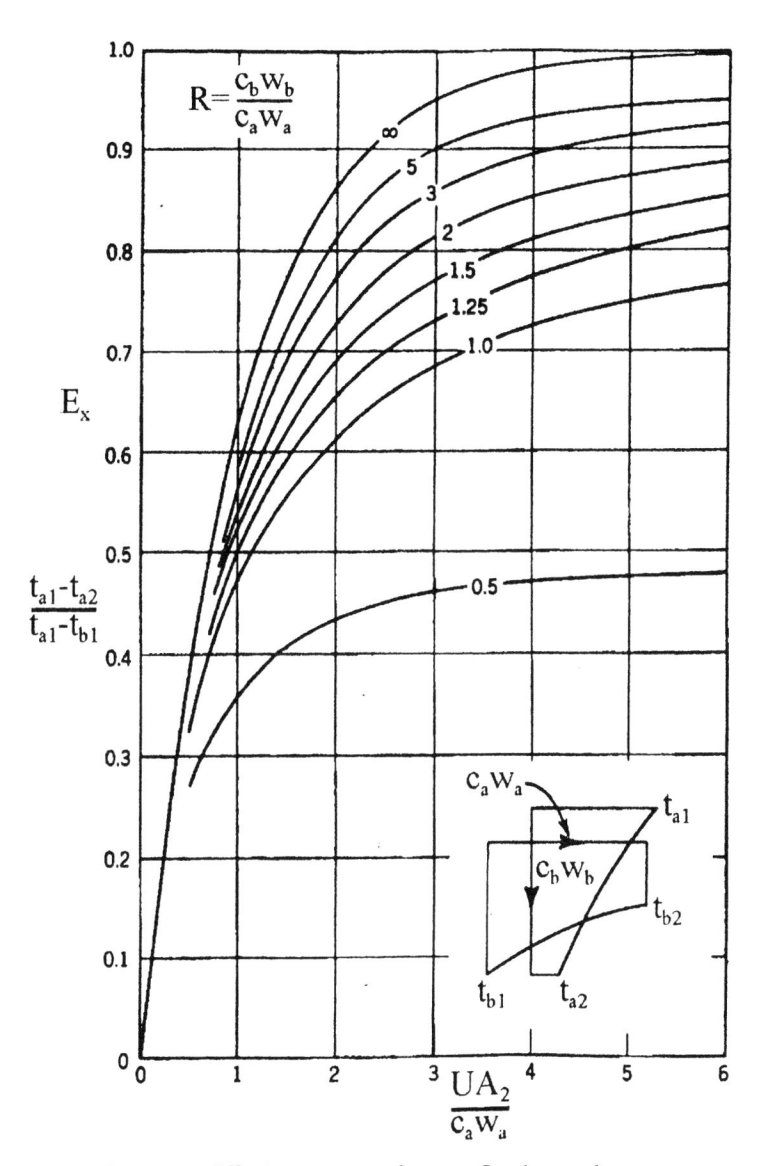

Figure 7.8. Effectiveness curves for cross-flow heat exchangers.

$$w_a c_a \Delta t_a = w_b c_b \Delta t_b$$

$$\Delta t_b = (w_a c_a/(w_b c_b))\Delta t_a = \Delta t_a/R$$

$$= (60 - 35) \text{ K}/1.333 = 18.8 \text{ K}$$

$$t_{b1} = t_{b2} + 18.8 \text{ K} = 10 \text{ C} + 18.8 \text{ C} = 28.8 \text{ C}$$

7.3.3 Crossflow Heat Exchanger Effectiveness Ratio

For the cross-flow exchanger, the analytic solution is more involved, since temperatures of both fluids vary with both the x- and y-directions. Fortunately the effectiveness of the crossflow exchanger, E_x, can be expressed in terms of the same dimensionless parameters, NTU and R, as the simpler heat exchangers. Values for crossflow heat exchangers as given by Saban (1952) are plotted in figure 7.8.

Nomenclature

A	area (m^2)
B	buoyancy (N/m^3)
Bo	Biot Number (dimensionless)
c_p	specific heat (J/kg K)
E	heat exchanger effectiveness ratio (dimensionless)
g	acceleration due to gravity (m/s^2)
Gr	Grashof Number (dimensionless)
h	convective heat transfer coefficient (W/m^{2}K)
k	thermal conductivity (W/mK)
l	length (m)
Nu	Nusselt Number (dimensionless)
NTU	number of transfer units (dimensionless)
Pr	Prandtl Number (dimensionless)
q	heat flow (W)
q'	heat transfer per unit length (W/m)
q'	heat flux (W/m^2)
Re	Reynolds Number (dimensionless)
t	temperature (C)
T	absolute temperature (K)
U	overall heat transfer coefficient (W/m^{2}K)
v	velocity (m/s)
w	mass flow rate (kg/s)
β	coefficient of thermal expansion (K^{-1})
μ	dynamic viscosity (Pa s)
ν	kinematic viscosity (m^2/s)
ρ	mass density (kg/m^3)
Θ	temperature difference (K)

References

Am. Soc. Heating, Refrigerating and Air Conditioning Engrs. 1997. *ASHRAE Handbook. 1997 Fundamentals.* New York: ASHRAE.

Jakob, M. and G. A. Hawkins. 1957. *Elements of Heat Transfer.* New York: John Wiley & Sons, Inc.

Kreith, F. 1958. *Principles of Heat Transfer.* Scranton, Pa.: International Textbook Co.

Saban, R. 1952. Personal communication. University of California.

Problems

1. It has been found that the friction head, F, for a fluid flowing through a smooth straight pipe depends upon the average velocity v of the fluid in the pipe, the inside diameter D of the pipe, the density ρ, the length of pipe l, the acceleration due to gravity g, the roughness factor ε, and the viscosity of the fluid μ. Develop an equation for the friction head by dimensional analysis, in which the Reynolds number appears.

2. Estimate the heat loss by convection from the top of a blancher box, 1.8 m wide and 11.0 m long, in a room at 20 C. The cover plates are uninsulated aluminum-painted galvanized iron. The box is direct steam-heated at 99 C.

3. Estimate the convective heat transfer coefficient, h, for air flowing normal to a single cylinder with a diameter of 2.5 cm if the average air temperature is 37 C and the average air velocity is 0.4 m/s.

4. Air at 60 C is flowing normal to a cylindrical copper tube with a diameter of 15 cm at a velocity of 2 m/s. Estimate the convective heat transfer coefficient at the surface.

5. Find the length of 2.24-cm-inside-diameter tube required for heating 1 770 kg/h of tomato pulp from 15 to 75 C in a counter-flow exchanger with hot water entering at 85 C and leaving at 68 C. The overall heat transfer coefficient U is estimated at 1 420 W/m²K. Properties of tomato pulp are: density, $\rho = 1\,025$ kg/m³; specific heat, $c_p = 3.98$ J/g K.

6. Air is to be cooled while passing through a 1.2-cm-inside-diameter copper tube maintained at a surface temperature of 4 C by a water bath. The air will pass through the tube at the rate of 0.09 kg/min. How long will the tube need to be in order to cool the air from 50 to 10 C?

7. Find the temperature to which milk with an initial temperature of 28 C will be cooled at a rate of 6.0 kg/min by a counter-flow surface cooler with 1.0 m² of surface, supplied with chilled water at 2 C and a flow rate three times the milk mass flow rate. An overall heat transfer coefficient U of 620 W/m²K is expected. The specific heat of milk is 3.89 kJ/kgK. What will be the exit temperature of the water?

8. Work problem 7 for a parallel-flow heat exchanger.

9. In a counterflow heat exchanger, 5 000 kg/h of water at 15 C is to cool 10 000 kg/h of an oil having a specific heat of 2.09 kJ/kgK from 95 to 65 C. If the overall heat transfer coefficient is 285 W/m^{2}K, determine the surface area required.

10. Water flowing at a rate of 50 000 kg/h is to be cooled from 95 to 65 C by means of an equal flow of water entering at 40 C. The water velocity will be such that the overall coefficient of heat transfer U is 2 270 W/m^{2}K. Calculate the area of heat-exchanger surface needed for each of the following arrangements: (a) parallel-flow or (b) counter-flow.

11. Prove that the coefficient of thermal expansion for an ideal gas at constant pressure is the inverse of its absolute temperature.

8 Heat Transfer—Radiation

8.1 The Electromagnetic Wave Spectrum

The transfer of heat by radiation is only one of a number of electromagnetic phenomena. The term "radiation" is applied generally to all kinds of processes which transmit energy by means of electromagnetic waves. The entire spectrum of such waves is shown in figure 8.1. The wavelength of radiation is defined as the ratio of the propagation velocity to the frequency:

$$\lambda = c/\nu \qquad (8.1)$$

where

λ = wavelength (m)
c = speed of light (m/s)
ν = radiation frequency (s^{-1})

The speed of light is approximately 299 000 km/s. Wavelengths are often expressed in microns. One micron is 10^{-6} m. Table 8-1 tabulates the wavelength ranges for some of the types of radiation shown in figure 8.1. For practical purposes the radiation of importance in heat-transfer calculations is limited to wavelengths ranging from 0.1 to 100 microns (1 × 10^{-7} to 1 × 10^{-4} m).

Radiant energy from a perfect emitter, known technically as a black body, is emitted to a continuous spectrum of wavelengths according to Planck's Law:

$$W_{b\lambda} = (3.66 \times 10^8 \text{ W micron}^4/\text{m}^2)\lambda^{-5}/(\exp((14\,300 \text{ micron K})/\lambda T) - 1) \qquad (8.2)$$

where

$W_{b\lambda}$ = the monochromatic or spectral emissive power of a black body (W/m^2 micron)
λ = wavelength (microns)
T = absolute temperature (K)

Figure 8.2 gives plots of the black body emissive power at five temperatures. The maximum value of $W_{b\lambda}$ is given by:

$$W_{b\lambda \text{ max}} = (129.8 \times 10^{-13} \text{ W/m}^2 \text{ micron K}^5)T^5 \qquad (8.3)$$

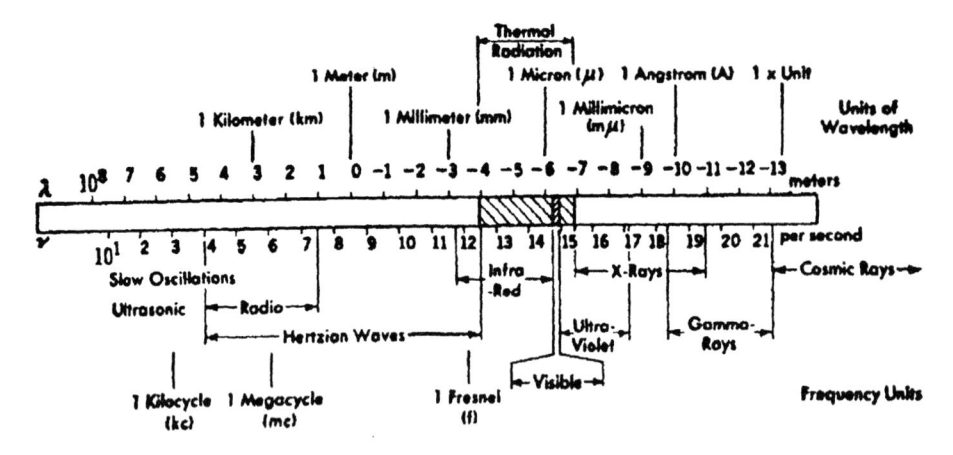

Figure 8.1. Electromagnetic-wave spectrum.

and occurs at a wavelength given by,

$$\lambda_{max} = (2\,880 \text{ microns} * K)/T \tag{8.4}$$

The integral of $W_{b\lambda}$ from wavelength 0 to infinity gives the black-body emissive power per unit area, W_b:

$$W_b = \int_0^\infty W_{b\lambda}\, d\lambda = \sigma T^4 \tag{8.5}$$

where

W_b = black-body emissive power (W/m^2)
σ = Stefan-Boltzmann constant = 5.73×10^{-8} W/m^{2}K^4
T = absolute temperature (K)

Equation 8.5 is known as the Stefan-Boltzmann equation.

8.2 Emissivity

Monochromatic emissivity ε_m is defined as the ratio of the emissive power of a non-black radiator in a given wavelength to black body emissive power in the same

Table 8-1. Electromagnetic waves

Name	Wavelength Range (m)
Cosmic rays	Up to 1×10^{-12}
Gamma rays	1×10^{-12} to 1.4×10^{-10}
X-rays	6×10^{-12} to 1×10^{-7}
Ultraviolet rays	1.4×10^{-8} to 4.0×10^{-7}
Visible or light rays	4.0×10^{-7} to 8.0×10^{-7}
Infrared or heat rays	8.0×10^{-7} to 4.0×10^{-4}
Radio	1×10 to 3.0×10^4

Heat Transfer—Radiation

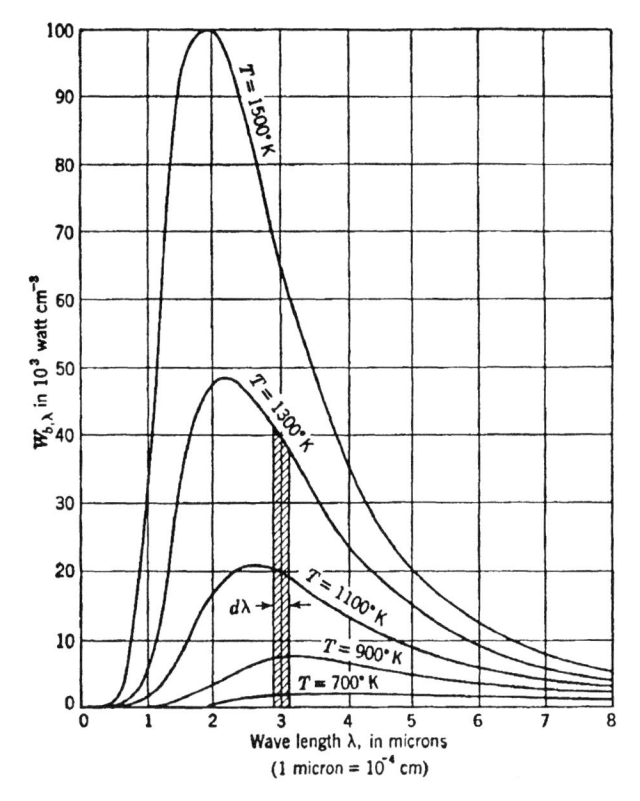

Figure 8.2. Monochromatic intensity of radiation for a black body at various absolute temperatures (Planck's law).

wavelength:

$$\varepsilon_\lambda = W_\lambda / W_{b\lambda} \tag{8.6}$$

where

ε_λ = monochromatic emissivity (dimensionless)
W_λ = monochromatic emissive power of body (W/m² micron)
$W_{b\lambda}$ = monochromatic emissive power of a black body (W/m² micron)

For most surfaces, ε_λ varies with wavelength. However, a surface having a constant monochromatic emissivity is called a gray body.

A mean emissivity can be defined by:

$$\varepsilon = (1/W_b) \int_0^\infty \varepsilon_\lambda W_{b\lambda} \, d\lambda \tag{8.7}$$

These mean values depend upon the temperature. Thus, the temperature should be stated when a value of emissivity is given. Emissivities do not change appreciably with temperature below 590 K. Table 8-2 gives emissivities for a number of surfaces at approximately 38 C and at solar temperatures.

Table 8-2. Emissivities of various surfaces

Material	Wavelength and Avg. Temp. 9.3 μ 38 C	0.6 μ Solar
Metals		
Aluminum		
Polished	0.04	0.3
Oxidized	0.11	
24-ST weathered	0.4	
Surface roofing	0.22	
Anodized (at 538 C)	0.94	
Brass		
Polished	0.10	
Oxidized	0.61	
Chromium		
Polished	0.08	0.49
Copper		
Polished	0.04	
Oxidized	0.87	
Iron		
Polished	0.06	0.45
Cast, oxidized	0.63	
Galvanized, new	0.23	0.66
Galvanized, dirty	0.28	0.89
Steel plate, rough	0.94	
Oxide	0.96	0.74
Magnesium	0.07	0.30
Molybdenum filament		0.2
Silver		
Polished	0.01	0.11
Stainless steel		
18-8, polished	0.15	
18-8, weathered	0.85	
Tungsten filament	0.03	0.35
Zinc		
Polished	0.02	0.46
Galvanized sheet	0.25	
Building and Insulating Materials		
Asbestos paper	0.93	
Asphalt	0.93	0.93
Brick		
Red	0.93	0.7
Fire clay	0.9	
Silica	0.9	
Magnesite refractory	0.9	
Enamel, white	0.9	
Marble, white	0.95	0.47
Paper, white	0.95	0.28
Plaster	0.91	

Heat Transfer—Radiation

Table 8-2. (Continued)

Material	Wavelength and Avg. Temp. 9.3 μ 38 C	0.6 μ Solar
Roofing Board	0.93	
Enameled steel, white		0.47
Asbestos cement, red		0.66
Paints		
Aluminized lacquer	0.65	
Cream paints	0.95	0.35
Lacquer, black	0.96	
Lampblack paint	0.96	0.97
Red paint	0.96	0.74
Yellow paint	0.95	0.30
Oil paints (all colors)	0.94	
White ZnO	0.95	0.18
Miscellaneous		
Ice	0.97	
Water	0.96	
Carbon		
T-carbon, 0.9% ash	0.82	
Filament	0.72	
Wood	0.93	
Glass	0.90	Low

Example 8.1. A gray body having an emissivity of 0.8 has a temperature of 50 C. (a) Determine the peak monochromatic emissive power of the body and the wavelength at which it occurs. (b) What is the monochromatic emissive power at 8 microns? (c) Determine the total emissive power of the body.

Solution

(a) $W_{\lambda\ max} = \varepsilon_\lambda W_{b\lambda\ max}$

Since for a gray body ε = constant, the monochromatic emissivity $\varepsilon_\lambda = \varepsilon = 0.8$. Then,

$$W_{\lambda\ max} = \varepsilon(129.8 \times 10^{-13})T^5$$
$$= 0.8(129.8 \times 10^{-13})(273.16 + 50)^5$$
$$= 36.1 \text{ W/m}^2 \text{ micron}$$

$$\lambda_{max} = 2\,880/T = 2\,880/(273.16 + 50) = 8.91 \text{ micron}$$

(b) $W_{\lambda 8} = \varepsilon W_{b\lambda 8}$

$$= (0.8)(3.66 \times 10^8)(8)^{-5}/(\exp(14\,300/((8)(50 + 273.16))) - 1)$$
$$= 35.5 \text{ W/m}^2 \text{ micron}$$

(c)
$$W = \varepsilon W_b = \varepsilon \sigma T^4$$
$$= 0.8(5.73 \times 10^{-8} \text{ W/m}^2\text{K}^4)(50 + 273.16)^4 \text{ K}^4$$
$$= 499.9 \text{ W/m}^2$$

8.3 Absorption, Reflection, and Transmission of Radiation

When radiation impinges on a body, it is partially absorbed, partially reflected, and partially transmitted as follows:

$$\alpha + \rho + \tau = 1 \tag{8.8}$$

where

α = absorptivity, i.e., the fraction of the incident radiation absorbed by the body
ρ = reflectivity, the fraction of the incident radiation reflected from the body
τ = transmissivity, the fraction of incident radiation transmitted through the body

Figure 8.3 illustrates the division of the incident radiation into the absorbed, transmitted, and reflected portions.

For the majority of the opaque solid materials encountered in engineering, except for extremely thin layers, practically none of the radiant energy is transmitted through the body. If the discussion is limited to opaque bodies, equation 8.8 becomes:

$$\alpha + \rho = 1 \tag{8.9}$$

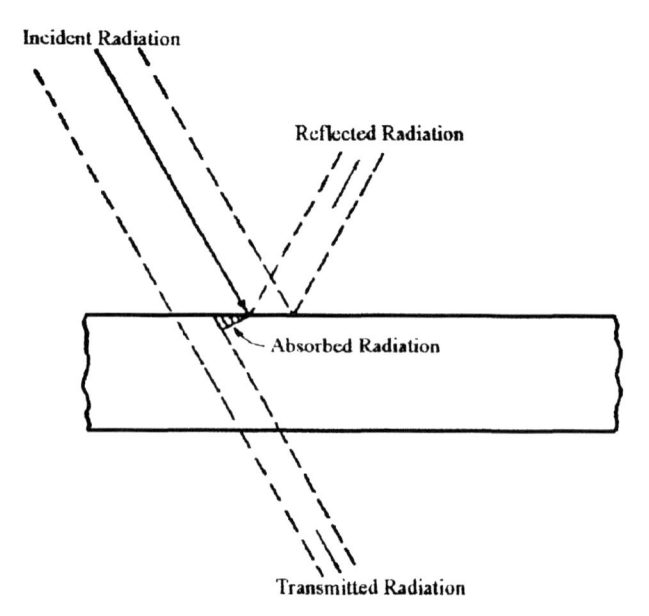

Figure 8.3. Reflection, absorption, and transmission of radiation.

Heat Transfer—Radiation

Regular Reflection $\alpha=\beta$ **Diffuse Reflection**

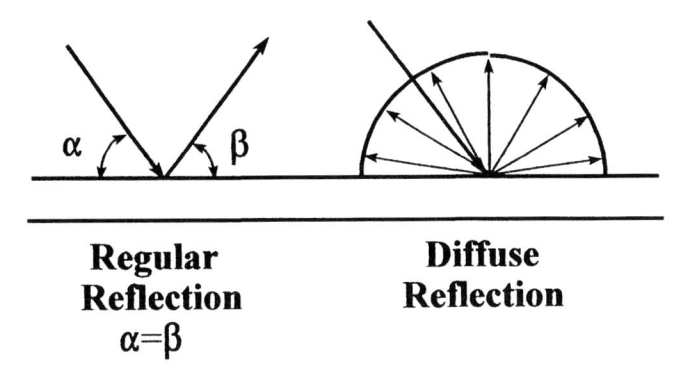

Figure 8.4. Types of reflection of radiation.

The reflection of radiation may be of two types as shown in figure 8.4. They are: (1) regular and (2) diffuse. If a surface is highly polished and smooth, the reflection of radiation will be similar to the reflection of a light beam, that is, the angle of incidence will be equal to the angle of reflectance. This is called regular reflection. Most materials are "rough" because their surfaces have asperities which are large compared with one wavelength. The reflection of radiation from a rough surface occurs practically indiscriminately in all directions and is called diffuse.

8.4 Kirchoff's Law

Suppose two small bodies B_1 and B_2 of surface areas A_1 and A_2 are placed in a large evacuated enclosure which is perfectly insulated from its surroundings. Radiation will be exchanged between the bodies and the walls of the enclosure until equilibrium is attained. Then the rate at which each body emits radiation of a given wavelength must equal the rate at which it absorbs radiation of that wavelength. If Γ_λ is the rate at which radiant energy of wavelength λ falls on each of the bodies, $\alpha_{\lambda 1}$ and $\alpha_{\lambda 2}$ are the absorptivities, and $W_{\lambda 1}$ and $W_{\lambda 2}$ the emissive powers of B_1 and B_2, respectively, an energy balance yields:

$$A_1 \Gamma_\lambda \alpha_{\lambda 1} = A_1 W_{\lambda 1} = A_1 \varepsilon_{\lambda 1} W_{b\lambda} \qquad (8.10)$$

or

$$\alpha_{\lambda 1}/\varepsilon_{\lambda 1} = W_{b\lambda}/\Gamma_\lambda \qquad (8.11)$$

Similarly,

$$\alpha_{\lambda 2}/\varepsilon_{\lambda 2} = W_{b\lambda}/\Gamma_\lambda \qquad (8.12)$$

$$\alpha_{\lambda 1}/\varepsilon_{\lambda 1} = \alpha_{\lambda 2}/\varepsilon_{\lambda 2} \qquad (8.13)$$

Since by definition the emissivity and the absorptivity of a black body are one:

$$\alpha_{\lambda 1}/\varepsilon_{\lambda 1} = 1 \qquad (8.14)$$

or

$$\alpha_\lambda = \varepsilon_\lambda \qquad (8.15)$$

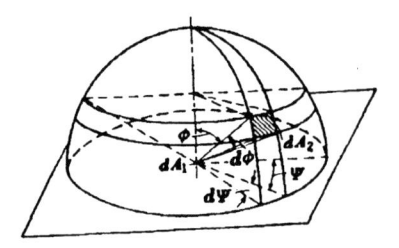

Figure 8.5. Nomenclature for intensity of radiation.

Thus at thermal equilibrium, the absorptivity and the emissivity of a body are equal. Thus, for gray and black bodies:

$$\alpha = \varepsilon \tag{8.16}$$

8.5 Radiation Intensity and Total Emissive Power

Previously we have only considered the total radiation emitted by a black body in all directions. This condition may be visualized by placing a hemispherical surface over a surface element dA_1, as shown in figure 8.5. The hemisphere will then intercept all of the radiation beams emitted from the surface element, but only from directly above it will dA_1 be seen without distortion. When viewed from a point on the hemisphere displaced by the angle ϕ from the normal to the surface, the element dA_1, will appear as the projected area $dA_1 \cos \phi$.

To determine the radiant heat per unit time emitted by dA_1 which reaches an area dA_2 on the surrounding hemisphere of radius r, we introduce a term called "intensity". The intensity of radiation I from dA_1 in space is defined as the radiant energy propagated in a particular direction per unit solid angle and per unit of area dA_1 as projected on a plane perpendicular to the direction of propagation. Then the rate of radiant heat flow from dA_1 to dA_2 is:

$$dq_{1-2} = I \cos \phi \, dA_1 \, dA_2 / r^2 \tag{8.17}$$

where

dq_{1-2} = rate of radiant heat flow from dA_1 to dA_2 (W)
I = intensity (W/m² steradian)
ϕ = angle between normal to surface and line connecting dA_1 to dA_2 (radians)
r = radius of hemisphere (m)

Notice that $\cos \phi \, dA_1$ is the projected surface area of dA_1 and dA_2/r^2 is the solid angle subtended by dA_2. Thus,

$$I = dq_{1-2}/dA_{1p} \, d\omega_1 \tag{8.18}$$

where dA_{1p} = projected area of surface 1 (m²).

For a diffuse surface the intensity is constant and does not vary with the emission angle ϕ. Then we can relate the intensity to the emissive power by integrating the radiation intercepted by an elemental area dA_2 over the half space represented by the surface of

Heat Transfer—Radiation

the hemisphere, or
$$dW = dq_{1-2}/dA_1 = I\cos\phi\, dA_2/r^2 \tag{8.19}$$

$$W = \iint_{A_2} I\cos\phi\, dA_2/r^2 \tag{8.20}$$

From figure 8.5:
$$dA_2 = (r\,d\phi)(r\sin\phi\, d\psi) \tag{8.21}$$

Thus,
$$W = \iint_{A_2} I\cos\phi(r^2\sin\phi\, d\phi\, d\psi/r^2) \tag{8.22}$$

Since I is constant,
$$W = I\int_0^{2\pi}\left[\int_0^{\pi/2}\cos\phi\,\sin\phi\, d\phi\right]d\psi \tag{8.23}$$

Then by integration over ϕ,
$$W = I\int_0^{2\pi}\left(\frac{1}{2}\right)\sin^2\phi\bigg|_0^{\pi/2} d\psi \tag{8.24}$$

or
$$W = I\int_0^{2\pi}\left(\frac{1}{2}\right)d\psi \tag{8.25}$$

Then,
$$W = \left(\frac{1}{2}\right)I\psi\bigg|_0^{2\pi} = \left(\frac{1}{2}\right)I(2\pi - 0) = \pi I \tag{8.26}$$

Thus, the emissive power of a diffuse surface is π times its intensity.

8.6 Heat Exchange by Radiation Between Black Surfaces

Consider two black bodies (fig. 8.6) separated by a medium which does not absorb radiation appreciably (e.g., air). Then dq_{1-2}, the rate at which radiation from dA_1 is received by dA_2 is:
$$dq_{1-2} = I_1\cos\phi_1\, dA_1\, d\omega_{1-2} \tag{8.27}$$

but
$$d\omega_{1-2} = \cos\phi_2\, dA_2/r^2 \tag{8.28}$$
$$I_1 = W_1/\pi \tag{8.29}$$

Then,
$$dq_{1-2} = W_1\, dA_1(\cos\phi_1\,\cos\phi_2\, dA_2/\pi r^2) \tag{8.30}$$

where the term in the parenthesis is equal to the fraction of the total radiation emitted

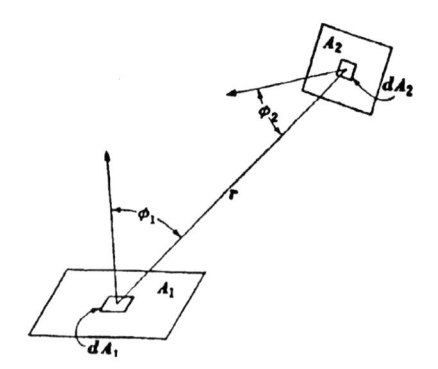

Figure 8.6. Geometrical shape-factor notation.

from dA_1 that is intercepted by dA_2. Similarly, the rate at which radiation from dA_2 is received by dA_1 is:

$$dq_{2-1} = W_2 dA_2(\cos\phi_1 \cos\phi_2 \, dA_1/\pi r^2) \tag{8.31}$$

Thus, the net rate of radiant heat transfer between dA_1 and dA_2 is:

$$dq_{1 \rightleftarrows 2} = dq_{1-2} - dq_{2-1}$$
$$= (W_1 - W_2)(\cos\phi_1 \cos\phi_2 \, dA_1 \, dA_2/\pi r^2) \tag{8.32}$$

To determine $q_{1 \rightleftarrows 2}$, the net rate of radiation between the entire surfaces A_1 and A_2, we simply integrate the fraction in the preceeding equation over both surfaces and obtain:

$$q_{1 \rightleftarrows 2} = (W_1 - W_2) \int_{A_1} \int_{A_2} (\cos\phi_1 \cos\phi_2 \, dA_1 \, dA_2/\pi r^2) \tag{8.33}$$

The double integral is conveniently written in shorthand notation either as $A_1 F_{1-2}$ or $A_2 F_{2-1}$, where F_{1-2} is called the shape factor evaluated on the basis of area A_1 and F_{2-1} is called the shape factor evaluated on the basis of area A_2. Physically F_{1-2} represents the fraction of the total radiant energy leaving A_1 which is intercepted by A_2 and F_{2-1} the fraction of energy leaving A_2 which is intercepted by A_1.

The equality,

$$A_1 F_{1-2} = A_2 F_{2-1} \tag{8.34}$$

is known as the reciprocity theorem. Substituting equation 8.34 for the double integral shows that the basic relation for the net rate of heat transfer by radiation between any two black bodies may be written as:

$$q_{1 \rightleftarrows 2} = (W_1 - W_2)F_{1-2}A_1 = (W_1 - W_2)F_{2-1}A_2 \tag{8.35}$$

The determination of a shape factor by evaluating the double integral of equation 8.33 is generally very tedious. Fortunately the shape factors for a large number of geometrical arrangements have been calculated and may be found in the literature. Many of the shape

Heat Transfer—Radiation

Table 8-3. Geometric shape factors

Surfaces Between Which Radiant Energy is Transferred	Shape Factor, F_{1-2}
1. Infinite parallel plates	1
2. Body A_1 completely enclosed by another body, A_2. Body A_1 cannot see any part of itself.	1
3. Element dA (A_1) and parallel circular disk (A_2) with its center directly above dA.	$a^2/(a^2 + L^2)$ a = radius of disk L = perpendicular distance between dA and the disk.
4. Two parallel disks of unequal diameter, distance L apart with centers on same normal to their planes, smaller disk A_1 of radius a, larger disk of radius b.	$(1/2a^2)[L^2 + a^2 + b^2 - \sqrt{(L^2 + a^2 + b^2)^2 - 4a^2b^2}]$

factors are given by Hottel (1930–1931), Hamilton et al. (1952), and Hutchinson (1952). A selected group of shape factors is summarized in table 8-3.

8.7 Heat Flow by Radiation Between Gray Surfaces

The radiation from gray surfaces can be treated conveniently in terms of the radiosity J, which is defined as the rate at which radiation leaves a given surface per unit area. The radiosity is the sum of radiation emitted, reflected, and transmitted, but for opaque bodies which transmit no radiation, the radiosity can be defined symbolically as:

$$J = \rho \Gamma + \varepsilon W_b \qquad (8.36)$$

where

J = radiosity (W/m^2)
Γ = irradiation or radiation per unit time incident on unit surface area (W/m^2)
W_b = black body emissive power (W/m^2)
ρ = reflectivity (dimensionless)
ε = emissivity (dimensionless)

The net rate at which radiation is leaving a gray surface per unit area and time is equal to the difference between the radiosity and the irradiation, that is,

$$dq_{\text{net}}/dA = J - \Gamma \qquad (8.37)$$

For an opaque gray surface, ρ is constant and $\rho + \varepsilon = 1$. The irradiation can therefore be eliminated from equation 8.37.

$$dq_{\text{net}}/dA = J - (J - \varepsilon W_b)/\rho \qquad (8.38)$$
$$dq_{\text{net}}/dA = J(1 - 1/\rho) + (\varepsilon/\rho)W_b \qquad (8.39)$$
$$dq_{\text{net}}/dA = (\varepsilon/\rho)(W_b - J) \qquad (8.40)$$

If the irradiation is uniformly distributed over the surface, the net rate of radiation leaving a surface A is obtained by integrating equation 8.40. Then,

$$q_{\text{net}} = (\varepsilon/\rho)A(W_b - J) \qquad (8.41)$$

The effect of the system geometry on the net radiation between two gray surfaces A_i and A_j with radiosities J_i and J_j, respectively, is the same as for similar black surfaces. Therefore,

$$q_{i \rightleftarrows j} = (J_i - J_j)A_i F_{i-j} = (J_i - J_j)A_j F_{j-i} \tag{8.42}$$

where

$q_{i \rightleftarrows j}$ = net heat transferred from A_i to A_j (W)
F_{i-j} & F_{j-i} = shape factors as defined by equations 8.34 and 8.33

Equation 8.41 can be written for both A_i and A_j as follows:

$$q_i = (\varepsilon_i/\rho_i)A_i(W_{bi} - J_i) \tag{8.43}$$
$$q_j = (\varepsilon_j/\rho_j)A_j(W_{bj} - J_j) \tag{8.44}$$

where

q_i = net heat loss by A_i to A_j (W)
q_j = net heat loss by A_j to A_i (W)
J_i = radiosity of A_i (W/m^2)
J_j = radiosity of A_j (W/m^2)

Then,

$$q_i = q_{i \rightleftarrows j} = -q_j \tag{8.45}$$

Therefore,

$$W_{bi} - J_i = (\rho_i/\varepsilon_i)(q_{i \rightleftarrows j}/A_i) \tag{8.46}$$
$$W_{bj} - J_j = -(\rho_j/\varepsilon_j)(q_{i \rightleftarrows j}/A_j) \tag{8.47}$$

or

$$J_i = W_{bi} - (\rho_i/\varepsilon_i)(q_{i \rightleftarrows j}/A_i) \tag{8.48}$$
$$J_j = W_{bj} + (\rho_j/\varepsilon_j)(q_{i \rightleftarrows j}/A_j) \tag{8.49}$$

Then substituting equations 8.48 and 8.49 into equation 8.42 results in:

$$q_{i \rightleftarrows j} = (W_{bi} - (\rho_i/\varepsilon_i)(q_{i \rightleftarrows j}/A_i) - W_{bj} - (\rho_j/\varepsilon_j)(q_{i \rightleftarrows j}/A_j))(A_i F_{i-j}) \tag{8.50}$$

Equation 8.50 can be rearranged to obtain:

$$q_{i \rightleftarrows j} = (W_{bi} - W_{bj})A_i/(1/F_{i-j} + \rho_i/\varepsilon_i + A_i\rho_j/A_j\varepsilon_j) \tag{8.51}$$

By substitution of equation 8.5, the general equation for the net rate of heat transfer between two gray bodies may be obtained:

$$q_{i \rightleftarrows j} = \sigma A_i(T_i^4 - T_j^4)/(1/F_{i-j} + \rho_i/\varepsilon_i + A_i\rho_j/A_j\varepsilon_j) \tag{8.52}$$

If we consider the special case of A_i to be a body completely enclosed within A_j such as a concentric sphere, we have $F_{i-j} = 1$ and

$$q_{i \rightleftarrows j} = \sigma A_i(T_i^4 - T_j^4)/(1/\varepsilon_i + (A_i/A_j)(1/\varepsilon_j - 1)) \tag{8.53}$$

Heat Transfer—Radiation

If $A_i \ll A_j$ the equation becomes:

$$q_{i \rightleftarrows j} = \sigma \varepsilon_i A_i (T_i^4 - T_j^4) \tag{8.54}$$

If A_i approaches A_j the equation becomes the equation for infinite parallel planes:

$$q_{i \rightleftarrows j} = \sigma A_i (T_i^4 - T_j^4)/(1/\varepsilon_i + 1/\varepsilon_j - 1) \tag{8.55}$$

Example 8.2. A sphere having a diameter of 1 m is suspended in a small room having dimensions of $2.4 \times 2.0 \times 2.0$ m. The surface temperature of the sphere is 60 C while the temperature of the room surfaces is 20 C. The sphere is polished brass having an emissivity of 0.10 while the room surfaces are coated with a cream paint having an emissivity of 0.95. (a) Determine the net rate of heat transfer by radiation from the brass sphere to the room surfaces. (b) Determine the percent error if the surface area of the sphere is considered very small compared with the room surface area.

Solution
(a) Since the sphere is completely enclosed within the room, F_{1-2} is 1.0 and equation 8.53 is applicable.

$$q_{s \rightleftarrows r} = \sigma A_s (T_s^4 - T_r^4)/(1/\varepsilon_s + (A_s/A_r)(1/\varepsilon_r - 1))$$
$$A_s = 4\pi r^2 = 4\pi (0.5 \text{ m})^2 = 3.1416 \text{ m}^2$$
$$A_r = 4(2.4 \text{ m})(2.0 \text{ m}) + 2(2.0 \text{ m})(2.0 \text{ m}) = 27.2 \text{ m}^2$$
$$T_s = 333.16 \text{ K} \quad \text{and} \quad T_r = 293.16 \text{ K}$$
$$q_{s \rightleftarrows r} = (5.73 \times 10^{-8} \text{ W/m}^2\text{K}^4)(3.1416 \text{ m}^2)(333.16^4 - 293.16^4)/$$
$$(1/0.10 + (3.1416/27.2)(1/0.95 - 1))$$
$$= 888.16/(10 + 0.1155(1.0526 - 1)) \text{ W}$$
$$= 88.762 \text{ W}$$

(b) Assuming $A_s \ll A_r$,

$$q_{s \rightleftarrows r} = \sigma \varepsilon_s A_s (T_s^4 - T_r^4)$$
$$= 0.10(888.16) = 88.816 \text{ W}$$

% Error $= (100)(88.816 - 88.762)/88.762 = 0.06\%$.

8.8 Heat Flow Between Surfaces Separated by a Transparent Layer

Previous sections of this chapter have considered only opaque surfaces such that $\tau = 0$. However, there are many situations in which radiative heat transfer takes place through transparent materials such as glass or plastic films. The radiosity can again be used in the treatment of such problems. Consider figure 8.7 in which two parallel opaque surfaces (1 and 3) are separated by a transparent layer (2). The total energy leaving surface 1 (its radiosity) is the amount of energy striking the upper surface of 2. Likewise the radiosity of the upper surface of 2 is the irradiation striking surface 1. Following the procedure outlined by Delwiche et al. (1984), the radiosities may be written for each

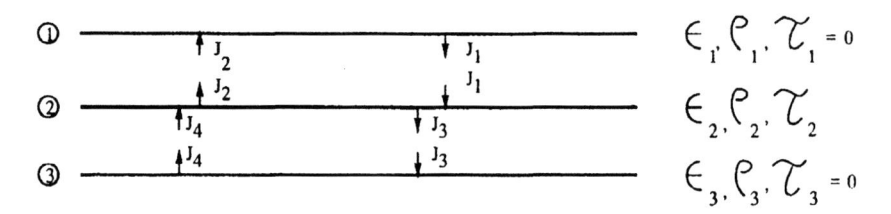

Figure 8.7. Radiation between two parallel opaque surfaces separated by a transparent layer.

surface as follows:

$$J_1 = \rho_1 J_2 + \varepsilon_1 W_{b1} \tag{8.56}$$
$$J_2 = \rho_2 J_1 + \varepsilon_2 W_{b2} + \tau_2 J_4 \tag{8.57}$$
$$J_3 = \rho_2 J_4 + \varepsilon_2 W_{b2} + \tau_2 J_1 \tag{8.58}$$
$$J_4 = \rho_3 J_3 + \varepsilon_3 W_{b3} \tag{8.59}$$

where

J_1 = radiosity of surface 1 (W/m^2)
J_2 = radiosity of upper surface of 2 (W/m^2)
J_3 = radiosity of lower surface of 2 (W/m^2)
J_4 = radiosity of surface 3 (W/m^2)
ρ_i = reflectivity of surface i (dimensionless)
ε_i = emissivity of surface i (dimensionless)
τ_i = transmissivity of surface i (dimensionless)
W_{bi} = black-body emissive power of surface i (W/m^2)

For surfaces having known property values, the set of four equations 8.56 through 8.59 contain seven unknowns (J_1, J_2, J_3, J_4, W_{b1}, W_{b2}, and W_{b3}). Thus, three additional equations or given conditions must be specified. If we consider the case where the temperatures of surfaces 1 and 3 are known, then W_{b1} and W_{b3} are fixed by:

$$W_{b1} = \sigma T_1^4 \tag{8.60}$$
$$W_{b3} = \sigma T_3^4 \tag{8.61}$$

where

σ = Stefan-Boltzmann constant = 5.73×10^{-8} W/m^{2}K^4
T_i = absolute temperature of surface i (K)

The final equation needed for solving for the various heat flux terms may be obtained by observing that at equilibrium an energy balance on layer 2 requires:

$$J_1 + J_4 = J_2 + J_3 \tag{8.62}$$

The set of equations 8.56 through 8.62 may now be solved simultaneously to evaluate all the heat fluxes. The equilibrium temperature of layer 2 may be obtained from the equation:

$$W_{b2} = \sigma T_2^4 \tag{8.63}$$

while the net heat flux to surface 1 is given by:

$$q_{net}'' = J_2 - J_1 \tag{8.64}$$

If additional transparent materials are placed between the opaque surfaces, three additional unknowns and three additional equations are introduced into the set of equations for each additional material in the system.

Example 8.3. Consider the system of figure 8.7 in which two opaque materials (surfaces 1 and 3) are separated by a transparent material (2). The temperatures of surfaces 1 and 3 are maintained at 30 and 60 C, respectively. Assume that surface 1 has an emissivity of 0.8 and surface 3 has an emissivity of 0.9. The transparent material (2) has an emissivity of 0.2 and a transmissivity of 0.7. (a) Determine the rate of heat transfer by radiation between surfaces 3 and 1 and determine the temperature of surface 2. (b) Determine the percentage reduction in heat transfer caused by the presence of the transparent layer.

Solution
(a) Equations 8.56 through 8.64 must be solved simultaneously. The reflectivities of surfaces 1 and 3 may be obtained from $\rho + \varepsilon = 1$ for opaque materials to obtain $\rho_1 = 0.2$ and $\rho_3 = 0.1$. The reflectivity of surface 2 is solved from $\rho + \varepsilon + \tau = 1$ to obtain $\rho_2 = 0.1$. The complete set of equations is then:

$$J_1 = 0.2 J_2 + 0.8 W_{b1}$$
$$J_2 = 0.1 J_1 + 0.2 W_{b2} + 0.7 J_4$$
$$J_3 = 0.1 J_4 + 0.2 W_{b2} + 0.7 J_1$$
$$J_4 = 0.1 J_3 + 0.9 W_{b3}$$
$$W_{b1} = \sigma(30 + 273.16)^4 \text{ K}^4$$
$$W_{b2} = \sigma T_2^4$$
$$W_{b3} = \sigma(60 + 273.16)^4 \text{ K}^4$$
$$J_1 + J_4 = J_2 + J_3$$
$$q_{net}'' = J_2 - J_1$$

Thus, the nine unknowns are J_1, J_2, J_3, J_4, W_{b1}, W_{b2}, W_{b3}, T_2, and q_{net}''. The nine equations may be solved for the unknowns using an iterative procedure. The following TK Solver rules were written to solve the problem:

```
J1 = rho1*J2 + eps1*Wb1
J2 = rho2*J1 + eps2*Wb2 + tau2*J4
J3 = rho2*J4 + eps2*Wb2 + tau2*J1
J4 = rho3*J3 + eps3*Wb3
J1+J4 = J2+J3
```

$$Wb1 = \text{sigma}*T1^4$$
$$Wb2 = \text{sigma}*T2^4$$
$$Wb3 = \text{sigma}*T3^4$$
$$rho1 + eps1 = 1$$
$$rho2 + eps2 + tau2 = 1$$
$$rho3 + eps3 = 1$$
$$qnet = J2-J1$$

The problem was then solved using the iterative solution feature of TK Solver to obtain: $J_1 = 518.44$ W/m^2, $J_2 = 656.19$ W/m^2, $J_3 = 552.87$ W/m^2, $J_4 = 690.63$ W/m^2, $W_{b1} = 484.00$ W/m^2, $W_{b2} = 604.53$ W/m^2, $W_{b3} = 705.94$ W/m^2, $T_2 = 320.49$ K $=$ 47.33 C, and $q''_{net} = 137.76$ W/m^2.

(b) The rate of heat transfer in the absence of the transparent layer may be calculated in either of two ways. First, the same solution technique may be used with τ_2 set to 1.0 ($\varepsilon_2 = \rho_2 = 0$). Secondly, equation 8.55 may be solved directly for the heat flow between the parallel surfaces.

Using method 1 and the TK Solver program given above, the net rate of heat transfer between the plates was resolved using tau2 $= 1.0$ to obtain 163.06 W/m^2. Using method 2,

$$q''_{net} = \sigma(T_3^4 - T_1^4)/(1/\varepsilon_3 + 1/\varepsilon_1 - 1)$$
$$= (5.73 \times 10^{-8} \text{ W/m}^2\text{K}^4)(333.16^4 - 303.16^4)\text{K}^4/(1/0.\rho + 1/0.8 - 1)$$
$$= 163.06 \text{ W/m}^2$$

The percent reduction caused by the transparent layer is then:

$$\% \text{ Reduction} = 100(163.06 - 137.76)/163.06 = 15.5\%$$

8.9 Solar Radiation and the Greenhouse Effect

For radiant heat transfer between surfaces at "normal" temperatures, the material properties which affect the proportions of irradiation reflected, absorbed, and transmitted may be considered the same as those which affect the amount of radiation emitted. However, for surfaces exposed to direct solar radiation (short wavelengths) the material properties affecting incoming radiation may vary greatly from the properties affecting outgoing radiation which is at a much longer wavelength. As shown by table 8-2, some materials have higher emissivities at short wavelengths (higher temperatures) while others have higher emissivities at long wavelengths (low temperatures). When solar radiation is present or when extremely high temperature differences are present, different emissivities must be used for incoming and outgoing radiation. If a transparent material is considered, its transmissivity may also vary greatly with wavelength. Table 8-4 presents transmissivity values for some materials at both long and short wavelengths.

Consider two flat surfaces exposed to solar irradiation at the rate of 700 W/m^2 (fig. 8.8). In order to illustrate the effect of differences in the emissivity values at short

Heat Transfer—Radiation

Table 8-4. Transmissivity of some glazing materials for long and short wavelength radiation

Material	Transmissivity Long Wavelength	Transmissivity Short Wavelength
Polyethylene, 101.6 μm UV Resistant	0.80	0.888
Flat Fiber Glass		
635 μm Regular	0.12	0.831
1 016 μm Premium	0.06	0.729
Polyester		
127 μm Weatherable	0.32	0.865
Corrugated Fiber Glass		
1 016 μm	0.075	0.787
Glass		
3 175 μm Double Strength	0.03	0.878
Polycarbonate, 1 588 μm	0.06	0.844
Polyvinylfluoride, 76.2 μm	0.43	0.91

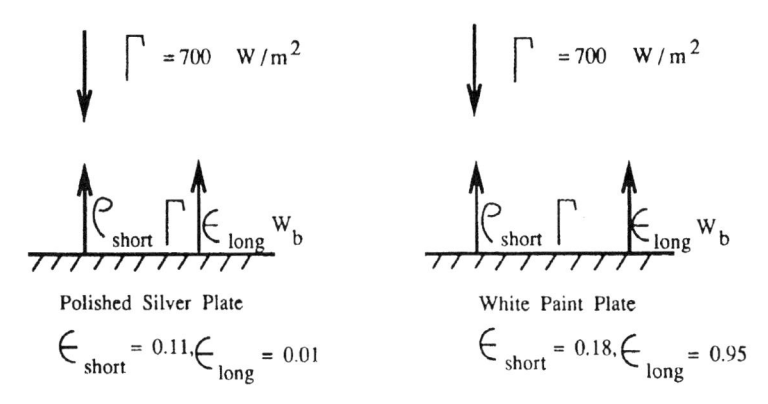

Figure 8.8. Two flat plates exposed to solar irradiation.

Polished Silver Plate: $\varepsilon_{short} = 0.11$, $\varepsilon_{long} = 0.01$

White Paint Plate: $\varepsilon_{short} = 0.18$, $\varepsilon_{long} = 0.95$

$\Gamma = 700$ W/m^2

and long wavelengths, let us assume that there is no heat transfer by conduction or convection between the plates and their environment. Then the total energy striking the plates must equal the radiosity of the plates and the radiosity is given by:

$$J = \rho_{short}\Gamma + \varepsilon_{long} W_b \tag{8.65}$$

where

J = radiosity (W/m^2)
Γ = solar irradiation (W/m^2)
ρ_{short} = reflectivity for short wavelength radiation (dimensionless)
ε_{long} = emissivity for long wavelength radiation (dimensionless)

Since $\Gamma = J$ at equilibrium, if other forms of energy transfer are neglected, and

$W_b = \sigma T^4$, then equation 8.65 may be solved for the equilibrium temperature of the plates:

$$T = (\Gamma \varepsilon_{short}/\varepsilon_{long} \sigma)^{0.25} \tag{8.66}$$

where

T = equilibrium absolute temperature of plate (K)
ε_{short} = emissivity for short wavelength radiation (dimensionless)
σ = Stefan-Boltzmann constant = 5.73×10^{-8} W/m²K⁴

For the situation of figure 8.8, the polished silver plate has an emissivity of 0.11 at solar wavelengths and an emissivity of 0.01 at long wavelengths. Inserting these values into equation 8.66 gives an equilibrium temperature of 605 K or 332 C. The white paint surface has an emissivity of 0.18 at solar wavelengths and an emissivity of 0.95 at long wavelengths giving an equilibrium temperature of 219 K or −54 C. Thus the equilibrium temperatures of the two plates differs by 386 K due to differences in the ratio of short wavelength to long wavelength emissivities. Of course if the plates are located in air, the convection to the surrounding air must be considered. Also, any conduction to surfaces below the plate must be considered.

If we place a partially transparent cover above one of the plates as illustrated in figure 8.9, we must account for the portion of all radiosity terms which is of short wavelength and the portion which is of long wavelength. Let us again neglect heat convection and heat conduction and also assume that the reflectivity of the transparent cover is zero. Then the heat fluxes become those shown in figure 8.9. An energy balance results in the following two equations:

$$\Gamma = \varepsilon_{1long} W_{b1} + \rho_{2short} \tau_{1short}{}^2 \Gamma + \tau_{1long} \rho_{2long} \varepsilon_{1long} W_{b1}$$
$$+ \tau_{1long} \varepsilon_{2long} W_{b2} \tag{8.67}$$

$$\tau_{1short} \Gamma + \varepsilon_{1long} W_{b1} = \rho_{2short} \tau_{1short} \Gamma + \rho_{2long} \varepsilon_{1long} W_{b1} + \varepsilon_{2long} W_{b2} \tag{8.68}$$

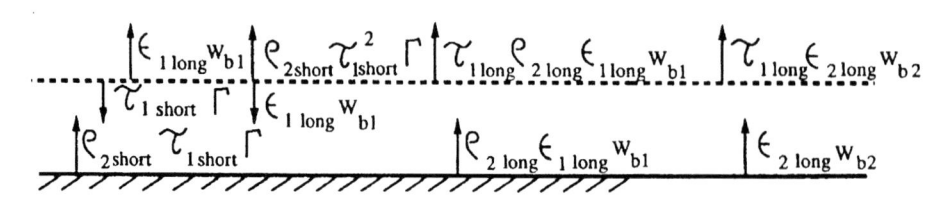

Figure 8.9. Plate covered by partially transparent cover and exposed to solar irradiataion.

Heat Transfer—Radiation

where

$\varepsilon_{i\,\text{long}}$ and $\varepsilon_{i\,\text{short}}$ = emissivities of surface i at long and short wavelengths (dimensionless)

$\rho_{i\,\text{long}}$ and $\rho_{i\,\text{short}}$ = reflectivities of surface i at long and short wavelengths (dimensionless)

$\tau_{i\,\text{long}}$ and $\tau_{i\,\text{short}}$ = transmissivities of surface i at long and short wavelengths (dimensionless)

W_{bi} = black-body emissive power of surface i (W/m^2)
W_{bi} = σT_i^4
T_i = absolute temperature of surface i (K)
σ = Stefan-Boltzmann constant = 5.73×10^{-8} W/m^{2}K^4

If we use the properties of white paint for surface 2, the emissivity of glass for surface 1 (with reflectivity assumed zero), and a solar irradiation of 700 W/m^2, then equations 8.67 and 8.68 may be solved for values of the emissive powers which may then be converted to temperatures. Solving this set of equations results in an equilibrium temperature of the white painted plate of 9 C and of the glass cover of -15 C. Thus, the effect of the glass cover is to raise the temperature of the plate by 63 C over the uncovered equilibrium temperature. This temperature rise is commonly referred to as the "greenhouse" effect.

Example 8.4. Determine the equilibrium temperature of the polished silver plate of figure 8.8 if convection to air at 20 C is considered and the convective heat transfer coefficient is assumed to be 5 W/m^{2}K.

Solution
At equilibrium the rate of solar irradiation must equal the sum of the plate radiosity and the heat transferred by convection to the air. Then,

$$\Gamma = J + q''_{\text{conv}}$$

where

$$J = \rho_{\text{short}}\Gamma + \varepsilon_{\text{long}} W_b$$

$$W_b = \sigma T^4$$

Then

$$\Gamma = \rho_{\text{short}}\Gamma + \varepsilon_{\text{long}}\sigma T^4 + h(T - T_a)$$

or

$$700\,\text{W/m}^2 = 0.89(700\,\text{W/m}^2) + 0.01(5.73 \times 10^{-8}\,\text{W/m}^2\text{K}^4)T^4$$
$$+ (5\,\text{W/m}^2\text{K})(T - 293.16)$$

An iterative solution yields an equilibrium plate temperature of 307.5 K or 34.4 C. Thus, the heat loss by convection greatly reduces the equilibrium temperature from that obtained when neglecting convection.

Nomenclature

- A area (m^2)
- c speed of light (m/s)
- h convective heat transfer coefficient (W/m^{2}K)
- I intensity of radiation (W/m^2 steradian)
- k thermal conductivity (W/mK)
- l length (m)
- q heat flow (W)
- q'' heat flux (W/m^2)
- t temperature (C)
- T absolute temperature (K)
- W the radiant energy emissive power of a body (W/m^2)
- α absorptivity (dimensionless)
- ε emissivity (dimensionless)
- Γ incident radiation (W/m^2)
- λ wavelength (m)
- ω solid angle (steradians)
- ν radiation frequency (s^{-1})
- ϕ angle from normal to surface (radians)
- ψ angle in plane tangent to surface (radians)
- ρ reflectivity (dimensionless)
- σ Stefan-Boltzmann constant $= 5.73 \times 10^{-8}$ W/m^{2}K^4
- τ transmissivity (dimensionless)

References

Delwiche, S. R. and D. H. Willits. 1984. The Effect of condensation on heat transfer through polyethylene film. *Transactions of the ASAE* 27:1476–1482.

Godbey, L. C., T. E. Bond and H. F. Zoring. 1979. Transmission of solar and long-wavelength energy by materials used as covers for solar collectors and greenhouses. *Transactions of the ASAE* 22:1137–1144.

Hamilton, D. C. and W. R. Morgan. 1952. Radiant-Interchange Configuration Factors. NACA TN2836.

Hottel, H. C. 1930. Radiant heat transmission. *Mechanical Engineering* 52:699–704.

———. 1931. Radiant heat transmission between surfaces separated by non-absorbing media. *Transactions of the ASME* FSP-53-19b. 53:265–271.

Hutchinson, F. W. 1952. *Industrial Heat Transfer*. New York: The Industrial Press.

Jakob, M. and G. A. Hawkins. 1957. *Elements of Heat Transfer*. New York: John Wiley & Sons, Inc.

Kreith, F. 1958. *Principles of Heat Transfer*. Scranton, Pa.: International Textbook Co.

Silverstein, S. D. 1976. Effect of infrared transparency on the heat transfer through windows: A clarification of the greenhouse effect. *Science* 193:229–231.

Heat Transfer—Radiation

Problems

1. Determine the temperature of light sources emitting their maximum amount of radiation at the following wavelengths:
   (a) 405 mμ — violet light
   (b) 480 mμ — blue light
   (c) 520 mμ — green light
   (d) 580 mμ — yellow light
   (e) 600 mμ — orange light
   (f) 640 mμ — red light

2. One steradian is the solid angle subtended at the center of a sphere by a portion of the surface equal to the square of the radius of the sphere. Thus:

$$d\omega = dA/r^2$$

   where

   ω = differential solid angle (steradians)
   dA = differential area on surface (m^2)
   r = radius of sphere (m)

   By integrating the above equation, determine the solid angle subtended by the whole sphere.

3. Estimate the rate of heat loss by radiation from the top of an aluminum painted galvanized iron blancher box, 1.8 m wide and 11.0 m long, in a room where the mean wall and ceiling temperatures are 24 C. The box is direct steam heated at 99 C.

4. Find the heat loss rate by radiation from a furnace tube, 1.0 m in diameter and 3 m long, which is in a concrete-block dehydrator furnace chamber, 2 m wide, 3 m long, and 2.4 m high. The tube-surface temperature is 230 C, and the chamber-wall temperature averages 80 C. Assume the tube is constructed of galvanized sheet iron having an emissivity of 0.276. Assume the concrete to have an emissivity of 0.91.

5. Calculate the equilibrium temperature of a thermocouple in a large air duct if the air temperature is 1 100 C, the duct-wall temperature is 260 C, the emissivity of the thermocouple is 0.4, and the convective heat transfer coefficient, h, is 110 W/m^{2}K.

6. A metal plate is placed in the sunlight. The incident radiant energy Γ is 80 W/m^2. The air and the surroundings are at 10 C. The heat transfer coefficient for free convection from the upper surface of the plate is 17.0 W/m^{2}K. The plate has an average emissivity of 0.9 at solar wavelengths and 0.1 at long wavelengths. Neglecting conduction losses on the lower surface, determine the equilibrium temperature of the plate.

9 Psychrometrics

Psychrometrics is the study of the physical and thermal properties of air and water vapor mixtures. Dry air at sea level has a percentage volumetric composition of: N_2, 78.03; O_2, 20.99; A, 0.94; CO_2, 0.03; H_2, 0.01; Ne, 0.00123; He, 0.0004; Kr, 0.00005; Xe, 0.000006. For engineering purposes, air is considered composed of nitrogen and oxygen. At sea level, atmospheric pressure, 101.325 kPa, air has the following composition.

Composition of Air

	By Volume	By Weight	Molecular Mass
Nitrogen	79%	76.8%	28.02
Oxygen	21%	23.2%	32.00
Air (dry)	—	—	28.97

9.1 Ideal Gas Law

The ideal gas law, which has been discovered experimentally as well as being derived theoretically from the kinetic theory of gases is:

$$pV = nRT \tag{9.1}$$

where

p = absolute pressure (Pa)
V = volume (m³)
n = number of kg-moles
R = universal gas constant = 8.314×10^3 Pa m³/kg-mole K
T = absolute temperature (K)

Values for R in other units are:

$$R = 1.987 \text{ cal/gm-mole K}$$
$$= 82.06 \text{ cm}^3 \text{ atm/gm-mole K}$$
$$= 8.314 \times 10^7 \text{ gm cm}^2/\text{s}^2 \text{ gm-mole K}$$
$$= 8.314 \times 10^3 \text{ kg m}^2/\text{s}^2 \text{ kg-mole K}$$

$$= 4.969 \times 10^4 \text{ lb}_m \text{ ft}^2/\text{s}^2 \text{ lb-mole R}$$
$$= 1.545 \times 10^3 \text{ ft lb}_f/\text{lb-mole R}$$

Then for a fixed number of moles of an ideal gas:

$$p_1 V_1 = nRT_1 \tag{9.2}$$
$$p_2 V_2 = nRT_2 \tag{9.3}$$
$$p_1 V_1 / T_1 = p_2 V_2 / T_2 \tag{9.4}$$

The ideal gas law does not hold perfectly for a real gas under extreme conditions. For high pressure (a number of atmospheres) and high temperatures, the deviation from the laws must be considered if accurate results are expected. It is entirely suitable for normal processing conditions, however.

9.2 Dalton's Law

Dalton's Law states that each component in a mixture of gases exerts the same pressure it would exert if it alone occupied the same volume at the same temperature and that the total pressure is equal to the sum of the partial pressures. Then:

$$p = p_1 + p_2 + p_3 + \cdots \tag{9.5}$$

The mass of the mixture is the sum of the masses of the components:

$$m = m_1 + m_2 + m_3 + \cdots \tag{9.6}$$

The mass of a particular component is:

$$m_1 = n_1 M_1 \tag{9.7}$$

where

m_1 = mass of component 1 (kg)
n_1 = number of moles of component 1 (kg-moles)
M_1 = molecular weight of 1 (kg/kg-mole)

Then

$$n_1 = m_1 / M_1 \tag{9.8}$$
$$p_1 V_1 = m_1 RT_1 / M_1 \tag{9.9}$$

or

$$m_1 = p_1 V_1 M_1 / RT_1 \tag{9.10}$$

Then,

$$m = p_1 V_1 M_1 / RT_1 + p_1 V_2 M_2 / RT_2 + p_3 V_3 M_3 / RT_3 + \cdots \tag{9.11}$$

For a mixture of gases, all the components occupy the same volume and are at the same

Psychrometrics

temperature. Therefore,

$$m = (p_1 M_1 + p_2 M_2 + p_3 M_3 + \cdots) V/RT \tag{9.12}$$

or

$$RTm/V = p_1 M_1 + p_2 M_2 + p_3 M_3 + \cdots \tag{9.13}$$

For a mixture,

$$pV = mRT/M \tag{9.14}$$

where M is the effective molecular weight for the mixture.

Then,

$$pM = mRT/V \tag{9.15}$$

$$pM = p_1 M_1 + p_2 M_2 + p_3 M_3 + \cdots \tag{9.16}$$

or

$$M = (p_1/p)M_1 + (p_2/p)M_2 + (p_3/p)M_3 + \cdots \tag{9.17}$$

9.3 Definitions of Psychrometric Terms

Normal atmospheric air is a mixture of dry air and water vapor, atmospheric air never being completely dry. A number of physical and thermal quantities are used to describe the state of the mixture.

9.3.1 Dry-Bulb Temperature

The dry-bulb temperature is the true temperature of the air-vapor mixture. The modifiers "dry-bulb" are added to distinguish this temperature from other psychrometric properties to be defined later.

9.3.2 Saturation Pressure

At any given temperature there is a water vapor pressure for which the vapor and the liquid forms of water are in equilibrium with each other. That is, the rate of evaporation of the liquid is exactly equal to the rate of condensation. This pressure is known as the saturation pressure. The saturation pressure is only very slightly affected by the total pressure of the atmosphere. Thus, the saturation vapor pressure can be taken directly from a standard steam table (table 9.1) or it may be calculated using empirical functions of temperature. Brooker (1967) gives the following relationship for saturation pressure in the temperature range from 255.38 to 273.16 K:

$$\ln p_{sat} = 31.9602 - 6270.3605/T - 0.46057 \ln T \tag{9.18}$$

where

p_{sat} = saturation vapor pressure (Pa)
T = temperature (K)

Table 9-1. Steam table*

Temperature t (°C)	Pressure p (kPa)	Specific Volume Sat. Liquid v_f (m³/kg)	Specific Volume Sat. Vapor v_g (m³/kg)	Internal Energy Sat. Liquid u_f (J/g)	Internal Energy Evap. u_{fg} (J/g)	Internal Energy Sat. Vapor u_g (J/g)	Enthalpy Sat Liquid h_f (J/g)	Enthalpy Evap. h_{fg} (J/g)	Enthalpy Sat. Vapor h_g (J/g)	Entropy Sat. Liquid n_f (J/gk)	Entropy Evap. n_{fg} (J/gk)	Entropy Sat. Vapor n_g (J/gk)
0	0.6109	0.00100	206.278	−.03	2375.4	2375.3	−.02	2501.4	2501.3	−.0001	9.1566	9.1565
.01	0.6133	0.00100	206.136	.00	2375.3	2375.3	.01	2501.3	2501.4	.0000	9.1562	9.1562
1	0.6567	0.00100	192.577	4.15	2372.6	2376.7	4.16	2499.0	2503.2	.0152	9.1147	9.1299
5	0.8721	0.00100	147.120	20.97	2361.3	2382.3	20.98	2489.6	2510.6	.0761	8.9496	9.0257
10	1.2276	0.00100	106.379	42.00	2347.2	2389.2	42.01	2477.7	2519.8	.1510	8.7498	8.9008
15	1.7051	0.00100	77.926	62.99	2333.1	2396.1	62.99	2465.9	2528.9	.2245	8.5569	8.7814
20	2.339	0.00100	57.791	83.95	2319.0	2402.9	83.96	2454.1	2538.1	.2966	8.3706	8.6672
25	3.169	0.00100	43.360	104.88	2304.9	2409.8	104.89	2442.3	2547.2	.3674	8.1905	8.5580
30	4.246	0.00100	32.894	125.78	2290.8	2416.6	125.79	2430.5	2556.3	.4369	8.0164	8.4533
35	5.628	0.00101	25.216	146.67	2276.7	2423.4	146.68	2418.6	2565.3	.5053	7.8478	8.3531
40	7.384	0.00101	19.523	167.56	2262.6	2430.1	167.57	2406.7	2574.3	.5725	7.6845	8.2570
45	9.593	0.00101	15.258	188.44	2248.4	2436.8	188.45	2394.8	2583.2	.6387	7.5261	8.1648
50	12.349	0.00101	12.032	209.32	2234.2	2443.5	209.33	2382.7	2592.1	.7038	7.3725	8.0763
55	15.758	0.00101	9.568	230.21	2219.9	2450.1	230.23	2370.7	2600.9	.7679	7.2234	7.9913
60	19.940	0.00102	7.671	251.11	2205.5	2456.6	251.13	2358.5	2609.6	.8312	7.0784	7.9096
70	31.19	0.00102	5.042	292.95	2176.6	2469.6	292.98	2333.8	2626.8	.9549	6.8004	7.7553
80	47.39	0.00103	3.407	334.86	2147.4	2482.2	334.91	2308.8	2643.7	1.0753	6.5369	7.6122
90	70.14	0.00104	2.361	376.85	2117.7	2494.5	376.92	2283.2	2660.1	1.1925	6.2866	7.4791
100	101.35	0.00104	1.6729	418.94	2087.6	2506.5	419.04	2257.0	2676.1	1.3069	6.0480	7.3549
120	198.53	0.00106	0.8919	503.50	2025.8	2529.3	503.71	2202.6	2706.3	1.5276	5.6020	7.1296
140	361.3	0.00108	0.5089	588.74	1961.3	2550.0	589.13	2144.7	2733.9	1.7391	5.1908	6.9299
160	617.8	0.00110	0.3071	674.84	1893.5	2568.4	675.55	2082.6	2758.1	1.9427	4.8075	6.7502
180	1002.1	0.00113	0.19405	762.09	1821.6	2583.7	763.22	2015.0	2778.2	2.1396	4.4461	6.5857
200	1553.8	0.00116	0.12736	850.65	1744.7	2595.3	852.45	1940.7	2793.2	2.3309	4.1014	6.4323
250	3973	0.00125	0.05013	1080.39	1522.0	2602.4	1085.36	1716.2	2801.5	2.7927	3.2802	6.0730
300	8581	0.00140	0.02167	1332.0	1231.0	2563.0	1344.0	1404.9	2749.0	3.2534	2.4511	5.7045
350	16513	0.00174	0.00881	1641.9	776.5	2418.4	1670.6	893.4	2563.9	3.7777	1.4335	5.2112
374	22050	0.00288	0.00332	1985.9	67.8	2053.7	2049.4	77.5	2127.0	4.3529	.1198	4.4727
374.136	22090	0.00316	0.00316	2029.6	0	2029.6	2099.3	0	2099.3	4.4298	0	4.4298

Psychrometrics

An equation valid in the temperature range from 273.16 to 533.16 K has been adapted from Keenan and Keyes (1936):

$$\ln(p_{sat}/R') = (A + BT + CT^2 + DT^3 + ET^4)/(FT - GT^2) \qquad (9.19)$$

where

$R' = 22\,105\,649.25$ $\qquad D = 0.12558 \times 10^{-3}$
$A = -27\,405.526$ $\qquad E = -0.48502 \times 10^{-7}$
$B = 97.5413$ $\qquad F = 4.34903$
$C = -0.146244$ $\qquad G = 0.39381 \times 10^{-2}$

Example 9.1. Determine the saturation pressure at a temperature of 20 C.

Solution
The absolute temperature is 293.16 K and equation 9.19 gives:

$$\ln(p_{sat}/R') = [A + B(293.16) + C(293.16)^2 + D(293.16)^3 + E(293.16)^4]/$$
$$[F(293.16) - G(293.16)^2]$$
$$= -8573.192/936.5104 = -9.1544017$$
$$p_{sat}/R' = 0.000105753$$
$$p_{sat} = 0.000105753(22105649.25) = 2337.8 \text{ Pa} = 2.338 \text{ kPa}$$

Note that this calculated value compares with a value of 2.339 kPa in the steam table (table 9-1).

9.3.3 Humidity Ratio

The humidity ratio of an air-water vapor mixture is defined as the mass of water vapor per unit mass of dry air.

$$H = m_v/m_a \qquad (9.20)$$

where

H = humidity ratio (kg_{H_2O}/kg_{air})
m_v = mass of water vapor (kg_{H_2O})
m_a = mass of dry air (kg_{air})

The ideal gas law may be used to convert equation 9.20 to:

$$H = p_v M_v / p_a M_a \qquad (9.21)$$

where

M_a = the effective molecular mass of air (28.97 kg/kg-mole)
M_v = molecular mass of water (18.02 kg/kg-mole)
p_v = water vapor pressure (Pa)
p_a = air pressure (Pa)

Then,

$$H = p_v/(1.605(p_{atm} - p_v)) \tag{9.22}$$

where p_{atm} = total atmospheric pressure (Pa).

Example 9.2. Determine the humidity ratio at saturation at 20 C assuming standard atmospheric pressure of 101.325 kPa.

Solution
Using the saturation pressure from example 9.1,

$$H_{sat} = 2.338 \text{ kPa}/[1.605(101.325 - 2.338)] \text{ kPa}$$
$$= 0.0147 \text{ kg}_{H_2O}/\text{kg}_{air}$$

9.3.4 Relative Humidity

Relative humidity is defined as the ratio of the actual pressure of the water vapor in the air-vapor mixture to the saturation pressure at the temperature of the mixture. Then,

$$rh = p_v/p_{sat} \tag{9.23}$$

where rh = relative humidity expressed as a decimal. The relative humidity is often multiplied by 100 and given as a percent.

Example 9.3. If the actual water vapor pressure in the mixture is 1.75 kPa and the dry-bulb temperature is 20 C, determine the relative humidity.

Solution
From example 9.1, $p_{sat} = 2.338$ kPa
Then,

$$rh = p_v/p_{sat} = 1.75 \text{ kPa}/2.338 \text{ kPa} = 0.7485 = 74.85\%$$

9.3.5 Specific Volume

The specific volume is the volume per unit mass of dry air at the state of the mixture. The term "dry air" is used to indicate that only the mass of the gases making up the air is included (i.e., the mass of the water vapor which is present in the mixture is not included). From the ideal gas law:

$$p_a V = m_a RT/M_a \tag{9.24}$$

Then, the specific volume is:

$$v = V/m_a = RT/p_a M_a \tag{9.25}$$

or

$$v = RT/((p_{atm} - p_v)M_a) \tag{9.26}$$

where v = specific volume, m^3/kg_{air}.

Example 9.4. Determine the specific volume of the air-water vapor mixture of example 9.3 if the total atmospheric pressure is 101.325 kPa.

Psychrometrics

Solution

$$v = RT/((p_{atm} - p_v)M_a)$$
$$= (8.314 \times 10^3 \text{ Pa m}^3/\text{kg-mole K})(293.16 \text{ K})/$$
$$((101.325 - 1.75) \text{ kPa}(1000 \text{ Pa/kPa})(28.97 \text{ kg/kg-mole}))$$
$$= 0.845 \text{ m}^3/\text{kg}_{air}$$

9.3.6 Dew-Point Temperature

The dew-point temperature is the temperature at which the saturation pressure of water is equal to the partial pressure of the water vapor existing in the air-water vapor mixture. Thus, if an air-water vapor mixture is cooled to its dew-point temperature, condensation of the water vapor will take place.

9.3.7 Enthalpy

The enthalpy of an air-water vapor mixture is the total energy content of the mixture. It would more accurately be defined as the "specific enthalpy" since it is expressed as an energy per unit mass of dry air. The enthalpy of moist air consists of the enthalpy of the dry air plus the enthalpy of the water vapor. All enthalpy values are measured relative to the energy at some datum state where the enthalpy is defined as zero. For the purposes of this book we will use a temperature of 0 C (273.16 K) as the datum for enthalpy calculations. Thus, a mixture of dry air and liquid water at 0 C would have an enthalpy of zero. The enthalpy of the dry air is the energy which would have to be added to 0 C air to obtain air of a specified temperature and is given by:

$$h_a = c_{pa}(t - t_0) \tag{9.27}$$

where

h_a = enthalpy of the air (kJ/kg$_{air}$)
c_{pa} = specific heat of air (kJ/kg K)
t = temperature (C)
t_0 = reference temperature = 0 C

The enthalpy of water vapor at low temperatures can be expressed as the sum of the energy required to vaporize liquid water at 0 C and the energy required to heat the resulting vapor from 0 C to the specified temperature or:

$$h_v = h_{g,0} + c_{pv}(t - t_0) \tag{9.28}$$

where

h_v = enthalpy of water vapor (kJ/kg$_{H_2O}$)
c_{pv} = specific heat of water vapor (kJ/kg$_{H_2O}$ K)
t = temperature (C)
t_0 = reference temperature (C)
$h_{g,0}$ = heat of vaporization of water at reference temperature of 0 C (273.16 K) (kJ/kg$_{air}$)

The total enthalpy of an air-water vapor mixture per kilogram of dry air is then given by:

$$h = h_a + H h_v \tag{9.29}$$

where

h = total enthalpy (kJ/kg$_{air}$)
h_a = enthalpy of air (kJ/kg$_{air}$)
H = humidity ratio (kg$_{H_2O}$/kg$_{air}$)
h_v = enthalpy of water vapor (kJ/kg$_{H_2O}$)

Then,

$$h = c_{pa}(t - t_0) + H[h_{g,0} + c_{pv}(t - t_0)] \tag{9.30}$$

At normal temperatures and pressures the values of c_{pa}, c_{pv}, and $h_{g,0}$ are:

$$c_{pa} = 1.0069254 \text{ kJ/kg}_{air} \text{ K}$$
$$c_{pv} = 1.8756864 \text{ kJ/kg}_{H_2O}$$
$$h_{g,0} = 2.502535259 \times 10^3 \text{ kJ/kg}_{H_2O}$$

When the air is saturated with water vapor and also contains water in liquid form, the enthalpy of the liquid water must be added. This is the energy required to heat the liquid water from 0 C to the specified temperature:

$$h_f = c_{pf}(t - t_0) \tag{9.31}$$

where

h_f = enthalpy of liquid water (kJ/kg$_{H_2O}$)
c_{pf} = specific heat of liquid water = 4.1868 kJ/kg K

Then,

$$h = c_{pa}(t - t_0) + H_{sat}[h_{g,0} + c_{pv}(t - t_0)] + (H - H_{sat})c_{pf}(t - t_0) \tag{9.32}$$

If the air is saturated with water vapor and contains an ice fog, the enthalpy of the ice must be added. Ice has a negative enthalpy since the enthalpy of liquid water at 0 C was chosen as the datum point of zero enthalpy. The enthalpy of the ice is less than zero by the sum of the energy which must be removed from liquid water at 0 C in order to freeze it and the energy which must be removed to cool the ice to the specified temperature:

$$h_i = -h_{s,0} + c_{pi}(t - t_0) \tag{9.33}$$

where

h_i = enthalpy of ice (kJ/kg)
$h_{s,0}$ = heat of fusion at 0 C = 333.3685 kJ/kg
c_{pi} = specific heat of ice = 2.0934 kJ/kg K

Psychrometrics

Then the total enthalpy for air containing an ice fog is:

$$h = c_{pa}(t - t_0) + H_{sat}[h_{g,0} + c_{pv}(t - t_0)] \\ + (H - H_{sat})[c_{pi}(t - t_0) - h_{s,0}] \quad (9.34)$$

Two alternative representations for the enthalpy of water vapor may be encountered. They are the result of assuming a different path by which the vapor is brought to a particular state. Whereas in equation 9.28 the liquid is assumed to be vaporized at 0 C and then heated to the specified temperature, there are many other paths by which the vapor may reach the same thermodynamic state. The first alternative which we will consider is to first heat the liquid water to the specified temperature and then vaporize the liquid at that temperature. The enthalpy of the vapor can then be expressed as:

$$h_v = c_{pf}(t - t_0) + h_{fg} \quad (9.35)$$

where h_{fg} = heat of vaporization at temperature t (kJ/kg).

A second alternative to be considered is to heat the liquid water to the dew-point temperature, vaporize the liquid at the dew-point temperature, and then heat the vapor to the specified dry-bulb temperature. Then:

$$h_v = c_{pf}(t_{dp} - t_0) + h''_{fg} + c_{pv}(t - t_{dp}) \quad (9.36)$$

where h''_{fg} = heat of vaporization at the dew-point temperature (kJ/kg).

Equations 9.28, 9.35, and 9.36 should give identical values of the water vapor enthalpy if the specific heat values are constant and the heat of vaporization can be accurately predicted at any given temperature.

Equations for the heat of vaporization are given by Brooker (1967) and ASAE (1996) as:

$$h_{fg} = h_{g,0} - (2.38576424 \text{ kJ/kg K})(t - t_0) \quad (9.37)$$

for $0 < t < 65.56$ C and

$$h_{fg} = (7329156 - 15.99596408 T^2)^{1/2} \quad (9.38)$$

for $65.56 < t < 260$ C ($338.72 < T < 533.16$ K).

For temperatures below 0 C, the enthalpy of water vapor can be visualized by assuming that the liquid water is first frozen at 0 C, cooled to the dew-point temperature, sublimed at the dew-point temperature, and then heated to the final temperature. Thus,

$$h_v = -h_{s,0} - c_{pi}(t_0 - t_{dp}) + h''_{ig} + c_{pv}(t - t_{dp}) \quad (9.39)$$

where

$h_{s,0}$ = heat of fusion at 0 C (kJ/kg)
c_{pi} = specific heat of ice (kJ/kg$_{H_2O}$ K)
h''_{ig} = heat of sublimation at the dew-point temperature (kJ/kg$_{H_2O}$)

The heat of sublimation as a function of temperature is given by Brooker (1967) and ASAE (1996) as:

$$h_{ig} = 2839.683144 \text{ kJ/kg} - 0.21256384 \text{ kJ/kg K}(t + 17.78) \quad (9.40)$$

for $-17.78 < t < 0$ C.

Example 9.5. Determine the enthalpy of water vapor at 60 C.

Solution

$$\begin{aligned} h_v &= h_{g,0} + c_{pv}(t - t_0) \\ &= 2502.535259 \text{ kJ/kg}_{H_2O} + 1.8756864 \text{ kJ/kg}_{H_2O} \text{ K}(60 - 0) \text{ K} \\ &= 2615.1 \text{ kJ/kg}_{H_2O} \end{aligned}$$

Example 9.6. Determine the enthalpy of an air-water vapor mixture having a dry-bulb temperature of 60 C and a humidity ratio of 0.02 kg_{H_2O}/kg_{air}.

Solution

$h_v = 2615.1$ kJ/kg$_{H_2O}$ from example 9.5.

$$\begin{aligned} h &= c_{pa}(t - t_0) + H h_v \\ &= 1.0069254 \text{ kJ/kg}_{air} \text{ K}(60 - 0) \text{ K} + (0.02 \text{ kg}_{H_2O}/\text{kg}_{air})(2615.1 \text{ kJ/kg}_{H_2O}) \\ &= 112.7 \text{ kJ/kg}_{air} \end{aligned}$$

9.3.8 Adiabatic Saturation Temperature

The adiabatic saturation temperature is the temperature at which the air-water vapor mixture can be brought to saturation adiabatically by the evaporation of water (liquid or solid) at the adiabatic saturation temperature. This process can be better visualized by first considering the adiabatic process depicted by figure 9.1. Air moves through a perfectly insulated tube. Water (as either a liquid or a solid) is added and evaporates as it moves through the tube and leaves as a vapor. Since there is no external heat exchange, this is an adiabatic process. In general, the water can enter the system at a temperature t_3 which can be above, below, or equal to either t_1 or t_2. There is, however,

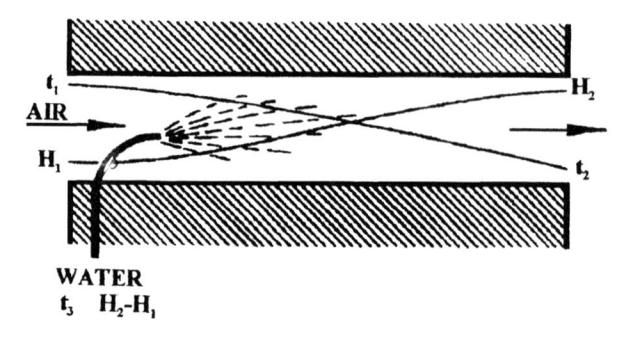

Figure 9.1. An adiabatic process.

Psychrometrics

a unique temperature (the adiabatic saturation temperature) such that if the water enters the system at that temperature the air-water vapor mixture may be brought to saturation at the same temperature. Referring again to figure 9.1, if $t_3 = t_{as}$ (the adiabatic saturation temperature) and $H_2 = H_{asat}$ (the saturation humidity ratio at the adiabatic saturation temperature) then t_2 is also equal to the adiabatic saturation temperature. For adiabatic saturation temperatures greater than or equal to 0 C, the energy balance per unit mass of "dry" air is:

$$c_{pa}(t_1 - t_0) + H_1[c_{pf}(t_{as} - t_0) + h'_{fg} + c_{pv}(t_1 - t_{as})] + c_{pf}(H_{asat} - H_1)(t_{as} - t_0)$$
$$= c_{pa}(t_{as} - t_0) + H_{asat}[c_{pf}(t_{as} - t_0) + h'_{fg}] \tag{9.41}$$

This energy balance evaluates the water vapor enthalpy by assuming that liquid water is heated to the adiabatic saturation temperature, vaporized, and then heated to the final temperature. Equation 9.41 can be solved for the adiabatic saturation temperature to obtain:

$$t_{as} = t_1 - h'_{fg}(H_{asat} - H_1)/(c_{pa} + c_{pv}H_1) \tag{9.42}$$

where

t_{as} = adiabatic saturation temperature (C)
h'_{fg} = heat of vaporization of water at the adiabatic saturation temperature (kJ/kg)
H_{asat} = saturation humidity ratio at the adiabatic saturation temperature (kg_{H_2O}/kg_{air})

Then for an air-water vapor mixture at dry-bulb temperature t_1 and humidity ratio H_1, equation 9.42 can be used to solve for the adiabatic saturation temperature. However, this must be solved by trial and error since H_{asat} is the saturation humidity ratio and h'_{fg} is the heat of vaporization at the unknown adiabatic saturation temperature. For adiabatic saturation temperatures below 0 C, the heat of vaporization term, h'_{fg}, is replaced by the heat of sublimation at the adiabatic saturation temperature, h'_{ig}.

Example 9.7. Determine the adiabatic saturation temperature of an air-water vapor mixture having a dry-bulb temperature of 50 C and a humidity ratio of 0.02 kg_{H_2O}/kg_{air}.

Solution
The solution of equation 9.42 must be determined by trial and error. The adiabatic saturation temperature will always be less than or equal to the dry-bulb temperature. Therefore, for a first guess, assume that $t_{as} = 50$ C. Then, the saturation humidity ratio and heat of vaporization of water at 50 C must be determined.
From equation 9.37:

$$h'_{fg} = 2502.535259 \text{ kJ/kg}_{H_2O} - (2.38576424 \text{ kJ/kg}_{H_2O} \text{ K})(50 - 0) \text{ K}$$
$$= 2383.2 \text{ kJ/kg}_{H_2O}$$

From table 9-1:

$$p_{sat} = 12.349 \text{ kPa}$$

(Note: This could have been calculated using equation 9.19)

From equation (9.22):

$$H_{asat} = 12.349 \text{ kPa}/(1.605(101.325 - 12.349) \text{ kPa})$$
$$= 0.0865 \text{ kg}_{H_2O}/\text{kg}_{air}$$

Then, from equation 9.42:

$$t_{as} = 50 \text{ C} - (2383.2 \text{ kJ/kg}_{H_2O})(0.0865 - 0.02) \text{ kg}_{H_2O}/\text{kg}_{air}/$$
$$[1.0069254 \text{ kJ/kg}_{air} + 1.8756864 \text{ kJ/kg}_{H_2O}(0.02 \text{ kg}_{H_2O}/\text{kg}_{air})]$$
$$t_{as} = 50 \text{ C} - 151.7 \text{ C} = -101.7 \text{ C}$$

Thus, the first guess for t_{as} was too high. As a second guess assume that $t_{as} = 30$ C.

$$h'_{fg} = 2502.535259 \text{ kJ/kg}_{H_2O} - (2.38576425 \text{ kJ/kg}_{H_2O})(30 - 0) \text{ K}$$
$$= 2431.0 \text{ kJ/kg}_{H_2O}$$
$$p_{sat} = 4.246 \text{ kPa}$$
$$H_{asat} = 4.246 \text{ kPa}/[1.605(101.325 - 4.246) \text{ kPa}]$$
$$= 0.0273 \text{ kg}_{H_2O}/\text{kg}_{air}$$
$$t_{as} = 50 \text{ C} - (2431.0 \text{ kJ/kg}_{H_2O})(0.0273 - 0.02) \text{ kg}_{H_2O}/\text{kg}_{air}/$$
$$[1.0069254 \text{ kJ/kg}_{air} + 1.8756864 \text{ kJ/kg}_{H_2O}(0.02 \text{ kg}_{H_2O}/\text{kg}_{air})]$$
$$= 50 \text{ C} - 17.0 \text{ C} = 33.0 \text{ C}$$

The guess of 30 C is somewhat low as evidenced by the calculated value of 33 C. Continued iterations would obtain a value of approximately 30.5 C. Computer programs may easily be written to make the tedious calculations necessary for the iterative solution.

9.3.9 Wet-Bulb Temperature

The wet-bulb temperature is the temperature of a wetted wick thermometer with sufficient air velocity past the wick and without radiation effects to approximate the adiabatic saturation temperature.

The evaporation of the water from the wet bulb attains a steady state in which heat is transferred just rapidly enough from the surroundings to provide energy for evaporation as shown in figure 9.2.

The quantity of air passing the bulb is so great that little change in the surrounding air temperature results. The wet bulb cools by evaporation of the water from the bulb, the rate being:

$$w = f_v A(p_{sat} - p_v) \tag{9.43}$$

where

$w =$ evaporation rate (kg/s)
$f_v =$ surface vapor transfer coefficient (kg/s N)

Psychrometrics

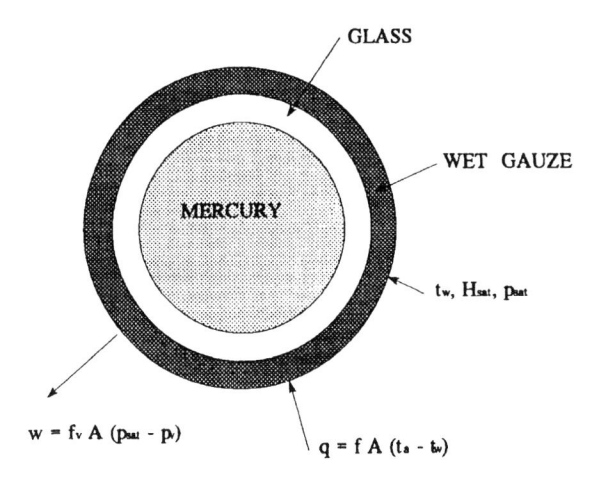

Figure 9.2. Cross section of a thermometer wet bulb showing the heat and water vapor transfer equations.

A = surface area of bulb (m²)
p_{sat} = saturation pressure of water at bulb temperature (Pa)
p_v = water vapor pressure in air (Pa)

The rate at which energy must be supplied for the evaporation to take place is:

$$q = w h_{fg} \tag{9.44}$$

where

q = rate of heat transfer to bulb (W)
h_{fg} = latent heat of evaporation of water at temperature of bulb (J/kg).

The heat required for the evaporation of the water must be obtained from the air stream. The heat transferred from the air is:

$$q = f A (t_a - t_w) \tag{9.45}$$

where

q = rate of heat transfer (W)
f = convective heat transfer coefficient (W/m²K)
A = area of bulb (m²)
t_a = air dry-bulb temperature (C)
t_w = wet-bulb temperature (C)

Then,

$$w h_{fg} = f_v A (p_{sat} - p_v) h_{fg} = f A (t_a - t_w) \tag{9.46}$$

$$p_{sat} - p_v = f(t_a - t_w)/(f_v h_{fg}) \tag{9.47}$$

When the water vapor pressure is small compared to the total air pressure, the humidity ratio is approximately:

$$H_v = M_v p_v/(M_a p_{atm}) \tag{9.48}$$

and

$$H_{sat} = M_v p_{sat}/(M_a p_{atm}) \tag{9.49}$$

Then substituting into equation 9.47,

$$(M_a/M_v) p_{atm}(H_{sat} - H_v) = f(t_a - t_w)/(f_v h_{fg}) \tag{9.50}$$

or

$$(H_{sat} - H_v)/(t_a - t_w) = f M_v/(f_v h_{fg} M_a p_{atm}) \tag{9.51}$$

Consider the heat and mass balance for an adiabatic process, figure 9.1. For the case where $t_2 = t_3 = t_{as}$, $H_2 = H_{sat} = H_{asat}$, $t_1 = t_a$, and $H_1 = H_v$, equation 9.42 becomes:

$$t_{as} = t_a - h'_{fg}(H_{sat} - H_v)/(c_{pa} + c_{pv} H_v) \tag{9.52}$$

or

$$(H_{sat} - H_v)/(t_a - t_{as}) = (c_{pa} + c_{pv} H_v)/h'_{fg} \tag{9.53}$$

Thus, comparing equations 9.51 and 9.53 shows that the wet-bulb temperature and the adiabatic saturation temperature are the same if and only if:

$$f M_v/(f_v h'_{fg} M_a p_{atm}) = (c_{pa} + c_{pv} H_v)/h'_{fg} \tag{9.54}$$

or

$$f M_v/(M_a f_v p'_{atm}) = c_{pa} + c_{pv} H_v \tag{9.55}$$

Fortunately, equation 9.55 is valid for engineering problems if the following operation factors are recognized. The rate of air past the bulb affects the value of the coefficients f and f_v. The error will be minimized for an air rate of 2.5 to 5.1 m/s. Radiant heat exchange between the wet bulb and the surroundings may be significant unless the following precautions are observed:
 (1) The bulb should be as small as practical to minimize the projected area that a radiant source or sink "sees".
 (2) The air rate past the bulb should be high so that the difference in temperature between the air and the bulb needed to compensate for radiant heat will be small.
 (3) Shielding the wet bulb will eliminate the radiant heat exchange.
Precaution (3) alone will eliminate the radiation problem, but shielding complicates the construction of a unit.

The validity of equation 9.55 is a result of fortuitous circumstance. It does not hold for systems composed of other materials. In other systems, air-benzene, for example, the adiabatic saturation temperature and the wet-bulb temperature are not the same.

Psychrometrics

9.3.10 Density

The density of the air-water vapor mixture is defined mathematically by:

$$\rho = (m_a + m_v)/V \tag{9.56}$$

or

$$\rho = m_a/V + m_v/V \tag{9.57}$$

where

ρ = density of air-water vapor mixture (kg/m^3)
m_a = mass of air (kg)
m_v = mass of water vapor (kg)
V = volume of mixture (m^3)

From the ideal gas law, $m_a/V = p_a M_a/RT$ and $m_v/V = p_v M_v/RT$.
Then,

$$\rho = (p_a M_a + p_v M_v)/RT \tag{9.58}$$

or

$$\rho = (1 + p_v M_v/p_a M_a)(p_a M_a/RT) \tag{9.59}$$

Substituting equations 9.21 and 9.25 for humidity ratio and specific volume, respectively, give:

$$\rho = (1 + H)/v \tag{9.60}$$

where

H = humidity ratio (kg$_{H_2O}$/kg$_{air}$)
v = specific volume (m^3/kg$_{air}$)

Example 9.8. Determine the density of the air-water vapor mixture of examples 9.3 and 9.4.

Solution

The specific volume is 0.845 m^3/kg$_{air}$ from example 9.4.

$$H = p_v/(1.605(p_{atm} - p_v))$$
$$= 1.75 \text{ kPa}/(1.605(101.325 - 1.75) \text{ kPa})$$
$$= 0.01095 \text{ kg}_{H_2O}/\text{kg}_{air}$$

Then,

$$\rho = (1 + H)/v$$
$$= (1 + 0.01195) \text{ kg/kg}_{air}/(0.845 \text{ m}^3/\text{kg}_{air})$$
$$= 1.196 \text{ kg/m}^3$$

9.4 The Psychrometric Chart

The psychrometric chart is a graphic representation of the physical and thermal properties of the air-water vapor mixture. Most psychrometric charts are drawn for standard atmospheric pressure. A Mollier-type psychrometric chart will be considered here due to its convenience in graphical solution of many problems with a minimum of thermodynamic approximations. The Mollier-type chart uses enthalpy and humidity ratio as the coordinates while the more traditional psychrometric chart uses dry-bulb temperature and humidity ratio. Eckert (1959) lists the advantages of a Mollier-type chart as:

(1) The adiabatic saturation temperature lines are straight
(2) The chart may be used to determine the resulting state when moisture at any state is added to the air
(3) The chart can be generalized for mixtures of substances other than air and water.

The charts currently produced by the American Society of Heating, Refrigeration, and Air-Conditioning Engineers (ASHRAE, 1997) are of the Mollier-type.

If the humidity ratio H is used as ordinate and the enthalpy h as the abscissa, figure 9.3 results. Note that the constant dry-bulb temperature lines are inclined to the right. The chart has the disadvantage that the region of unsaturated air is relatively small. This can be avoided by using a system of oblique-angled coordinates. Mollier (1923 & 1929) proposed giving the h axis such a direction that the 0 C isotherm is vertical in the unsaturated range. A diagram with an arbitrary temperature line vertical is shown schematically in figure 9.4.

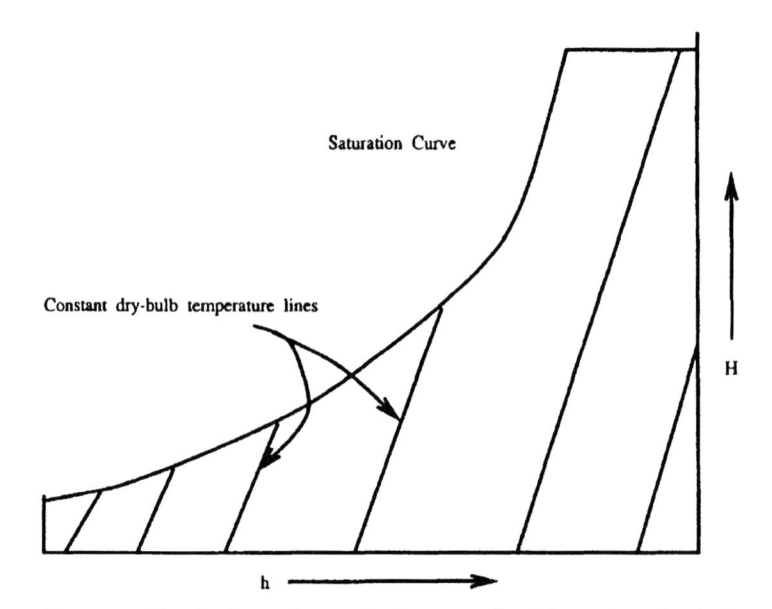

Figure 9.3. Sketch of a psychrometric chart using humidity ratio and enthalpy as the axis.

Psychrometrics

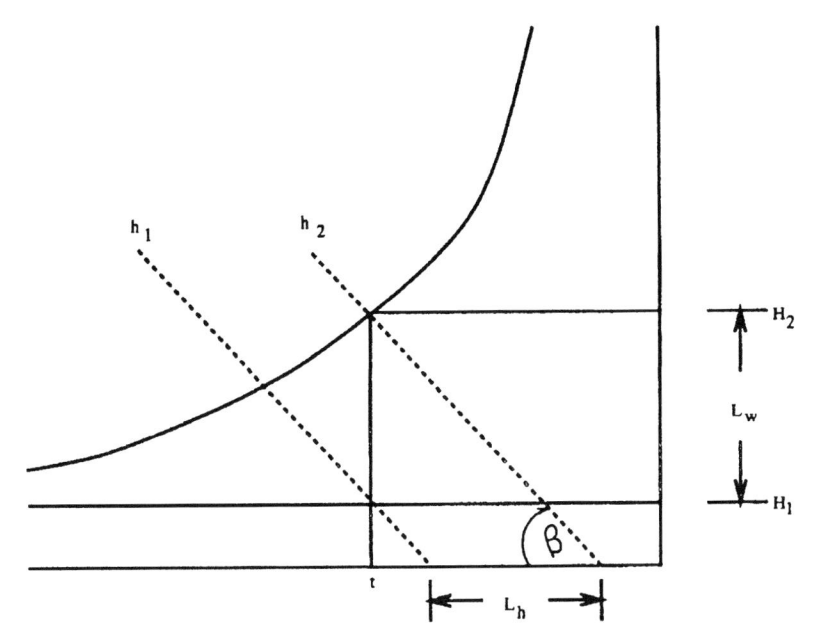

Figure 9.4. Mollier-type psychrometric chart with an arbitray temperature line vertical.

The angle β at which the enthalpy lines are to be inclined in order to make a given temperature line vertical in the unsaturated region is calculated as shown by Threlkeld (1962) by:

$$\tan \beta = s_w / \{[c_{pv}(t - t_0) + h_{g,0}]s_h\} \tag{9.61}$$

where

β = angle of inclination of enthalpy lines (radians)
s_w = humidity ratio scale factor (m kg$_{air}$/kg$_{H_2O}$)
c_{pv} = specific heat of water vapor (kJ/kg$_{H_2O}$ K)
t = dry-bulb temperature line which is to be vertical in the unsaturated region (C)
t_0 = reference temperature at which enthalpy is zero (C)
$h_{g,0}$ = heat of vaporization of water at 0 C (kJ/kg$_{H_2O}$)
s_h = enthalpy scale factor (m kg$_{air}$/kJ)

The scale factors s_w and s_h are defined by (fig. 9.4):

$$s_w = L_w/(H_2 - H_1) \tag{9.62}$$

and

$$s_h = L_h/(h_2 - h_1) \tag{9.63}$$

Example 9.9. If 2.5 cm vertically on the chart is to represent 0.01 kg_{H_2O}/kg_{air}, 2.5 cm horizontally is to represent 20 kg_{H_2O}/kg_{air}, and the 40 C dry-bulb temperature line is to be vertical in the region of unsaturated air, determine (a) the humidity ratio scale factor, (b) the enthalpy scale factor, and (c) the angle of inclination of the enthalpy lines.

Solution
(a) $\quad s_w = L_w/(H_2 - H_1)$
$\quad\quad = (.025 \text{ m})/(0.01 \text{ kg}_{H_2O}/\text{kg}_{air}) = 2.5 \text{ m kg}_{air}/\text{kg}_{H_2O}$
(b) $\quad s_h = L_h/(h_2 - h_1)$
$\quad\quad = (.025 \text{ m})/(20 \text{ kJ/kg}_{air}) = 0.00125 \text{ m kg}_{air}/\text{kJ}$
(c) $\quad \tan \beta = s_w/[c_{pv}(t - t_0) + h_{g,0}]s_h$
$\quad\quad = (2.5 \text{ m kg}_{air}/\text{kg}_{H_2O})/\{[1.8756864 \text{ kJ/kg}_{H_2O}\text{K} (40 - 0) \text{ K}$
$\quad\quad\quad + 2502.535259 \text{ kJ/kg}_{H_2O}](0.00125 \text{ m kg}_{air}/\text{kJ})\}$
$\quad \tan \beta = 0.7759$
$\quad \beta = \tan^{-1}(0.7759) = 0.6599 \text{ radians} = 37.81°$

Figure 9.5 is a Mollier-type psychrometric chart prepared for ASHRAE (1997) covering the dry-bulb temperature range from 0 to 50 C with an atmospheric pressure of 101.325 kPa. The angle of inclination of the enthalpy lines is such that the 50 C dry-bulb temperature line is vertical. The curved lines on the chart are lines of constant relative humidity. The dashed lines which have slopes similar to the enthalpy lines are lines of constant adiabatic saturation temperature. The lines of constant adiabatic saturation temperature are labeled as wet-bulb temperature lines. This nomenclature is convenient for the casual user, but it is erroneous as regards the true properties of the lines. Note that the dry-bulb temperature lines intersect the adiabatic saturation temperature lines of the same value at the saturation line (100% relative humidity line).

The specific volume lines are not linear, but the deviation from linearity is slight so that the lines appear straight on the psychrometric chart. They are solid lines on the chart of figure 9.5 and have steeper slopes than the enthalpy and adiabatic saturation temperature lines. Figure 9.5 shows specific volume lines between 0.78 and 0.96 m^3/kg_{air}.

The dew-point temperature can be determined from the psychrometric chart by following a constant humidity ratio line horizontally to the left until the saturation line is intersected. The temperature at this point is the dew-point temperature for the given air-water vapor mixture.

It may be noted that the crossing of any two independent property lines on the chart establishes a state point from which all other values can be secured.

Figure 9.6 is a Mollier-type chart which was computer generated by Young and Day (1988) for the same temperature range and atmospheric pressure as that of figure 9.5. However, the region of super-saturated air is also included on figure 9.6. Note that in the super-saturated region, dry-bulb temperatures and adiabatic saturation temperatures are equal. Graphical solutions of psychrometric problems which are discussed in the following sections are valid in both the regions of unsaturated and saturated air.

Psychrometrics

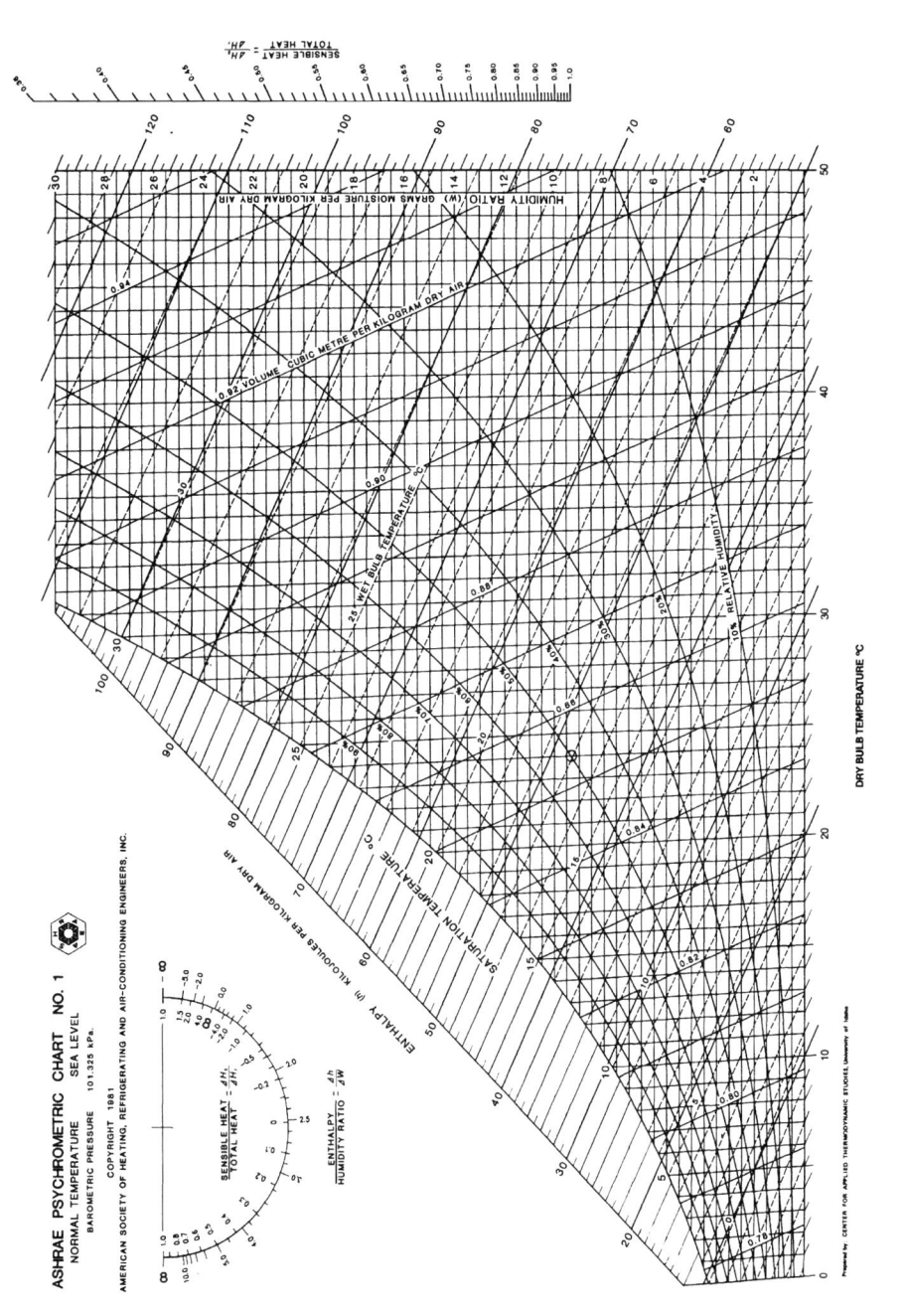

Figure 9.5. ASHRAE Mollier-type psychrometric chart.

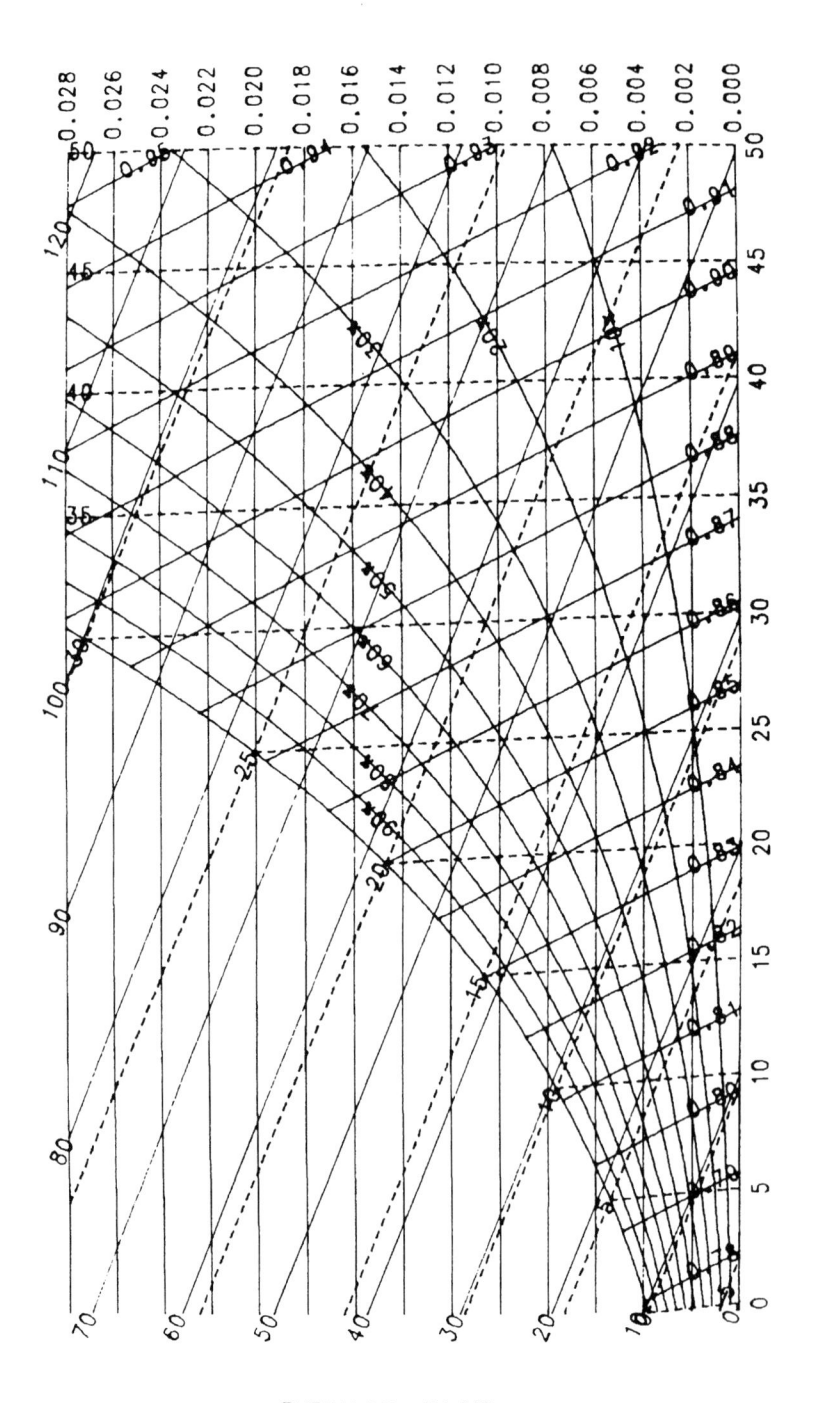

Figure 9.6. Mollier-type psychrometric chart for a 0 to 50 C.

Psychrometrics

9.5 Uses of the Psychrometric Chart

The psychrometric chart may be used to determine thermal and physical properties of air-water vapor mixtures and to solve psychrometric problems concerning processes which take place at constant pressure (i.e., the pressure for which the chart is prepared).

9.5.1 Determination of State Factors

If any two properties of the air-vapor mixture are known, any of the other properties can be determined directly from the psychrometric chart. For example, 20 C air having an adiabatic saturation temperature (wet-bulb temperature) of 15 C has a humidity ratio H of 0.0086 kg$_{H_2O}$/kg$_{air}$. The dew-point temperature is 11.7 C; relative humidity, 59%; and specific volume, 0.842 m^3/kg$_{air}$.

9.5.2 Cooling and Heating

Cooling or heating an air-water vapor mixture without changing the amount of moisture in the mixture takes place along a horizontal line on the chart as shown in figure 9.7. The heat involved per kilogram of "dry" air is $h_2 - h_1$ where h_1 and h_2 are the enthalpies before and after the heating or cooling process. In cooling, the temperature of the cooling medium must be above the dew-point temperature or dehumidification will result.

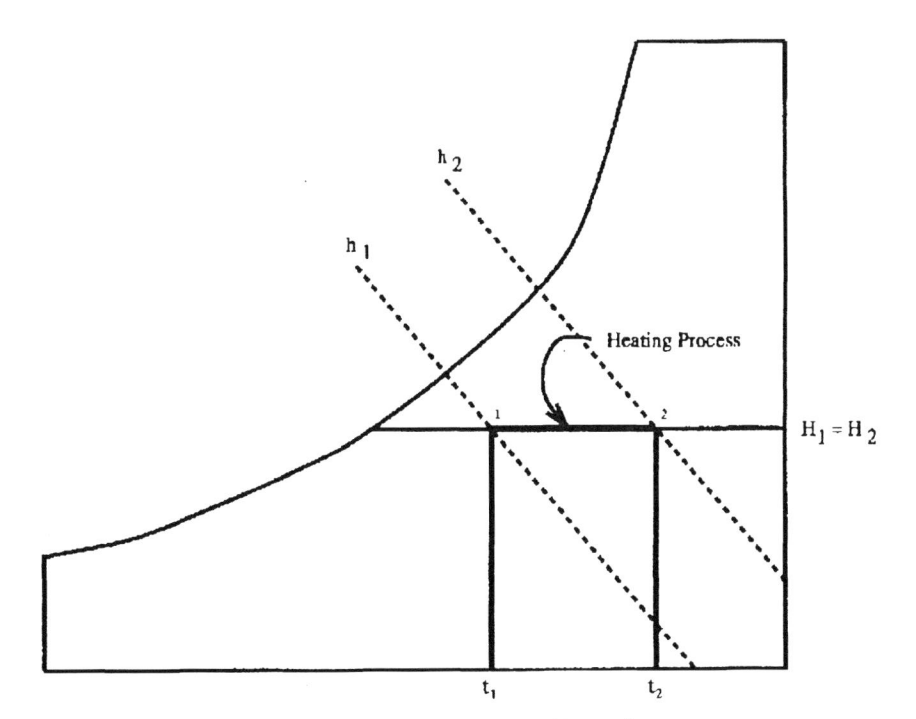

Figure 9.7. Heating process on a psychrometric process.

Mass and energy balances for the heating process depicted in figure 9.7 are:
Mass of Air

$$\dot{m}_1 = \dot{m}_2 = \dot{m} \tag{9.64}$$

Mass of Water

$$\dot{m}_1 H_1 = \dot{m}_2 H_2 = \dot{m} H \tag{9.65}$$

Energy Balance

$$\dot{m} h_1 + q = \dot{m} h_2 \tag{9.66}$$

$$q = \dot{m}(h_2 - h_1) \tag{9.67}$$

where

$\dot{m}_i$ = mass flow rate of air at state i (kg$_{air}$/s)
q = rate of heat transfer to the air-vapor mixture (W)

The mass flow rate of the air can be expressed in terms of its volumetric flow rate by:

$$\dot{m} = Q/v \tag{9.68}$$

where

Q = volumetric flow rate of air (m^3/s)
v = specific volume of air (m^3/kg$_{air}$)

Then,

$$q = (Q/v)\Delta h \tag{9.69}$$

Example 9.10. An air stream having a volumetric flow rate of 5 m^3/s, a dry-bulb temperature of 30 C and a humidity ratio of 0.01 kg$_{H_2O}$/kg$_{air}$ is to be heated to 50 C with electrical resistance heaters. What is the power requirement for the heaters?

Solution

$$q = (Q/v)\Delta h$$

From the psychrometric chart, $v = 0.8725$ m^3/kg$_{air}$, $h_1 = 55.8$ kJ/kg$_{air}$, and $h_2 = 76.3$ kJ/kg$_{air}$.
Then,

$$q = [(5 \text{ m}^3/\text{s})/(0.8725 \text{ m}^3/\text{kg}_{air})](76.3 - 55.8) \text{ kJ/kg}_{air}$$
$$= 117.5 \text{ kJ/s} = 117.5 \text{ kW}$$

9.5.3 Mixtures

The state point of an air-water vapor mixture resulting from mixing air streams of different state points falls on a straight line connecting the two initial states, figure 9.8. Proof of this procedure follows from the following mass and energy balances:

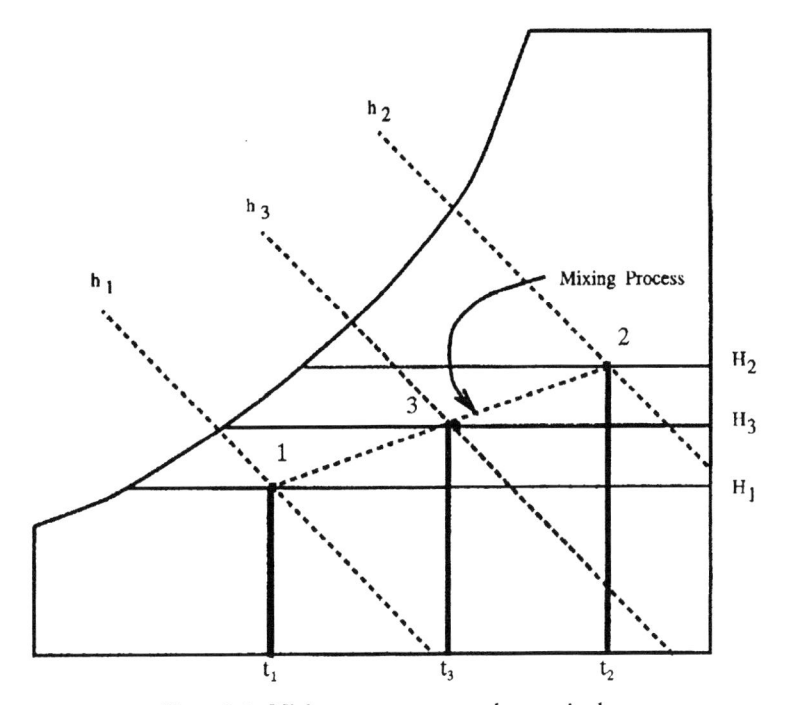

Psychrometrics

![Figure 9.8. Mixing process on a psychrometric chart.]

Figure 9.8. Mixing process on a psychrometric chart.

Mass of Air
$$\dot{m}_1 + \dot{m}_2 = \dot{m}_3 \tag{9.70}$$

Mass of Water
$$\dot{m}_1 H_1 + \dot{m}_2 H_2 = \dot{m}_3 H_3 \tag{9.71}$$

Energy Balance
$$\dot{m}_1 h_1 + \dot{m}_2 h_2 = \dot{m}_3 h_3 \tag{9.72}$$

Equations 9.70 and 9.71 may be solved for H_3 to obtain:
$$H_3 = (\dot{m}_1 H_1 + \dot{m}_2 H_2)/(\dot{m}_1 + \dot{m}_2) \tag{9.73}$$

while equations 9.70 and 9.72 yield:
$$h_3 = (\dot{m}_1 h_1 + \dot{m}_2 h_2)/(\dot{m}_1 + \dot{m}_2) \tag{9.74}$$

Equations 9.73 and 9.74 may be rearranged to obtain:
$$(H_3 - H_1)/(H_2 - H_3) = \dot{m}_2/\dot{m}_1 \tag{9.75}$$
$$(h_3 - h_1)/(h_2 - h_3) = \dot{m}_2/\dot{m}_1 \tag{9.76}$$

Then,
$$(H_3 - H_1)/(H_2 - H_3) = (h_3 - h_1)/(h_2 - h_3) \tag{9.77}$$

and by similarity of the triangles, state 3 must lie on the line connecting state 1 and state 2.

Example 9.11. An air stream having a temperature of 20 C and a relative humidity of 90% mixes with a stream having a temperature of 40 C and a relative humidity of 20%. If the flow rate of the 20 C stream is 5 m^3/s and the flow rate of the 40 C stream is 10 m^3/s, determine the state of the mixture.

Solution
From the psychrometric chart,

$v_1 = 0.848$ m^3/kg$_{air}$, $h_1 = 53.6$ kJ/kg$_{air}$,   and $H_1 = .0133$ kg$_{H_2O}$/kg$_{air}$
$v_2 = 0.900$ m^3/kg$_{air}$, $h_2 = 64.0$ kJ/kg$_{air}$,   and $H_2 = .0093$ kg$_{H_2O}$/kg$_{air}$

Then,

$$\dot{m}_1 = Q_1/v_1 = (5 \text{ m}^3/\text{s})/(0.848 \text{ m}^3/\text{kg}_{air}) = 5.896 \text{ kg}_{air}/\text{s}$$
$$\dot{m}_2 = Q_2/v_2 = (10 \text{ m}^3/\text{s})/(0.9 \text{ m}^3/\text{kg}_{air}) = 11.111 \text{ kg}_{air}/\text{s}$$
$$H_3 = (\dot{m}_1 H_1 + \dot{m}_2 H_2)/(\dot{m}_1 + \dot{m}_2)$$
$$= [(5.896)(0.0133) + (11.111)(0.0093)]/(5.896 + 11.111)$$
$$= 0.0107 \text{ kg}_{H_2O}/\text{kg}_{air}$$
$$h_3 = [(5.896)(53.6) + (11.111)(64.0)]/(5.896 + 11.111)$$
$$= 60.4 \text{ kJ/kg}_{air}$$

From the psychrometric chart, $t^3 = 32.8$ C and $rh^3 = 34\%$

9.5.4 Cooling and Dehumidifying

When a stream of air at a temperature t_1 comes in contact with a heat-removing sink with an effective surface temperature, t_c, below the dew-point temperature, both the temperature and the humidity ratio of the stream are reduced. This process is best represented by the cooling and dehumidification which results when air is passed through a finned-type unit refrigeration evaporator.

An approximate representation of this process may be obtained using the following three assumptions:

1. Assume that a portion of the air is cooled to a temperature of t_c, the remainder staying at the t_1 condition. The portion of the air which is cooled is dehumidified such that its humidity ratio is the saturation humidity ratio for the effective temperature of the coil, t_c. A process involving mixtures then exists and the principles of the previous section apply.
2. Assume that the enthalpy of the water vapor in the air stream is small compared to the enthalpy of the air so that a change in enthalpy of the mixture can be considered proportional to a change in temperature of the mixture.
3. Assume that the effect of the phase change due to condensation of water vapor is small enough so that the heat exchanger equations of Chapter 7 may be considered valid.

Psychrometrics

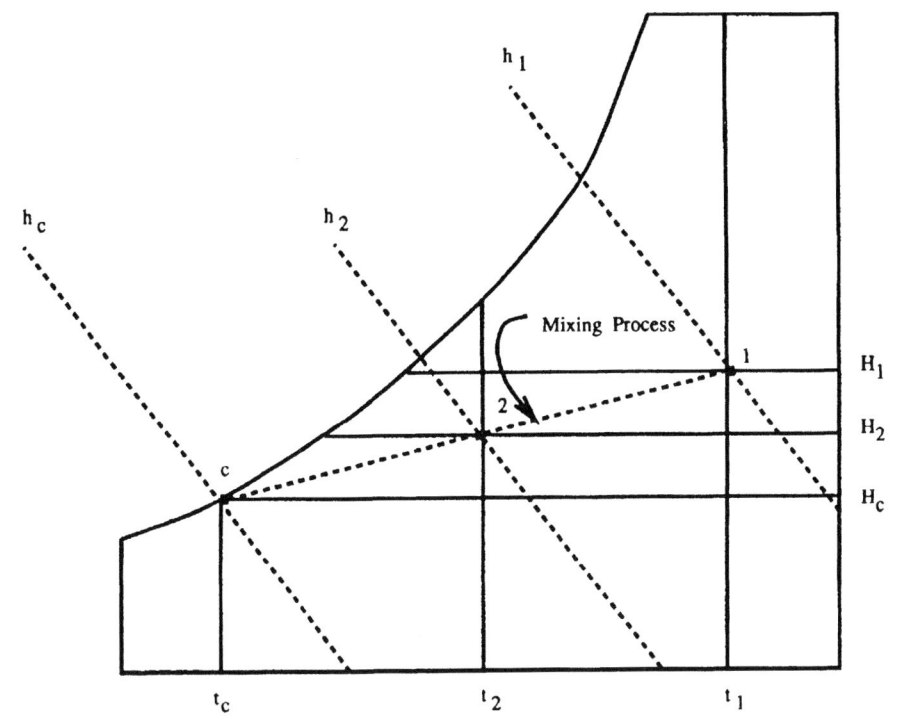

Figure 9.9. Cooling and dehumidifying process on the psychrometric chart.

Assumption 1 allows the representation of the process as a mixing process depicted in figure 9.9. The state of the cooled air stream falls approximately on the straight line connecting the initial state of the air stream with the point on the saturation line at the effective temperature of the coil. The portion of the air stream which is assumed to bypass the cooling coil, B, is defined as the *bypass factor*. Mass and energy balance equations for the mixing process may then be written as follows:

Mass of Air

$$\dot{m}_c + \dot{m}_1 = \dot{m}_2 \tag{9.78}$$

$$(1 - B)\dot{m} + B\dot{m} = \dot{m} \tag{9.79}$$

where

$\dot{m}_c$ = mass flow rate of air cooled to the coil temperature (kg$_{air}$/s)
$\dot{m}_1$ = mass flow rate of air assumed to bypass the coil (kg$_{air}$/s)
$\dot{m} = \dot{m}_2$ = total mass flow rate of air (kg$_{air}$/s)
B = bypass factor (decimal fraction)

Mass of Water

$$(1 - B)\dot{m} H_c + B\dot{m} H_1 = H_2 \tag{9.80}$$

Energy Balance

$$(1 - B)\dot{m}h_c + B\dot{m}h_1 = \dot{m}h_2 \tag{9.81}$$

Using the nomenclature of Figure 9.9 in equation 9.75 for a mixing process we obtain:

$$(H_2 - H_c)/(H_1 - H_2) = B\dot{m}/((1 - B)\dot{m}) = B/(1 - B) \tag{9.82}$$

or

$$B = (H_2 - H_c)/(H_1 - H_c) \tag{9.83}$$

and solving for the final humidity ratio gives:

$$H_2 = H_c + B(H_1 - H_c) \tag{9.84}$$

Likewise,

$$h_2 = h_c + B(h_1 - h_c) \tag{9.85}$$

Assumption 2 allows the substitution of dry-bulb temperatures for the enthalpies of equation 9.85. Then:

$$t_2 = t_c + B(t_1 - t_c) \tag{9.86}$$

Using assumption 3 and equation 7.101, the effectiveness ratio for the cooling coil in which one fluid is at a constant temperature is given by:

$$E = 1 - \exp(-\text{NTU}) \tag{9.87}$$

where

E = heat exchanger effectiveness ratio (dimensionless)
NTU = number of transfer units (dimensionless)

The number of transfer units is given by:

$$\text{NTU} = UA/\dot{m}c_p \tag{9.88}$$

where

U = overall heat transfer coefficient for the coil (W/m²K)
A = total external surface area of coil (m²)
$\dot{m}$ = mass flow rate of air (kg$_{\text{air}}$/s)
$c_p = c_{\text{pa}} + c_{\text{pv}} H_1$ = specific heat of mixture at state 1 (kJ/kg)

Then,

$$(t_1 - t_2)/(t_1 - t_c) = E = 1 - \exp(-\text{NTU}) \tag{9.89}$$

$$t_2 = t_c + \exp(-\text{NTU})(t_1 - t_c) \tag{9.90}$$

where t_2 = exit air temperature (C).

Psychrometrics

Comparison of equations 9.86 and 9.90 indicate that:

$$B = \exp(-\text{NTU}) = 1 - E \tag{9.91}$$

Thus, the bypass factor B is one minus the effectiveness ratio of the coil or we could call it the "ineffectiveness ratio" of the coil.

9.5.5 Direct Heating by Combustion of a Fuel

When an air-water vapor mixture is heated by the combustion of a fuel within the mixture, both energy and water are added to the mixture by the combustion process. Consider the combustion of propane as given by:

$$C_3H_8 + 5O_2 \rightarrow 3CO_2 + 4H_2O + \text{heat} \tag{9.92}$$

For each molecule of propane burned in the process, four molecules of water are given off along with the heat of combustion. Since the molecular mass of propane is 44 (i.e., $3(12) + 8(1) = 44$) and the molecular mass of water is 18 (i.e., $2(1) + 1(16) = 18$), the combustion of 1 kg of propane will release $72/44 = 1.636$ kg of water. The heat of combustion of propane is given in table 9-2 as 50 150 kJ/kg$_{\text{propane}}$. Thus, during the combustion of propane, the ratio of energy released to water released is 30 654 kJ/kg$_{H_2O}$. The mass and energy balances for the process of heating by combustion are:

Mass of Water

$$\dot{m}_a H_1 + \dot{m}_{\text{fuel}} W = \dot{m}_a H_2 \tag{9.93}$$

where

$\dot{m}_a$ = mass flow rate of air (kg/s)
$\dot{m}_{\text{fuel}}$ = rate of combustion of fuel (kg/s)

Table 9-2. Heats of combustion and densities of some fuels*

Fuel	High[†] (kJ/kg)	Low[‡] (kJ/kg)	Density (kg/m³)
Gasoline	48 260	45 360	736
Kerosene	46 080	43 030	817
Fuel Oil	43 810	41 300	954
Butane	49 260	45 730	580
Propane	50 150	46 360	508
Methane (nat. gas)	37 090*	33 390*	—
Coal	29 075	—	—

* Per cubic meter at 20 C and 101.325 kPa.
[†] High heat of combustion refers to total energy released by combustion of fuel (i.e., Δh_{total}).
[‡] Low heat of combustion refers to sensible energy released by combustion of fuel (i.e., $\Delta h_{\text{total}} - h_{\text{latent}}$).

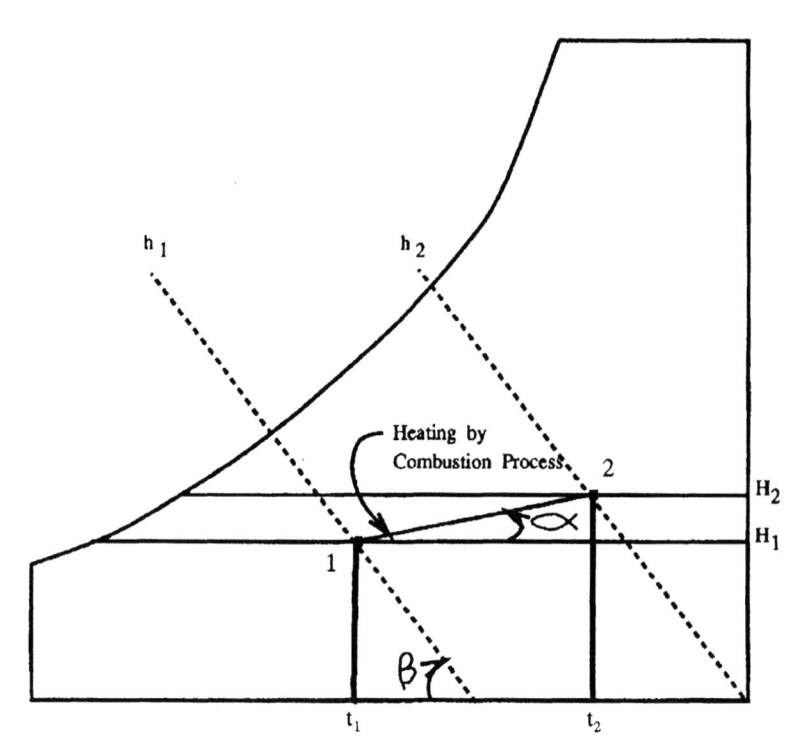

Figure 9.10. Direct heating by combustion of a fuel.

W = mass of water released per unit mass of fuel burned (kg_{H_2O}/kg_{fuel})
H_1 = initial humidity ratio of mixture (kg_{H_2O}/kg_{air})
H_2 = final humidity ratio of mixture (kg_{H_2O}/kg_{air})

Energy Balance

$$\dot{m}_a h_1 + \dot{m}_{fuel} L = \dot{m}_a h_2 \qquad (9.94)$$

where L = heat of combustion of fuel (kJ/kg_{fuel}).
Equations 9.93 and 9.94 may be combined to obtain:

$$\Delta h / \Delta H = (h_2 - h_1)/(H_2 - H_1) = L/W \qquad (9.95)$$

Equation 9.95 defines the direction on the psychrometric chart which the heating by combustion process must follow. This direction is illustrated in figure 9.10. The horizontal and vertical distances on the chart are given by:

$$x_h = s_h \Delta h - \Delta H s_w / \tan \beta \qquad (9.96)$$

$$x_w = s_w \Delta H \qquad (9.97)$$

Psychrometrics

The angle from the horizontal at which the process proceeds is then:

$$\alpha = \tan^{-1}\{\tan\beta/[(\Delta hs_h/\Delta Hs_w)\tan\beta - 1]\} \qquad (9.98)$$

$$\alpha = \tan^{-1}\{\tan\beta/[(s_h L/s_w W)\tan\beta - 1]\} \qquad (9.99)$$

Example 9.12. Determine the angle from the horizontal at which the process for direct heating by combustion of propane will take place on the chart specified in example 9.9.

Solution
From example 9.9, $s_h = 0.00125$ m kg$_{air}$/kJ, $s_w = 2.5$ m kg$_{air}$/kg$_{H_2O}$ and $\beta = 37.81°$.
Then,

$$\alpha = \tan^{-1}\{\tan 37.81°/\{[(0.00125 \text{ m kg}_{air}/\text{kJ})(30\,654 \text{ kJ/kg}_{H_2O})/$$
$$(2.5 \text{ m kg}_{air}/\text{kg}_{H_2O})]\tan 37.81° - 1\}\}$$
$$\alpha = \tan^{-1}\{0.7759/[(15.327)(0.7759) - 1]\}$$
$$= \tan^{-1}\{0.7759/10.892\} = \tan^{-1}(0.07123)$$
$$= 0.07111 \text{ radians} = 4.07°$$

9.5.6 Drying Processes

A drying process in psychrometrics refers to a process in which the air is passed through or over a moist material and evaporates water from the material. Thus, the material (not the air) is dried. Drying systems in which heat energy is supplied only by the air, with heat content of the dry matter small in proportion to the latent heat of vaporization of the water, may be treated as cases of adiabatic humidification. As air passes over or through the material being dried, its temperature drops and its humidity rises so that the adiabatic saturation temperature remains constant. This process is illustrated in figure 9.11. It will be discussed in more detail in the following chapter.

Example 9.13. If air at 50 C and 20% relative humidity passes through a wet material at the rate of 10 m³/s, determine the maximum rate of moisture removal which can be achieved.

Solution
The maximum amount of moisture removal would occur if the air exited in a saturated condition. The process moves along an adiabatic saturation temperature line.

$$H_2O \text{ Removal Rate} = (Q/v)\Delta H$$

From the psychrometric chart,

$$H_1 = 0.0156 \text{ kg}_{H_2O}/\text{kg}_{air}$$
$$t_{as} = 28.3 \text{ C}$$
$$H_2 = H_{asat} = 0.0248 \text{ kg}_{H_2O}/\text{kg}_{air}$$
$$v = 0.938 \text{ m}^3/\text{kg}_{air}$$

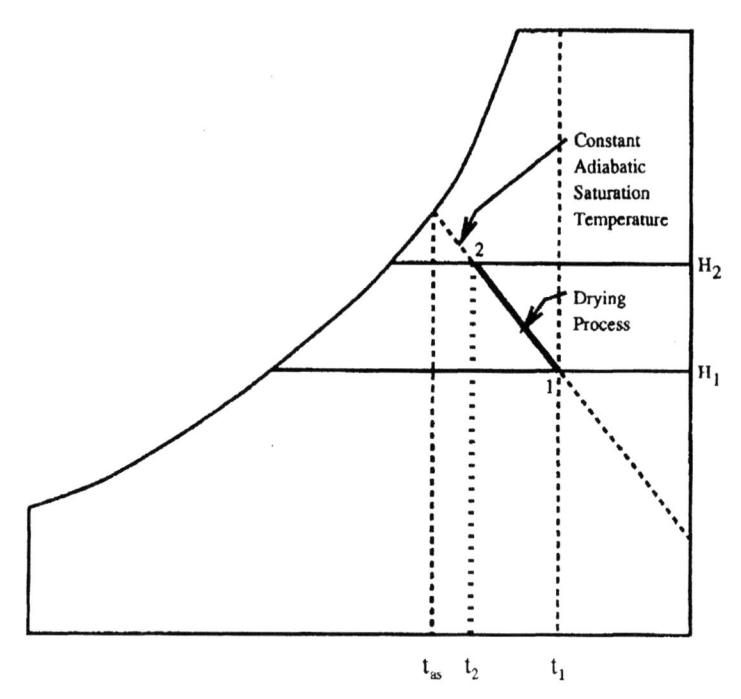

Figure 9.11. Drying process on psychrometric chart.

Then,

$$\text{H}_2\text{O Removal Rate} = [(10\text{m}^3/\text{s})/(0.938\ \text{m}^3/\text{kg}_{air})](.0248 - .0156)\ \text{kg}_{\text{H}_2\text{O}}/\text{kg}_{air}$$
$$= 0.0981\ \text{kg}_{\text{H}_2\text{O}}/\text{s} = 5.88\ \text{kg}_{\text{H}_2\text{O}}/\text{min}$$

9.5.7 Addition of Water at any State

The addition of water at any state may be described by the following mass and energy balances:

Mass of Water

$$\dot{m}_a H_1 + \dot{m}_w = \dot{m}_a H_2 \tag{9.100}$$

Energy Balance

$$\dot{m}_a h_1 + \dot{m}_w h_w = \dot{m}_a h_2 \tag{9.101}$$

where

$\dot{m}_a$ = mass flow rate of air (kg/s)
$\dot{m}_w$ = mass flow rate of water added (kg/s)
h_1 = initial enthalpy of mixture (kJ/kg$_{air}$)
h_2 = final enthalpy of mixture (kJ/kg$_{air}$)
H_1 = initial humidity ratio of mixture (kg$_{air}$/kg$_{air}$)

Psychrometrics

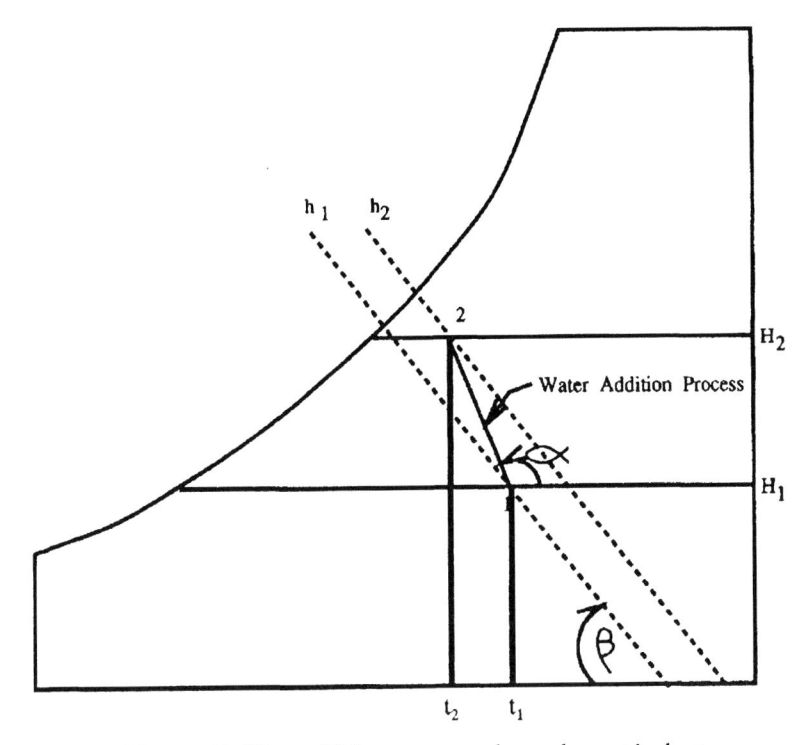

Figure 9.12. Water addition process on the psychrometric chart.

H_2 = final humidity ratio of mixture (kg_{H_2O}/kg_{air})
h_w = enthalpy of water added (kJ/kg_{H_2O})

Dividing equation 9.101 by equation 9.100 gives:

$$(h_2 - h_1)/(H_2 - H_1) = \Delta h/\Delta H = h_w \quad (9.102)$$

Thus, the direction in which the state of the air moves on the psychrometric chart is fixed by the enthalpy of the water which is added. Figure 9.12 illustrates the process. Equation 9.98 gives the angle from the horizontal at which the process proceeds. Substitution of equation 9.102 gives:

$$\alpha = \tan^{-1}\{\tan\beta/[(h_w s_h/s_w)\tan\beta - 1]\} \quad (9.103)$$

Since the direction on the psychrometric chart which describes the process is a function only of the enthalpy of the water added, a scale can be added to the chart to aid in graphical solutions. Figure 9.5 (ASHRAE psychrometric chart no. 1) contains a semicircle in the upper left portion of the chart which has rays representing the direction to move on the chart to represent processes in which water at different states is added to the mixture. The scale around the outer edge of the semicircle is the enthalpy of the water that is added in kilojoules per gram. (Note that this chart has humidity ratio units of grams of water per kilogram of air). The scale on the inner side of the semicircle gives an approximate ratio of sensible to total heat (or enthalpy) for a process which proceeds along a given direction on the chart. The sensible heat is the "dry" heat which raises the temperature of the mixture at a constant humidity ratio. Total heat includes both sensible and latent heat. Latent heat is the heat (or enthalpy) content of the water which is added to the mixture. The chart of figure 9.5 also contains rays at the right side of the chart which originate from a point at the intersection of the 24 C temperature line and the 50% relative humidity curve. These rays also give the direction to be followed for processes which have given sensible to total heat ratios.

Example 9.14. An air stream having a temperature of 40 C and a humidity ratio of 0.01 kg_{H_2O}/kg_{air} is injected with steam at 160 C. If the mass flow rate of the air is 1000 kg/min and the mass flow rate of the steam is 5 kg/min, determine the state of the mixture.

Solution
From the psychrometric chart, $h_1 = 66.0$ kJ/kg_{air}
From table 9.1, $h_w = 2758.1$ kJ/kg_{H_2O}

$$\Delta H = H_2 - H_1 = \dot{m}_w/\dot{m}_a = (5 \text{ kg}_{H_2O}/\text{min})/(1\,000 \text{ kg}_{air}\text{min})$$
$$= 0.005 \text{ kg}_{H_2O}/\text{kg}_{air}$$
$$\Delta h = h_2 - h_1 = \Delta H h_w = (0.005 \text{ kg}_{H_2O}/\text{kg}_{air})(2758.1 \text{ kJ/kg}_{H_2O})$$
$$= 13.8 \text{ kJ/kg}_{air}$$
$$h_2 = h_1 + \Delta h = 66.0 + 13.8 = 79.8 \text{ kJ/kg}_{air}$$
$$H_2 = H_1 + \Delta H = 0.010 + 0.005 = 0.015 \text{ kg}_{H_2O}/\text{kg}_{air}$$

From the psychrometric chart,

$$t_2 = 40.9 \text{ C and } rh_2 = 31\%$$

9.5.8 Dehumidification Using an Absorbent

The process by which moisture is removed from the air-water vapor mixture by passing the mixture through or over an absorbent such as silica-gel is the reverse of the drying process described in section 9.5.6. Thus, the process moves approximately along an adiabatic saturation temperature line in the direction toward lower humidity ratios.

Nomenclature

A	area (m^2)
B	bypass factor of cooling coil (decimal)
c_{pa}	specific heat of air (1.0069254 kJ/kg_{air} K)
c_{pv}	specific heat of water vapor (1.8756864 kJ/kg_{H_2O} K)
c_{pf}	specific heat of liquid water (4.1868 kJ/kg_{H_2O} K)
c_{pi}	specific heat of ice (2.0934 kJ/kg_{H_2O} K)
E	effectiveness ratio (decimal)

Psychrometrics

f	convective heat transfer coefficient (W/m²K)
f_v	surface vapor transfer coefficient (kg/h N)
h	enthalpy of air-water vapor mixture (kJ/kg$_{air}$)
h_a	enthalpy of air (kJ/kg$_{air}$)
h_v	enthalpy of water vapor (kJ/kg$_{H_2O}$)
h_{fg}	heat of vaporization of water (kJ/kg$_{H_2O}$)
h'_{fg}	heat of vaporization at the adiabatic saturation temperature (kJ/kg$_{H_2O}$)
h''_{fg}	heat of vaporization at the dew-point temperature (kJ/kg$_{H_2O}$)
$h_{g,0}$	heat of vaporization of water at 0 C (2502.535259 (kJ/kg$_{H_2O}$)
h_{ig}	heat of sublimation of ice (kJ/kg$_{H_2O}$)
h'_{ig}	heat of sublimation at the adiabatic saturation temperature (kJ/kg$_{H_2O}$)
h''_{ig}	heat of sublimation at the dew-point temperature (kJ/kg$_{H_2O}$)
$h_{s,0}$	heat of fusion of water at 0 C (333.3685 kJ/kg$_{H_2O}$)
H	humidity ratio (kg$_{H_2O}$/kg$_{air}$)
H_{asat}	saturation humidity ratio at the adiabatic saturation temperature (kg$_{H_2O}$/kg$_{air}$)
H_{sat}	humidity ratio at saturation (kg$_{H_2O}$/kg$_{air}$)
L	heat of combustion of fuel (kJ/kg$_{fuel}$)
m_a	mass of air (kg)
m_v	mass of water vapor (kg)
$\dot{m}_i$	mass flow rate of air at state i (kg$_{air}$/s)
M	molecular mass (kg/kg-mole)
M_a	molecular mass of air (kg/kg-mole)
M_v	molecular mass of water vapor (kg/kg-mole)
n	number of kg-moles
p	absolute pressure (Pa)
p_a	air pressure (Pa)
p_{atm}	total atmospheric pressure (Pa)
p_{sat}	saturation vapor pressure (Pa)
p_v	water vapor pressure (Pa)
rh	relative humidity (dimensionless)
R	universal gas constant (8.314 × 10³ Pa m³/kg-mole K)
s_h	enthalpy scale factor (m kg$_{air}$/kJ)
s_w	humidity ratio scale factor (m kg$_{air}$/kg$_{H_2O}$)
t	temperature (C)
t_0	reference temperature at which enthalpy is zero (C)
t_{as}	adiabatic saturation temperature (C)
t_{dp}	dew-point temperature (C)
t_w	wet-bulb temperature (C)
T	absolute temperature (K)
v	specific volume (m³/kg$_{air}$)
V	volume (m³)
W	mass of water released per unit mass of fuel burned (kg$_{H_2O}$/kg$_{fuel}$)
α	angle from horizontal at which process for direct heating by combustion takes place (radians)

β   angle of inclination of enthalpy lines (radians)
ρ   density of air-water vapor mixture (kg/m^3)

References

ASAE Standards, 43rd Ed. 1996. Psychrometric Data. D271.2. St. Joseph, Mich.: ASAE.

Am. Soc. Heating, Refrigerating and Air Conditioning Engrs. *ASHRAE Handbook. 1997 Fundamentals*. Atlanta, Ga.: ASHRAE.

Brooker, D. B. 1967. Mathematical model of the psychrometric chart. *Transactions of the ASAE*. 10(4):558–560, 563.

Brunt, D. 1941. *Physical and Dynamical Meteorology*, 85–87. Cambridge, CONN.: Cambridge University Press.

Eckert, E. R. G. and R. M. Drake Jr. 1959. *Heat and Mass Transfer*. New York: McGraw-Hill Book Co., Inc.

Keenan, J. H. and F. G. Keyes. 1936. *Thermodynamic Properties of Steam*. New York: John Wiley and Sons, Inc.

Mollier, R. 1923. *ZVDI* 67:869–872.

Mollier, R. 1929. *ZVDI* 73:1009–1013.

Seltz, W. G. and G. J. Silvestri. 1958. The formulation of steam properties for digital computer application. *Transactions of the ASME* 80:967.

Threlkeld, J. L. 1962. *Thermal Environmental Engineering*. Englewood Cliffs, N.J.: Prentice-Hall, Inc.

Young, J. H. and B. H. Day. 1988. Computer generation of Mollier-type psychrometric charts. *Transactions of the ASAE* 31:1224–1232.

Problems

1. The partial pressure of water vapor in an air-water vapor mixture is 1.7051 kPa, the temperature is 35 C, and the total atmospheric pressure is 55.0 kPa. Determine the value of the following psychrometric terms: (1) saturation pressure, (2) humidity ratio, (3) relative humidity, and (4) specific volume. Assume ideal gas laws to apply.

2. If a unit volume of the air-water vapor mixture given in problem 1 is isothermally compressed to one-half its original volume, what will be the values of the psychrometric terms calculated in problem 1?

3. Determine the following terms for the air-water vapor mixture in problem 1: (1) enthalpy, (2) dew-point temperature, (3) adiabatic saturation temperature, and (4) density.

4. Atmospheric air at 20 C and 50% relative humidity is compressed at constant temperature to one-third its original volume. Which of the following psychrometric factors will change in value? Determine their initial and final values. (a) saturation pressure, (b) relative humidity, (c) humidity ratio, (d) specific volume, (e) dew-point temperature, and (f) density.

Psychrometrics

5. Dry- and wet-bulb temperatures are 30 and 20 C, respectively. Find the specific volume, relative humidity, dew-point temperature, and humidity ratio using the psychrometric chart.

6. Atmospheric air has a temperature of 30 C and a dew-point of 11 C. What is the relative humidity when cooled at constant pressure and without dehumidification to 15 C? How many kilograms of water are removed from 400 m^3 if cooled at constant pressure to a saturated condition at 2 C?

7. A grain drier requires 300 m^3/min of 46 C air. The atmospheric air is at 24 C and 68% relative humidity. How much power must be supplied to heat the air?

8. Air at 35 C and 25% relative humidity passes over a cooling coil with an effective surface temperature of 2 C and is cooled to 24 C. What is the relative humidity of the cooled air? Determine the bypass factor, the number of transfer units, and the effectiveness ratio for the cooling coil.

9. An air "conditioning" unit is to be designed to "condition" 14 m^3/min of air at 20 C and 50% relative humidity to any dry-bulb temperature between 2 and 135 C and any dew-point temperature between 2 and 57 C. Determine the following:
   (a) maximum power requirement for heating
   (b) maximum cooling load
   (c) maximum humidification load
   (d) maximum dehumidification load

10. Five kilograms per minute of water at 20 C is added to an air stream having a flow rate of 1 000 kg$_{air}$/min, a dry-bulb temperature of 45 C, and a humidity ratio of 0.005 kg$_{H_2O}$/kg$_{air}$. What is the state of the mixture?

11. Solve problem 10 for the case where the water is added as steam at 200 C.

12. A stream of air at 50 C and 10% relative humidity merges with a stream at 20 C and 90% relative humidity. The temperature of the mixture is 25 C.
    (a) What is the relative humidity of the mixture?
    (b) If the higher temperature air stream has a volumetric flow rate of 300 m^3/min, what is the rate of the lower temperature stream?

10 Drying

The drying of farm crops is important from the following standpoints:
1. Permits early harvest which reduces field losses due to storms and natural shattering and facilitates earlier fall tillage operations.
2. Permits planning the harvest season to make better use of labor.
3. Permits long-time storage without deterioration.
4. Permits farmer to take advantage of higher prices early in the season or by holding in storage until prices rise again.
5. Permits maintenance of the viability of seeds.
6. Permits the farmer to sell a better quality product.
7. Permits use of waste products. Example: livestock feed from fruit pulp and almond hulls.

10.1 Moisture Content

The moisture content of a material may be expressed in either of two different ways: (1) the wet-basis moisture content or (2) the dry-basis moisture content. The wet-basis moisture content is given by the following equation:

$$m = W_m/(W_m + W_d) \tag{10.1}$$

where

m = wet basis moisture content expressed as a decimal
W_m = mass of moisture (kg)
W_d = mass of dry matter (kg)

The dry-basis moisture content is given by:

$$M = W_m/W_d \tag{10.2}$$

where M = dry-basis moisture content expressed as a decimal.

The wet-basis and dry-basis moisture contents are thus related by the following equations:

$$m = M/(1 + M) \tag{10.3}$$

$$M = m/(1 - m) \tag{10.4}$$

Both wet-basis and dry-basis moisture contents are often expressed as a percent by multiplying by 100.

Example 10.1. Find the mass of moisture to be removed in drying grain having an initial mass of 1 000 kg and initial moisture content of 25% wet basis to a final moisture content of 14% wet basis. Also, determine the total mass of the grain after drying to 14% wet basis.

Solution

$$m_1 = W_{m_1}/(W_{m_1} + W_d) = W_{m_1}/W_{tot1} = 0.25$$

$$W_{m_1} = 0.25(1\,000 \text{ kg}) = 250 \text{ kg}$$

$$W_d = W_{tot1} - W_{m_1} = 1\,000 \text{ kg} - 250 \text{ kg} = 750 \text{ kg}$$

$$m_2 = W_{m_2}/(W_{m_2} + W_d) = .14$$

$$W_{m_2} = .14(W_{m_2} + W_d)$$

$$(1 - .14)W_{m_2} = .14W_d$$

$$W_{m_2} = (.14/.86)W_d = (.14/.86)(750 \text{ kg}) = 122.1 \text{ kg}$$

$$\text{Water Removed} = W_{m_1} - W_{m_2} = 250 \text{ kg} - 122.1 \text{ kg} = 127.9 \text{ kg}$$

$$\text{Final Total Mass} = W_{m_2} + W_d = 122.1 \text{ kg} + 750 \text{ kg} = 872.1 \text{ kg}$$

10.2 Interaction of Moisture and Materials

The absorption of water vapor by an organic, chemically inert material is a complex process which is not entirely understood. The complexity becomes much greater when biological materials are involved. This complexity is due principally to the fact that water may be present in several different forms.

The first form in which water may be present is water of hydration. This moisture is chemically bound to the constituents of the material and in most cases would not be considered in moisture content determinations. It is considered to be an integral part of the material.

The second form of moisture which may be present is physically bound moisture which is held within the material by surface forces in excess of the forces which act on normally condensed water. This is commonly referred to as "bound" moisture. The heat of absorption of this moisture is a measure of the forces binding it to the material. It is generally believed to be present in a unimolecular layer on the surface of the material.

The third form of water which may be present is normally condensed moisture, sometimes called "free" water. The heat of adsorption of this moisture is equal to the normal heat of vaporization of water at the same temperature.

A suggested fourth form of moisture in biological materials is "absorbed" moisture or moisture which has passed through the cell walls and entered the cytoplasm of the cell. This fourth form of moisture is believed to account for the hysteresis between sorption and desorption equilibrium moisture contents.

Drying

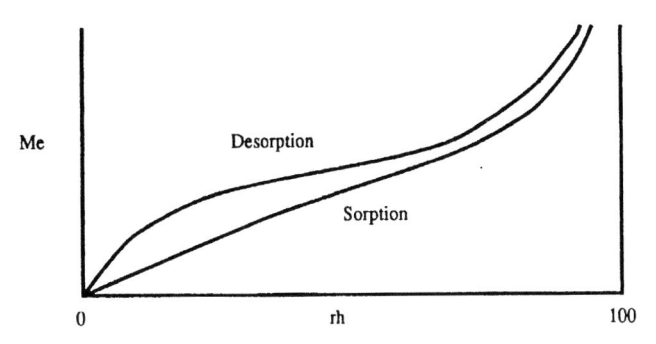

Figure 10.1. Typical equilibrium moisture content isotherms.

10.3 Equilibrium Moisture Content

Moisture is exchanged between a material and its surroundings until the material reaches some equilibrium moisture content at which there is no net gain or loss of moisture. The equilibrium moisture content is known to be a function of temperature, relative humidity, the physical properties of the material, and the previous moisture history of the material. A plot of the equilibrium moisture content versus relative humidity at a constant temperature is called an equilibrium moisture content isotherm. A sorption isotherm indicates that the equilibrium was reached in a wetting environment, and a desorption isotherm indicates that the equilibrium was reached in a drying environment.

Equilibrium moisture content isotherms for most materials are sigmoidal in shape (figure 10.1) with the desorption isotherm being above the sorption isotherm. Numerous theories have been suggested for describing the equilibrium moisture content isotherms.

10.3.1 Langmuir Equation

The Langmuir Equation (Langmuir, 1918) for describing adsorption of vapors was developed under the assumption of a monomolecular layer of the adsorbed gas developing on the surface of the material. The equation is:

$$v = v_m bp/(1 + bp) \tag{10.5}$$

where

$v\ \ $ = volume of gas adsorbed (m^3)
v_m = volume of a unimolecular layer of the adsorbate (m^3)
$p\ \ $ = vapor pressure of the gas (Pa)
$b\ \ $ = parameter dependent upon the material and the temperature (Pa^{-1})

This equation gives an isotherm of the type shown in figure 10.2 for the adsorption of water vapor. Thus, the moisture content approaches a maximum value asymptotically as the relative humidity approaches 100%. This type of equation describes the adsorption of some gases by some inorganic materials but does not correctly describe the sorption of water vapor by biological materials.

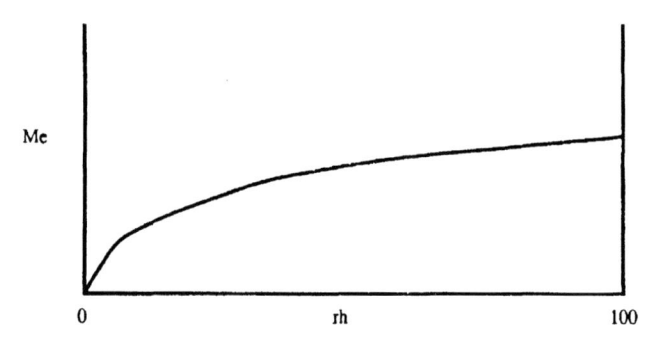

Figure 10.2. Langmuir equilibrium moisture content equation.

10.3.2 BET Equation

The BET Equation (named for Brunauer, Emmett, and Teller, 1938) was developed under the assumption of multimolecular layers of adsorbed gas on the surface of the adsorbent. The equation is:

$$v = cv_m p / \{(p_{sat} - p)[1 + (c - 1)(p/p_{sat})]\} \quad (10.6)$$

where

- v = volume of gas adsorbed (m³)
- v_m = volume of a unimolecular layer of the gas (m³)
- p = vapor pressure of gas (Pa)
- p_{sat} = saturation vapor pressure at the same temperature (Pa)
- c = parameter dependent upon the adsorbent material and the temperature (dimensionless)

The BET equation has been widely used to describe the adsorption of gases by inorganic materials and is used to determine the surface area of porous materials such as silica gel. A plot of the BET equation gives an isotherm such as the one shown in figure 10.3. Thus, a sigmoidal isotherm is obtained which is of the same general shape found experimentally for water vapor adsorption by biological materials. However, the

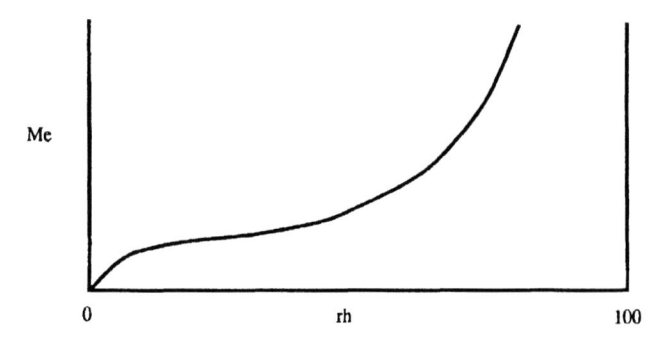

Figure 10.3. BET equilibrium moisture content equation.

Drying

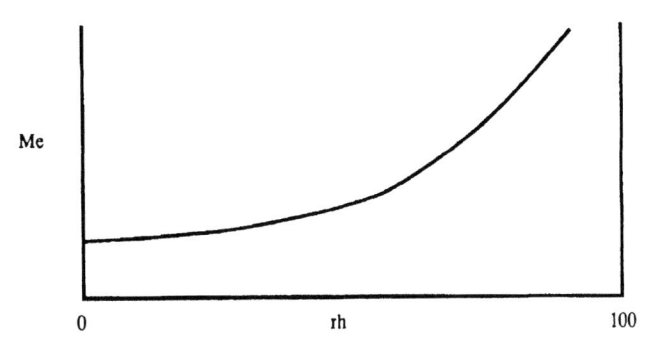

Figure 10.4. Smith equilibrium moisture content equation.

BET equation approaches infinite moisture contents too rapidly at relative humidities above approximately 50%.

10.3.3 Smith Equation

The Smith Equation (Smith, 1947) was developed specifically for the case of water vapor sorption at high relative humidities. The equation is:

$$v = v_b - v_m \ln(1 - rh) \tag{10.7}$$

where

- v = volume of water adsorbed (m^3)
- v_b = volume of bound water adsorbed (m^3)
- v_m = volume of unimolecular layer of normally condensed moisture (m^3)
- rh = relative humidity (expressed as a decimal)

This equation gives an isotherm of the form shown in figure 10.4. The equation has been found to fit experimental data quite accurately at high relative humidities. However, it could not be used at low relative humidities.

10.3.4 Henderson Equation

The best known and most widely used equation for the equilibrium moisture content of biological materials is an empirical equation developed by Henderson (1952):

$$1 - rh = \exp\left[-cT M_e^n\right] \tag{10.8}$$

where

- rh = relative humidity (decimal)
- T = absolute temperature (K)
- M_e = equilibrium moisture content, dry-basis (percent)
- c, n = empirical constants given for some materials in table 10-1

Although this relationship must be considered empirical, it has been found quite useful in describing the equilibrium moisture isotherms of many biological materials (especially if the c and n values are determined for the temperature of interest). Figure 10.5 gives experimental desorption curves for several of the materials listed in table 10-1.

Table 10-1. Values of equilibrium constants c and n for some materials

Material	c	n
Shelled corn	1.98×10^{-5}	1.90
Wheat	10.06×10^{-7}	3.03
Sorghum	6.12×10^{-6}	2.31
Soybeans	5.78×10^{-5}	1.52
Flaxseed	12.40×10^{-6}	2.02
Raisins	12.83×10^{-5}	1.02
Dried peaches	7.40×10^{-4}	0.564
Dried prunes	2.25×10^{-4}	0.865
Cotton	8.84×10^{-5}	1.70
Wood	9.61×10^{-5}	1.41
Spray-dried eggs	5.31×10^{-5}	2.00
Natural clay	13.55×10^{-5}	1.72

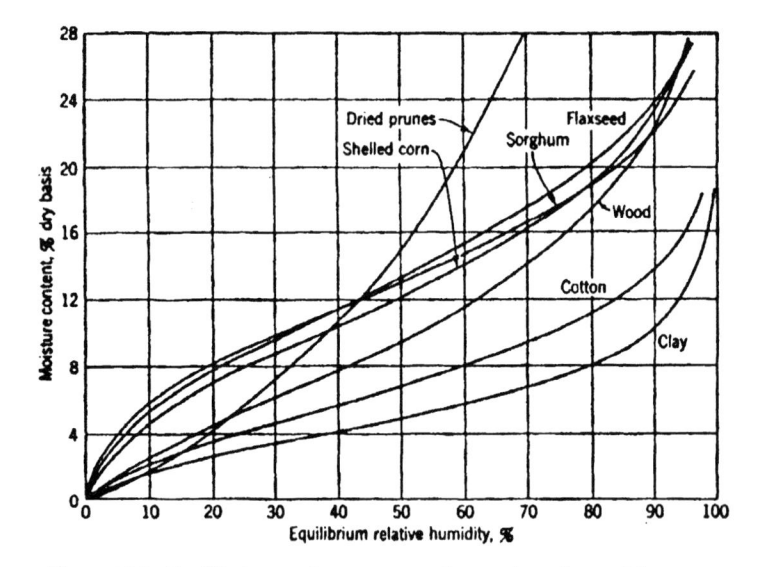

Figure 10.5. Equilibrium moisture curves of a number of materials at room temperature, approximately 25 C.

Example 10.2. Determine the equilibrium moisture content of soybeans at 20 C and 75% relative humidity using the Henderson equation.

Solution

$$1 - rh = \exp\left(-cT M_c^n\right)$$
$$\ln(1 - rh) = -cT M_c^n$$
$$M_c^n = -\ln(1 - rh)/(cT)$$
$$M_c = [-\ln(1 - rh)/(cT)]^{1/n}$$

Drying

$$M_e = [-\ln(1 - .75)/(5.78 \times 10^{-5})(20 + 273.16)]^{1/1.52} = 18.13\%$$
$$m_e = 0.1813/(1 + 0.1813) = 0.1535 = 15.35\%$$

10.3.5 Young Equations

Equations have been developed (Young, 1966) based on the theory of moisture occurring as either (1) bound, (2) free or normally condensed, or (3) absorbed moisture. Equations for each type of moisture were developed and the sum of these equations gives the total moisture content of the material. The equation for the bound moisture is:

$$\theta = rh/[rh + (1 - rh)E] \tag{10.9}$$

where

θ = fraction of surface covered by a layer of bound water molecules (decimal)
rh = relative humidity (decimal)
$E = \exp[-(q_1 - q_2)/kT]$
q_1 = heat of adsorption of the bound layer of water molecules (J/molecule)
q_2 = normal heat of vaporization of water molecules (J/molecule)
k = Boltzmann's constant (1.38×10^{-23} J/molecule K)
T = absolute temperature (K)

The equation for the "free" or normally condensed moisture is:

$$\alpha = -Erh/[E - (E - 1)rh] + [E^2/(E - 1)]\ln\{[E - (E - 1)rh]/E\}$$
$$- (E + 1)\ln(1 - rh) \tag{10.10}$$

where α = total amount of normally condensed moisture measured in molecular layers. The equation for the absorbed moisture during a sorption process is:

$$\gamma_s = \rho V \phi \tag{10.11}$$

where

γ_s = amount of absorbed moisture in sorption process (kg)
ρ = density of water (kg/m^3)
V = amount of absorbed moisture at saturation (m^3)
$\phi = rh\,\theta$ = fraction of surface covered by at least one layer of normally condensed moisture (decimal)

The equation for the absorbed moisture during a desorption process is:

$$\gamma_d = \rho V \theta\, rh_{\max} \tag{10.12}$$

where

γ_d = amount of absorbed moisture in a desorption process (kg)
$rh_{\max}$ = maximum relative humidity with which the material has been in equilibrium prior to desorption process

Then the total dry-basis moisture contents for sorption and desorption processes are given by:

$$M_s = (\rho v_m / W_d)(\theta + \alpha) + (\rho V / W_d)\phi \tag{10.13}$$

and

$$M_d = (\rho v_m / W_d)(\theta + \alpha) + (\rho V / W_d)\theta \, rh_{\max} \tag{10.14}$$

where

M_s = total dry-basis moisture content for a sorption process (decimal)
M_d = total dry-basis moisture content for a desorption process (decimal)
v_m = volume of moisture in a unimolecular layer of water molecules on the surface of the cells (m^3)
W_d = mass of dry material (kg)
ρ = density of water (kg/m^3)

These equations give equilibrium moisture content isotherms similar to those shown in figure 10.5. The equations approximate the BET equation at low relative humidities and the Smith equation at high relative humidities. This combination has been found to quite accurately describe experimental isotherms. The equations have the unique feature that they can describe either sorption or desorption isotherms. None of the previously discussed theories distinguish between sorption and desorption. The constants of the equations would have to be determined from material having a similar moisture history. Different values of the constants would be determined for sorption processes than for desorption processes.

10.4 Moisture Content Measurement

Although moisture measurement is often regarded as a simple analytical procedure, it is actually beset with many and often unexpected complexities. There are problems not only in indirect measurements based on some property of the material which varies with moisture content, but also, with direct methods which depend on the weight lost when moisture is driven off. Thus, it is difficult even to establish a standard by which other methods may be calibrated.

10.4.1 Oven Methods

The most widely recognized methods for determining moisture content are based on drying known weights of material in ovens and calculating moisture content from the loss in weight. These methods are rather simple and direct and usually constitute the "basic" methods against which rapid, practical moisture-testing equipment is calibrated. Actually, however, all oven methods are somewhat empirical in nature since the results depend upon the degree of subdivision of the material being tested and the time, temperature, and atmospheric pressure under which the drying is accomplished. In biological material, it is very difficult if not impossible to remove all moisture by the application of heat without at the same time driving off other volatile substances or causing decomposition of some of the constituents with the formation of moisture not originally present as such. In living material, increased temperatures may simply increase the rate

Drying

of respiration which forms water and carbon dioxide from the burning of sugars within the material.

The United States Department of Agriculture has specified certain standard procedures for determining the moisture content of agricultural commodities. All moisture meters are tested and calibrated by these standards. These methods usually consist of grinding the material and then drying it in an oven at a specific temperature for a specific period of time. The loss in weight is considered to be moisture and is used for the moisture content calculation.

The American Society of Agricultural Engineers publishes standards for a number of biological materials (ASAE, 1996). These standards are for forced convection ovens with whole grains. Table 10-2 gives the temperatures and drying times specified by ASAE for a number of grains.

Table 10-2. Oven temperature and heating period for moisture content determinations

Seed	Oven Temperature, ±1 C	Heating Time hr	min
Alfalfa	130	2	30
Barley	130	20	0
Beans, edible	103	72	0
Bentgrass	130	1	0
Bluegrass	130	1	0
Bluestem	100	1	0
Bromegrass	130	0	50
Cabbage	130	4	0
Carrot	100	0	40
Clover	130	2	30
Collard	130	4	0
Corn	103	72	0
Fescue	130	3	0
Flax	103	4	0
Kale	130	4	0
Mustard	130	4	0
Oats	130	22	0
Onion	130	0	50
Orchardgrass	130	1	0
Parsley	100	2	0
Parsnip	100	1	0
Peanuts	130	6	0
Radish	130	1	10
Rape	130	4	0
Rye	130	16	0
Ryegrass	130	3	0
Safflower	130	1	0
Sorghum	130	18	0
Soybeans	103	72	0
Sunflower	130	3	0
Timothy	130	1	40
Turnip	130	4	0
Wheat	130	19	0

The Association of Official Agricultural Chemists has adopted vacuum oven procedures for various agricultural commodities. The procedures are similar to those specified by the USDA except that vacuum ovens are used. Generally this allows the use of lower oven temperatures.

10.4.2 Desiccant Drying

In order to avoid the hazard of possible organic matter decomposition by heat, a material may be dried for moisture content determination by placing a weighed sample of the finely ground material in a closed container with a relatively large quantity of a good desiccant. The moisture in the material is gradually vaporized and then adsorbed by the desiccant. The drying time required by this method is too great for the method to be of widespread practical importance.

10.4.3 Distillation Methods

There are several distillation methods which may be used for moisture determination. They involve removing the moisture from the material by heating it in oil or some other non-aqueous liquid and measuring either the weight loss or the volume of water distilled from the material. Three such distillation methods are: (1) toluene distillation method, (2) Brown-Duvel distillation method, and the (3) Dexter oil distillation method.

10.4.4 Chemical Methods

Chemical methods of moisture determination are based on stoichiometric chemical reactions in which water is one of the reactants. From a theoretical standpoint these methods should be highly accurate. However, there are practical difficulties in their use. Since the water within materials is bound to some degree by physical forces, it may be questionable as to whether all of the water present enters into the reaction. If all the water does react, it may be difficult to determine the time required for a complete reaction.

Some examples of chemical methods are: (a) calcium carbide method, (b) calcium hydride method, (c) Karl Fischer method, and (d) dichromate method. The Karl Fischer method is probably one of the most accurate moisture measurement methods available. There is a commercial moisture instrument available which uses the calcium carbide method.

10.4.5 Gas Chromatography

The gas chromatography method of moisture measurement involves the use of methanol to extract the water from the sample. The amount of water is then measured by means of gas-liquid chromatography.

10.4.6 Electrical Methods

Electric moisture meters are essentially of two types. The first type measures the electrical conductance of the material placed between two electrodes while the second type measures the capacitance of a condenser in which the sample material acts as the dielectric. There are a number of moisture meters on the market which utilize one or the other of these two methods.

Drying

One of the principal disadvantages of conductance meters is that their accuracy is dependent on the distribution of the moisture within the material. The surface moisture has a great effect on the readings obtained.

Although there is much disagreement as to which type of electric moisture meter is capable of greater over-all accuracy, both the United States and Canadian grain inspectors now use a capacitance meter. The Motomco Moisture Meter was adopted by the Standardization and Testing Branch of the Grain Division, USDA, because in laboratory tests it has been less influenced by electrical variations and has been less subject to area-to-area or season-to-season variations than other meters tested for grain moisture measurement.

For a material other than grain, extensive calibration tests would be required to determine which type and brand of electrical moisture meter best suits the requirements of that particular material.

10.4.7 Spectrophotometric Methods

The moisture determination of a material by spectroscopic means involves relating the energy absorption at one of the water absorption bands to the amount of water present in the material. Spectroscopic moisture measurement techniques which have been investigated are (a) optical spectroscopy or colorimetry, (b) near infrared spectroscopy, and (c) microwave spectroscopy.

10.4.8 Nuclear Methods

There are two nuclear methods which may be used for moisture measurement of materials. They are (1) nuclear magnetic resonance and (2) nuclear scattering devices. Both of these methods actually measure the number of hydrogen nuclei present rather than the number of water molecules present. Therefore, the presence of non-water hydrogen will affect the reading of meters based on nuclear principles.

10.4.9 Equilibrium Relative Humidity

The equilibrium relative humidity method of moisture measurement involves measuring the relative humidity of the environment in equilibrium with the material and then converting the humidity reading to a moisture content by a known relationship between the two. This method is relatively simple and rapid. However, there are a number of disadvantages. First, the relationship between moisture content and equilibrium relative humidity must be known. Second, hysteresis in the material causes variations from calibration curves. Third, the method is inaccurate at high relative humidities due to the large change in moisture content associated with a small change in relative humidity. Fourth, the material must be in a state of equilibrium with its environment.

10.4.10 Suction Methods

Suction methods of moisture measurement are based on the Kelvin and height-of-capillary rise equations relating the reduction in the vapor pressure over a curved water meniscus in a capillary to the length of the water column this curved meniscus will

Table 10-3. Relationship between vapor pressure ratio (i.e., relative humidity), height of water column, and pF values

p/p_{sat} (%)	Height of H_2O column (cm)	pF
0.08	10^7	7
49.0	10^6	6
93.0	10^5	5
99.3	10^4	4
99.9	10^3	3
99.99	10^2	2
99.999	10	1
100.0	0	—

support. The relationship between the column height and the relative vapor pressure is:

$$h = -(RT/Mg)\ln(p/p_{sat}) \quad (10.15)$$

where

h = height of column of water in capillary (m)
R = universal gas constant (8.314×10^3 Pa m^3/kg-mole K)
T = absolute temperature (K)
M = molecular mass of water (kg/kg-mole)
g = gravitational acceleration (m/s^2)
p = vapor pressure over the meniscus (Pa)
p_{sat} = saturation vapor pressure for flat water surface (Pa)

The value calculated from this equation is commonly referred to as the suction. As a convenience, units of suction are sometimes expressed as a pF value defined by the following relationship:

$$pF = \log(100h) = \log[-(100RT/Mg)\ln(p/p_{sat})] \quad (10.16)$$

The relationship between vapor pressure ratio (i.e. relative humidity), height of water column, and the pF are given in table 10-3.

The greatly expanded scale for relative humidities above 99% is evident. This region is extremely important in soils since at lower moisture levels plants are unable to obtain moisture from the soil.

Suction may be measured directly by suction plates, tensiometers, or pressure membranes. These methods involve placing the sample material in contact with a porous plate or membrane across which a pressure differential is maintained. The moisture potential at which the sample is at equilibrium is determined by adjusting the pressure differential until the sample neither gains or loses moisture.

10.4.11 Summary of Advantages and Disadvantages of Various Moisture Measuring Methods

Table 10-4 gives a brief summary of the advantages and disadvantages of each of the previously discussed moisture measurement techniques as given by Young (1991).

Table 10-4. Advantages and disadvantages of various moisture measuring methods

	Method	Advantages	Disadvantages
1.	Oven Methods	a. Simple and direct.	a. Heat may cause decomposition of dry matter creating water not originally present. b. Other volatile materials may be driven off. c. Degree of grinding and time, temperature, and pressure may affect results. d. Moisture may be gained or lost during grinding.
2.	Desiccant drying	a. Heat not required for drying.	a. Lengthy time requirements. b. Moisture may be lost or gained during grinding.
3.	Distillation methods	a. Water may be determined directly rather than as a weight loss.	a. Other volatile substances which are soluble in water may cause errors. b. Distilling time to obtain complete distillation of water is uncertain. c. Moisture may be lost or gained during grinding. d. Sample is destroyed.
4.	Chemical methods	a. Chemical reactions based on water present should theoretically produce high accuracy.	a. Water may not all take part in the reaction. b. Moisture may be lost or gained during grinding. c. Sample is destroyed.
5.	Gas chromatography	a. Highly accurate determination of water in methanol extract.	a. Complete extraction of water is uncertain. b. Expensive equipment required. c. Moisture may be lost or gained during grinding. d. Sample is destroyed.
6.	Electrical methods	a. Rapid, simple, and may be non-destructive	a. Variations in electrical properties of the dry matter may cause errors. b. Moisture distribution may affect readings. c. Ratio of "free" to "bound" water must be constant at a given moisture content.
7.	Spectrophotometric methods	a. Measurement based on fundamental properties of the water molecule. b. May be nondestructive in some cases.	a. Water in different forms may give different results. b. Dry matter absorbs some energy and presents some calibration difficulty. c. Moisture distribution affects the readings. d. Expensive equipment required in some cases.

Table 10-4. (Continued)

8.	Nuclear methods	a. Rapid. b. High accuracy for properly calibrated NMR instruments. c. Nondestructive.	a. Senses total hydrogen rather than water. b. Different energy forms of water may cause problems in separation of water signals from nonwater hydrogen NMR signals. c. Other elements may affect neutron scattering readings. d. Expensive equipment required.		
9.	Equilibrium relative humidity methods	a. Relatively simple. b. Nondestructive.	a. Relationship between moisture content and equilibrium relative humidity must be known. b. Hysteresis in material causes variations from calibration curves. c. Inaccurate at high relative humidities. d. Material must be in a state of equilibrium with its environment. e. Relative humidity measurement is difficult.		
10.	Suction methods	a. Relatively simple. b. Nondestructive. c. Increased accuracy at very high moisture levels.	a. Relationship between moisture content and suction must be known. b. Hysteresis in material causes variations from calibration curve. c. Cannot be used at low moisture levels.		

10.5 Thin-Layer Drying

Thin-layer drying refers to the drying of individual particles or grains of material which are fully exposed to the drying air. The thin-layer drying process is often considered to be divided into two periods: (1) the constant drying-rate period, and (2) the falling drying-rate period.

10.5.1 Constant Drying-Rate Period

In the constant drying- rate period a material or mass of material containing so much water that liquid surfaces exist will dry in a manner comparable to an open-faced body of water. The water and its surroundings, not the solid, will determine the rate of drying. Wet sand, soil, pigments, and washed seed are examples of materials that initially dry at a constant rate.

The energy required for drying may be provided by radiation, conduction, or convection. The convection case is represented by the adiabatic evaporation of moisture from the surface of a wet- bulb thermometer. The equation representing the drying is:

$$dW/d\tau = f_v A(p_{sat} - p_v) = f A(t_a - t_s)/h_{fg} \qquad (10.17)$$

where

$dW/d\tau$ = drying rate (kg/s)

Drying

f = convective heat transfer coefficient at the water-air interface (W/m²K)
A = water surface area (m²)
h_{fg} = latent heat of water at water surface temperature (J/kg)
t_a = air temperature (K)
t_s = water surface temperature (K)
f_v = water vapor transfer coefficient at the water-air interface (kg/s m² Pa)
p_{sat} = saturation water vapor pressure at t_s (Pa)
p_v = water vapor pressure in the air (Pa)

Values of f_v and f were determined by Gamson et al. (1943) for drying by forcing air through beds of moist spherical or cylindrical pellets. For vapor transfer, with Reynolds number greater than 350,

$$(f_v D_p p_{atm}/D_v) = 0.989(D_p G/\mu)^{0.59}(\mu/D_v \rho)^{1/3} \tag{10.18}$$

and for heat transfer:

$$(f D_p/k) = 1.064(D_p G/\mu)^{0.59}(c_p \mu/k)^{1/3} \tag{10.19}$$

where

D_p = diameter of particle (m)
D_v = diffusivity of water vapor (m²/s)
k = thermal conductivity of air (W/mK)
G = mass velocity (kg/s m²)
ρ = density of air (kg/m³)
μ = viscosity of air (kg/s m)
c_p = specific heat of air (J/kg K)
p_{atm} = total pressure of the atmosphere (Pa)

Note that:

$$(D_p G/\mu) = Re = \text{Reynolds number} \tag{10.20}$$

$$(c_p \mu/k) = Pr = \text{Prandtl number} \tag{10.21}$$

$$(\mu/D_v \rho) = Sm = \text{Schmidt number} \tag{10.22}$$

If the energy for the evaporation of moisture were supplied by radiation rather than by convection the following equation would describe the process:

$$dW/d\tau = f_v A(p_{sat} - p_v) = \varepsilon A \sigma (T_w^4 - T_s^4)/h_{fg} \tag{10.23}$$

where

ε = emissivity of the water surface (dimensionless)
σ = Stefan-Boltzmann constant (5.73×10⁻⁸ W/m²K⁴)
T_w = absolute temperature of radiating walls (K)
T_s = absolute temperature of water surface (K)

10.5.2 Falling Rate Period

Practically all agricultural drying takes place in the falling-rate period. The falling-rate period is bounded by equilibrium moisture contents of an equilibrium moisture curve between zero and nearly 100% relative humidity. Moisture contents near the 100% level would approximate a constant rate drying period because as moisture evaporates the equilibrium relative humidity and thus the vapor pressure driving force change only very slightly.

Drying in the falling-rate period involves two processes: (1) the movement of moisture within the material to the surface and (2) removal of the moisture from the surface.

10.5.2.1 Sherwood's Exponential Drying Equation

An equation which has been most used for describing the drying of agricultural products was developed by Sherwood (1936). He began with the assumption that the drying rate was proportional to the difference in moisture content between the material being dried and the equilibrium moisture content at the drying air state or:

$$dM/d\tau = -k(M - M_e) \quad (10.24)$$

where

$dM/d\tau$ = drying rate (kg_{H_2O}/kg_{dm} h)
k = drying parameter (h^{-1})
M = moisture content of material, kg_{H_2O}/kg_{dm}
M_e = equilibrium moisture content of material at conditions of drying air (kg_{H_2O}/kg_{dm})

Integration of equation 10.24 and substitution of the boundary condition that $M = M_o$ at $\tau = 0$ yields:

$$(M - M_e)/(M_o - M_e) = \exp(-k\tau) \quad (10.25)$$

The exponential drying curve is analogous to Newton's Law of heating or cooling such that the internal resistance to moisture movement is considered to be negligible with the surface resistance controlling the drying process.

Also, if we accept the theory that the vapor pressure gradient is the driving force for moisture movement,

$$dM/d\tau = -k_1(p_{sat} - p_v) \quad (10.26)$$

In order for equation 10.26 to take the form of 10.24, the following relations must hold:

$$k(M - M_e) = k_1(p_{sat} - p_v) \quad (10.27)$$

or in general the equilibrium moisture content must be given by:

$$M_e = (k_1/k)p + c_1 \quad (10.28)$$

Thus, equation 10.25 is based on the assumptions that the equilibrium moisture content is a linear function of vapor pressure and that the limiting resistance to moisture flow is at the surface. The assumption of a linear relationship between equilibrium moisture

Drying

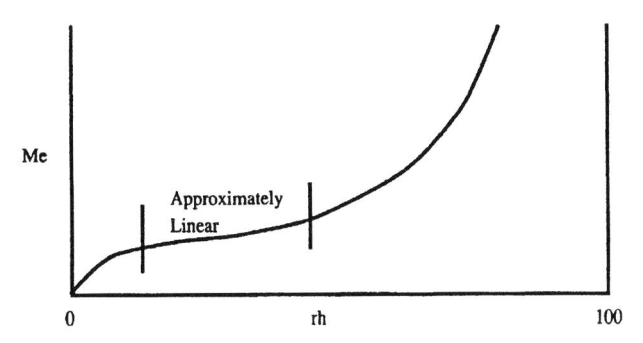

Figure 10.6. Assumption of linear relationship between equilibrium moisture content and vapor pressure at medium relative humidities.

content and vapor pressure is a fairly good approximation at medium relative humidities (fig. 10.6). However, the assumption of negligible internal resistance is questionable in most cases. The equation based on these assumptions has been found quite useful, however, in describing many drying processes.

10.5.2.2 Diffusion Equations

If the surface resistance to moisture transfer is negligible in comparison to the internal resistance, diffusion equations may be used to describe the drying process. These equations are based on a special case of Fick's first law of diffusion which states that:

$$J_x = -D(\partial C/\partial x) \tag{10.29}$$

where

J_x = flux of the diffusing medium, in this case water (kg/sm^2)
D = mass diffusivity of water (m^2/s)
C = concentration of water (kg/m^3)
x = distance (m)

This equation is completely analogous to the Fourier-Biot Law for heat conduction. Equation 10.29 states that the rate of movement of water within the material is proportional to the concentration gradient. However, the question arises as to whether the water moves as a liquid or as a vapor. If the movement is in the liquid form, then the concentration gradient to be used in equation 10.29 is the liquid concentration gradient. If the movement is in the vapor state, then the vapor concentration gradient should be used. It is also possible that water might be transferred by both liquid and vapor diffusion in which case the total flux of water would have to be the sum of the liquid flux and the vapor flux. Most drying work to date has assumed either liquid or vapor diffusion but not both. We shall consider these two cases.

10.5.2.2.1 Liquid Diffusion.
In the case of liquid diffusion we have:

$$J_x = -D_{liq}(\partial C_{liq}/\partial x) \tag{10.30}$$

where

C_{liq} = liquid concentration (kg/m³)
D_{liq} = mass diffusivity of liquid water (m²/s)

In a manner similar to the derivation of the Fourier-Poisson equation for heat conduction, Fick's 2nd Law of diffusion can be developed:

$$(\partial C_{liq}/\partial \tau) = D_{liq}((\partial^2 C_{liq}/\partial x^2) + (\partial^2 C_{liq}/\partial y^2) + (\partial^2 C_{liq}/\partial z^2)) \tag{10.31}$$

Since,

$$M = C_{liq}/\rho_{dm} \tag{10.32}$$

where

M = moisture content (dry basis)
ρ_{dm} = density of the dry matter (kg/m³)

equation 10.31 may be written as:

$$\partial M/\partial \tau = D_{liq}((\partial^2 M/\partial x^2) + (\partial^2 M/\partial y^2) + (\partial^2 M/\partial z^2)) \tag{10.33}$$

where $\partial M/\partial \tau$ = rate of change of moisture content with respect to time, (kg$_{H_2O}$/kg$_{dm}$ s).

If a plane sheet of material at an initial moisture content, M_o, is placed in an environment for which its equilibrium moisture content is M_e, the solution to equation 10.33 is:

$$(M - M_o)/(M_e - M_o) = 1 - (4/\pi) \sum_{n=0}^{\infty} ((-1^n/(2n+1))$$
$$\exp(-D_{liq}(2n+1)^2 \pi^2 \tau/(4L^2)) \cos((2n+1)\pi x/(2L)) \tag{10.34}$$

where

M = moisture content, dry basis (at time τ and point x)
L = half thickness of sheet (m)

The average moisture content at any time is given by:

$$(\overline{M} - M_e)/(M_o - M_e) = \sum_{n=0}^{\infty} \{8/[(2n+1)^2 \pi^2]\} \exp(-D_{liq}(2n+1)^2 \pi^2 \tau/(4L^2)) \tag{10.35}$$

where $\overline{M}$ = average moisture content, dry basis (at time τ). The analogous solutions for a sphere are:

$$(M - M_o)/(M_e - M_o) = 1 + (2a/\pi r) \sum_{n=1}^{\infty} ((-1)^n/n) \sin(n\pi r/a)$$
$$\exp(-D_{liq} n^2 \pi^2 \tau/a^2) \tag{10.36}$$

$$(\overline{M} - M_e)/(M_o - M_e) = (6/\pi^2) \sum_{n=0}^{\infty} (1/n^2) \exp(-D_{liq} n^2 \pi^2 \tau/a^2) \tag{10.37}$$

Drying

where

a = radius of the sphere (m)
r = radial distance from the center (m)

These equations assume a constant diffusivity, uniform initial concentrations, and constant surface concentrations. Under a few specialized cases, the differential equation can be solved for cases of variable diffusivities. Non-uniform initial moisture contents may also be analyzed.

For cases where the surface resistance to moisture movement is not negligible, the equations have also been solved considering the finite value of the surface conductance.

10.5.2.2.2 Vapor Diffusion. In the case of vapor diffusion we have:

$$J_x = -D_v(\partial C_v/\partial x) \qquad (10.38)$$

where

C_v = vapor concentration (kg/m³)
D_v = vapor mass diffusivity (m²/s)

The diffusion of the vapor must take place through the pores of the solid material and the liquid water present within the material provides a source of vapor due to evaporation within the body. The differential equation describing the vapor diffusion process is for a constant diffusivity:

$$D_v((\partial^2 C_v/\partial x^2) + (\partial^2 C_v/\partial y^2) + (\partial^2 C_v/\partial z^2)) = v(\partial C_v/\partial \tau) + (1-v)\rho_s(\partial M/\partial \tau) \qquad (10.39)$$

where

v = void fraction or porosity of body (dimensionless)
ρ_s = density of solid material (kg/m³)
τ = time (s)
M = moisture content, dry basis (decimal)

Equation 10.39 contains two dependent variables, c_v and M. Therefore, the relationship between the two must be known or assumed before the equation can be solved. Most attempts to solve equation 10.39 have begun with the assumption that the equilibrium moisture content is a linear function of the vapor concentration and temperature.

$$M = \alpha + \beta C_v - \gamma T \qquad (10.40)$$

where

T = absolute temperature (K)
α, β, γ = constants having appropriate dimensions for dimensional homogeneity

Then for a constant temperature:

$$\partial M/\partial \tau = \beta(\partial C_v/\partial \tau) \qquad (10.41)$$

Substituting equation 10.41 into 10.40 gives:

$$\partial C_v/\partial \tau = (D_v/(v + (1 - v)\rho_s\beta))((\partial^2 C_v/\partial x^2) + (\partial^2 C_v/\partial y^2) + (\partial^2 C_v/\partial z^2)) \quad (10.42)$$

Equation 10.42 is analogous to equation 10.31 for liquid diffusion with the $(v + (1 - v)\rho_s\beta)$ term being the primary difference. Again, if the moisture content is linearly related to the vapor concentration then:

$$\partial M/\partial \tau = (D_v/(v + (1 - v)\rho_s\beta))((\partial^2 M/\partial x^2) + (\partial^2 M/\partial y^2) + (\partial^2 M/\partial z^2)) \quad (10.43)$$

The only difference between equation 10.43 for vapor diffusion and equation 10.33 for liquid diffusion is the diffusivity function if the linear relationship between vapor concentration and moisture content is assumed. For equivalent drying relationships, the following identity must hold:

$$D_{\text{liq}} = D_v/(v + (1 - v)\rho_s\beta) \quad (10.44)$$

Thus, the assumption of liquid or vapor diffusion only affects the choice of the diffusivity to be used if the moisture content is assumed to vary linearly with vapor concentration and the material being dried is homogenous. However, the vapor diffusion equation is more appropriate for use with composite bodies composed of materials having different physical and hygroscopic properties.

Also, with the assumption of a vapor diffusion model, the heat transfer equation may be solved simultaneously for the case where the temperature is variable. The heat transfer equation for this case is:

$$k((\partial^2 T/\partial x^2) + (\partial^2 T/\partial y^2) + (\partial^2 T/\partial z^2)) = c_s(1 - v)\rho_s(\partial T/\partial \tau)$$
$$+ c_w(1 - v)\rho_s M(\partial T/\partial \tau) - h_{\text{fg}}(1 - v)\rho_s(\partial M/\partial \tau) - q''' \quad (10.45)$$

where

k = thermal conductivity of the body (W/mK)
T = temperature (K)
c_s = specific heat of solid (J/kg K)
c_w = specific heat of water (J/kg K)
h_{fg} = latent heat of vaporization of the moisture (J/kg)
q''' = rate of internal heat generation (W/m^3)

By a numerical solution of the vapor diffusion equation and the heat transfer equation simultaneously it has been shown that the heat transfer into the body is only a limiting factor if:

$$Le_m < 60 \quad (10.46)$$

where

$$Le_m = k(v + (1 - v)\rho_s\beta)/(D_v(1 - v)\rho_s(c_s + c_w M + h_{\text{fg}}\gamma)) \quad (10.47)$$
$$Le_m = \text{modified Lewis Number (dimensionless)}$$

Drying

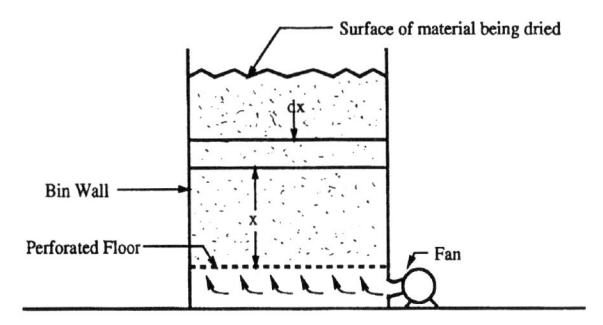

Figure 10.7. Sketch of bulk drying bin with differential layer of grain identified.

10.6 Bulk Drying Simulation

There are two fundamental tasks which must be performed by any drying system. First, the system must provide a means for supplying energy to the product to be dried in order to vaporize the moisture. Second, the system must provide a means for removing the vaporized moisture from the system.

Forced air drying systems are used to dry many agricultural products. In these systems, air (either heated or ambient) is forced by fans through the product to be dried. Energy is transferred from the air to the product to evaporate the water. The air movement through the system also "flushes" the vaporized moisture from the system.

We will next consider procedures for simulating the drying of a stationary bed of material by forcing air through the bed. A differential layer of the drying material (fig. 10.7) can be assumed to be fully exposed to a given temperature and relative humidity of the drying air. Under constant temperature and humidity of the supplied air, each successive layer is exposed to decreasing air temperatures and increasing relative humidities. In practical applications we are concerned primarily with the drying rate of the entire layer or some particular portion of it. By considering the energy and mass balance together with the information on fully exposed drying (thin-layer drying equations), physical relationships for design purposes can be developed.

10.6.1 Significance of Adiabatic Saturation Process

Consider a batch of moist grain of uniform initial moisture content in a bin with vertical sides. Drying air is forced into the grain through a perforated floor. The air moves upward at a constant rate, leaving through the upper portion of the grain. The temperature and humidity of the air entering the grain are assumed to be constant. As the air moves upward its humidity increases as a result of evaporation of moisture from the grain. The increase in air humidity is accompanied by a decrease in air temperature. If the change in sensible heat of the grain and the loss of energy to the surroundings is neglected, then the energy transferred from the air to the grain must be equal to the energy used to evaporate the moisture. Thus, this is an adiabatic saturation process. The state of the air follows approximately an adiabatic saturation temperature or wet-bulb temperature line on the psychrometric chart. The rate at which the air temperature drops depends on the rate at which moisture is evaporated from the grain.

10.6.2 Energy Balance

The energy that is used in evaporating moisture from the grain must be equal to the energy supplied by the drop in air temperature, plus any energy supplied by a change in grain temperature, plus any energy supplied by conduction or radiation from the surrounding grain or bin walls. For practical purposes, the energy lost or gained through the bin walls is small and may usually be neglected. It will be neglected in our discussion. For further simplicity, the case when sensible heat changes in the grain are negligible will be considered although under some conditions this factor may have an effect on drying rates.

This reduces the energy balance to one in which the heat of vaporization of the moisture evaporated from the grain is equal to the energy loss of the passing air.

Consider a differential thickness of grain dx at any height x of the bin (fig. 10.7). During a short time interval $d\tau$, the moisture content (M) decreases by dM. The quantity of energy required to decrease the moisture content by this amount is:

$$q_L = \rho A \, dx \, dM \, h_{fg} \qquad (10.48)$$

where

q_L = energy required for vaporization of moisture (J)
ρ = density of dry matter (kg/m^3)
A = cross-sectional area of bin (m^2)
dx = differential thickness of layer (m)
dM = change in moisture content, dry basis, expressed as a decimal (kg$_{H_2O}$/kg$_{dm}$)
h_{fg} = latent heat of vaporization of the water (J/kg$_{H_2O}$)

During the same time interval, the air moving through the thin layer undergoes a slight drop in temperature, dT. The energy loss associated with this drop in temperature is:

$$q_a = c_{pa} \dot{m} \, dT \, d\tau \qquad (10.49)$$

where

q_a = energy lost by air to the grain (J)
c_{pa} = specific heat of air (J/kg K)
$\dot{m}$ = mass flow rate of air (kg/h)
dT = change in temperature of air (K)
$d\tau$ = time interval (h)

Then,

$$q_L = q_a \qquad (10.50)$$

or

$$\rho A \, dx \, dM h_{fg} = c_{pa} \dot{m} \, dT \, d\tau \qquad (10.51)$$

Drying

Then,
$$dM/d\tau = (c_{pa}\dot{m}/(\rho A h_{fg}))(dT/dx) \tag{10.52}$$

or
$$dM/d\tau = P\, dT/dx \tag{10.53}$$

where
$$P = c_{pa}\dot{m}/(\rho A h_{fg}) \tag{10.54}$$

10.6.3 Hukill's Analysis

Hukill (1947) has developed a solution to equation 10.53 based on the exponential equation for thin-layer drying:
$$(M - M_e)/(M_o - M_e) = \exp(-k\tau) \tag{10.55}$$

and upon an approximate equation for the temperature distribution:
$$(T - T_G)/(T_o - T_G) = \exp(-cx) \tag{10.56}$$

where

T = temperature at x (K)
T_o = entering air temperature (K)
T_G = temperature at which, having cooled at a constant wet-bulb temperature, the relative humidity of the air is in equilibrium with the moisture content of the grain (fig. 10.8) (K)
x = distance from entrance to bed (m)

These two equations are used as the boundary conditions for the analysis by Hukill. Then:

when $\tau = 0$, $M = M_o$, and $T = T_G + (T_o - T_G)\exp(-cx)$ (10.57)
when $\tau = \infty$, $M = M_e$ (10.58)
when $x = 0$, $T = T_o$, and $M = M_e + (M_o - M_e)\exp(-k\tau)$ (10.59)
when $x = \infty$, $T = T_G$ (10.60)

Hukill then developed equations which satisfy the differential equation given by equation 10.53. They are:
$$(M - M_e)/(M_o - M_e) = \exp(cx)/(\exp(cx) + \exp(k\tau) - 1) \tag{10.61}$$

and
$$(T - T_G)/(T_o - T_G) = \exp(k\tau)/(\exp(cx) + \exp(k\tau) - 1) \tag{10.62}$$

in which
$$c = k(M_o - M_e)/(P(T_o - T_G)) \tag{10.63}$$

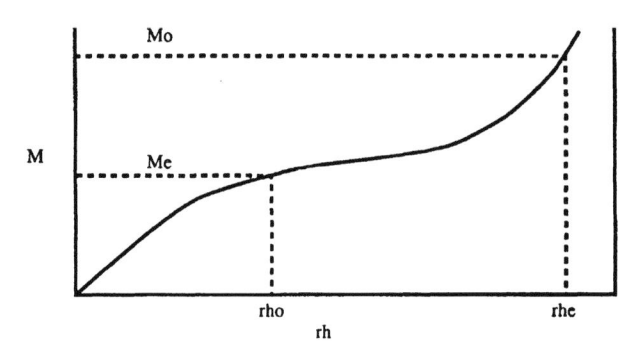

Figure 10.8a. Relationship between moisture contents and relative humidities.

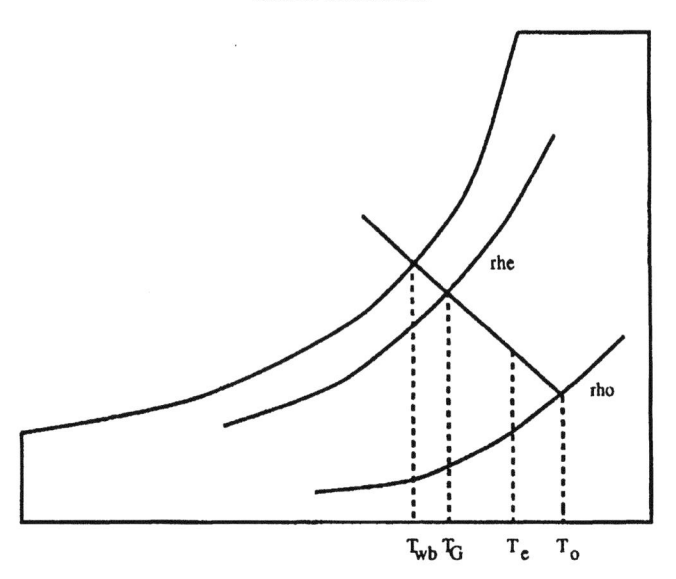

Figure 10.8b. Relationship between relative humidities and temperatures in bulk drying process.

The equations were apparently developed by trial and error by Hukill based on his vast experience in the area of bulk drying. Equation 10.63 is a necessary condition for equations 10.61 and 10.62 to satisfy the basic differential equation. However, there is no apparent physical reason why equation 10.63 must be true.

Hukill simplified equation 10.61 by introducing some new parameters. The first such parameter is the period of half response or the time required for the moisture ratio to reach 0.5 according to the exponential drying equation. Then, if H is one time unit:

$$(M - M_e)/(M_o - M_e) = \exp(-kH) = 0.5 \tag{10.64}$$

Drying

Then,
$$\exp(kH) = 1/0.5 = 2 \tag{10.65}$$

If Y is defined as the number of time units:
$$Y = \tau/H \tag{10.66}$$

then
$$MR = (M - M_e)/(M_o - M_e) = \exp(-k\tau) = \exp(-kHY) \tag{10.67}$$

and
$$\exp(k\tau) = \exp(kHY) = (\exp(kH))^Y = 2^Y \tag{10.68}$$

The unit of equivalent depth was defined as the depth which contains enough grain to make the heat requirement for evaporating its moisture, from an initial moisture content of M_o to a final moisture content of M_e equal to the energy which would be supplied by all the air passing through the bin in one unit of time if its temperature dropped from T_o to T_G. Then the number of depth units in a bin of depth x is given by:

$$D = x\rho A h_{fg}(M_o - M_e)/(c_{pa}\dot{m} H(T_o - T_G)) \tag{10.69}$$

Then
$$\exp(cx) = \exp(c(c_{pa}\dot{m}H)(T_o - T_G)D/(\rho A h_{fg}(M_o - M_e)))$$

By substitution of equation (10.63):
$$\exp(cx) = \exp(k c_{pa}\dot{m} H D A h_{fg}/(c_{pa}\dot{m} h_{fg} A))$$
$$= \exp(kHD) = \exp(kH))^D = 2^D \tag{10.70}$$

Then
$$MR = (M - M_e)/(M_o - M_e) = 2^D/(2^D + 2^Y - 1) \tag{10.71}$$

and
$$(T_e - T_G)/(T_o - T_G) = 2^Y/(2^D + 2^Y - 1) \tag{10.72}$$

Figure 10.9 plots the moisture ratio given by equation 10.71 as a function of the number of time factors (Y) and for various numbers of depth factors (D).

10.6.3.1 Mean or Average Moisture Content

The mean or average moisture content of a layer within the bin may be obtained by integration of equation 10.71 over the appropriate layer. In general, for a layer of grain between depths representing $D1$ and $D2$ depth factors, the average moisture ratio is given by:

$$\overline{MR} = (1/(\ln(2)(D2 - D1))) \ln((2^{D2} + 2^Y - 1)/(2^{D1} + 2^Y - 1)) \tag{10.73}$$

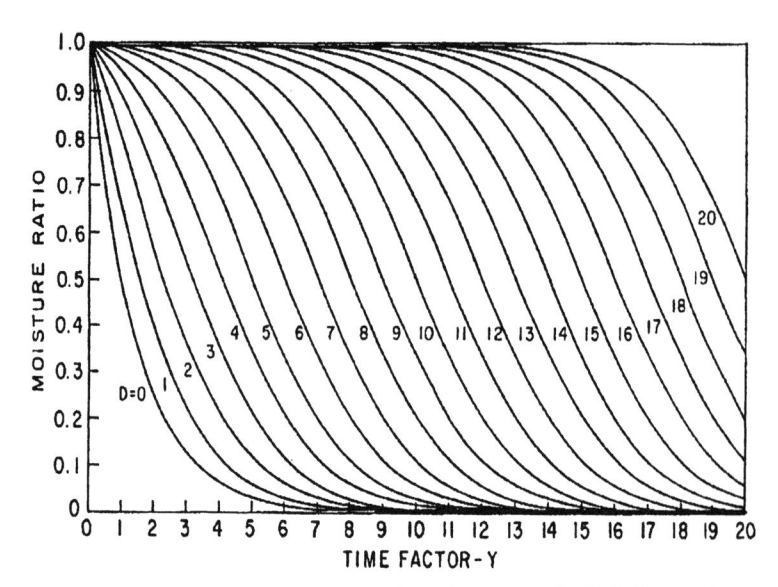

Figure 10.9. Moisture ratio vs. number of time units for Hukill analysis.

For a layer from the bottom (or entrance) of the bin to a depth corresponding to D depth units, equation 10.73 becomes:

$$\overline{MR} = (1/(D\ln(2)))\ln((2^D + 2^Y - 1)/2^Y) \quad (10.74)$$

Figure 10.10 plots the average moisture ratio as a function of number of time units and number of depth units as given by equation 10.74.

10.6.3.2 Drying Rate at Any Position

The non-dimensionalized drying rate at any position may be obtained by differentiating equation 10.71 with respect to the number of time units, Y. This gives:

$$d(MR)/dY = \dot{MR} = -2^D(\ln 2)2^Y/(2^D + 2^Y - 1)^2 \quad (10.75)$$

Figure 10.11 plots the rate of change in the moisture ratio with respect to number of time units as a function of number of depth factors and number of time factors as given by equation 10.75.

10.6.3.3 Mean Drying Rate of Grain for Any Layer

The mean drying rate of any layer may be determined by integrating equation 10.75 between the number of depth factors representing the lower and upper boundaries ($D1$ and $D2$) of the layer. Thus,

$$\overline{\dot{MR}} = (2^Y/(D2 - D1))[1/(2^{D2} + 2^Y - 1) - 1/(2^{D1} + 2^Y - 1)] \quad (10.76)$$

For the average drying rate from the bottom of the bin to a depth corresponding to D depth units, equation 10.76 becomes:

$$\overline{\dot{MR}} = (1/D)[2^Y/(2^D + 2^Y - 1) - 1] \quad (10.77)$$

Drying

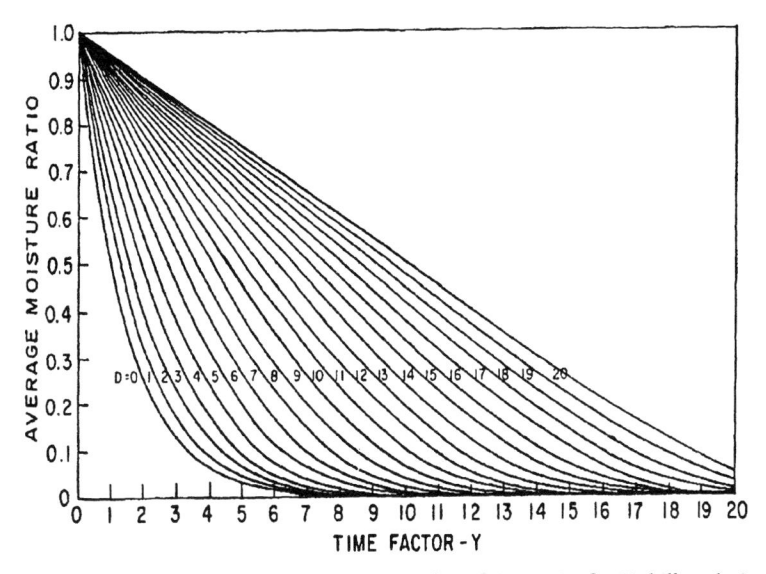

Figure 10.10. Average moisture ratio vs. number of time units for Hukill analysis.

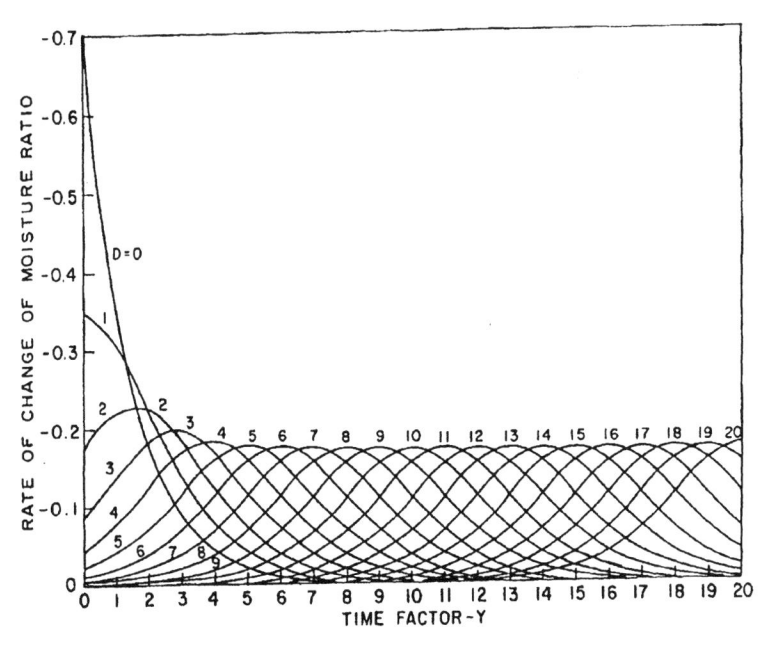

Figure 10.11. Rate of change of moisture ratio vs. number of time units for Hukill analysis.

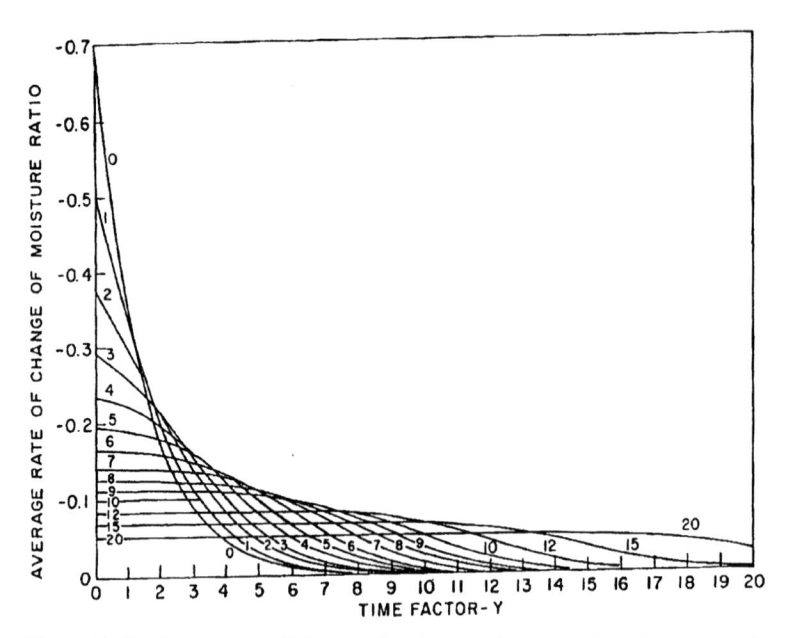

Figure 10.12. Average rate of change of moisture ratio vs. number of time units for Hukill analysis.

Figure 10.12 plots the non-dimensionalized average drying rate as a function of number of time units and depth factors as given by equation 10.77.

10.6.3.4 Drying Efficiency

Drying efficiency may be defined as the energy used for drying divided by the energy potentially available for drying. These energy terms are proportional to the changes in temperature of the air. Thus,

$$E = (T_o - T_e)/(T_o - T_G) \tag{10.78}$$

where T_e = exit temperature of air (K).

If we subtract one from each side of equation 10.72, we obtain:

$$(T_e - T_o)/(T_o - T_G) = 2^Y/(2^D + 2^Y - 1) - 1 \tag{10.79}$$

or

$$E = (T_o - T_e)/(T_o - T_G) = 1 - 2^Y/(2^D + 2^Y - 1) \tag{10.80}$$

Figure 10.13 plots the drying efficiency as a function of number of time units and number of depth factors as given by equation 10.80.

10.6.3.5 Mean Drying Efficiency

The mean drying efficiency from the beginning of the drying process until a given number of time units has elapsed may be found by integration of equation 10.80 over

Drying

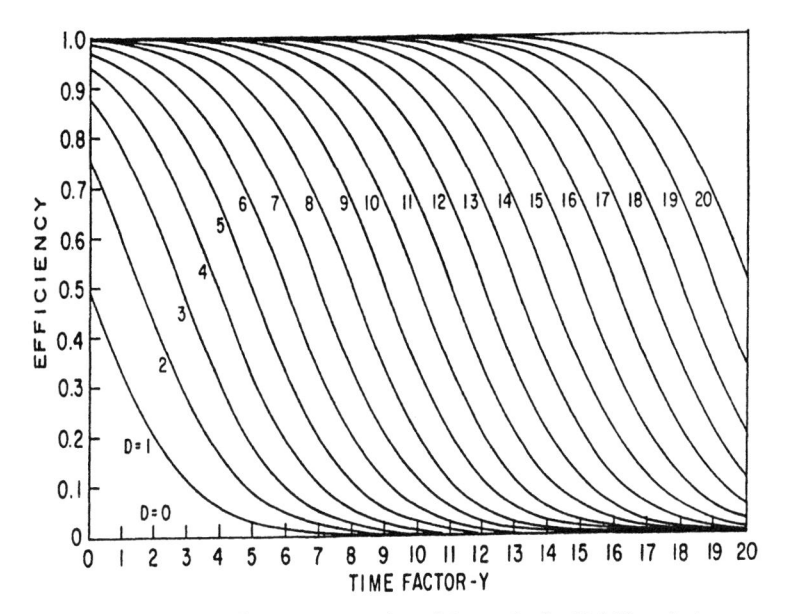

Figure 10.13. Efficiency vs. number of time units for Hukill analysis.

time units to obtain:

$$\overline{E} = 1 - (1/(Y \ln 2)) \ln[(2^D + 2^Y - 1)/2^D] \quad (10.81)$$

Figure 10.14 plots the mean drying efficiency as a function of number of time units and number of depth factors as given by equation 10.81.

Example 10.3. A circular metal bin 5.5 m in diameter is filled to a depth of 1.8 m with shelled corn at a moisture content of 22% w.b. It is to be dried with air at 32 C and 30% rh. Assume the time of half response to be 3.5 h. Air is to be supplied at a rate of 0.25 m³/s m³. Determine: (1) time required for average moisture content to reach 15% w.b., (2) moisture content of bottom layer when average moisture content is 15% w.b., (3) moisture content at top of bin when average moisture content is 15% w.b., (4) time required for grain at top of bin to reach 15% w.b., (5) moisture content at bottom of bin when top of bin reaches 15% w.b., (6) drying efficiency when top of bin reaches 15% w.b., (7) mean drying rate when top of bin reaches 15% w.b., (8) mean drying efficiency over the drying period until top of bin reaches 15% w.b. Assume $\rho = 605$ kg/m³ for shelled corn.

Solution
Moisture Contents
Initial

$$m_o = .22 \quad M_o = .22/(1 - .22) = .282$$

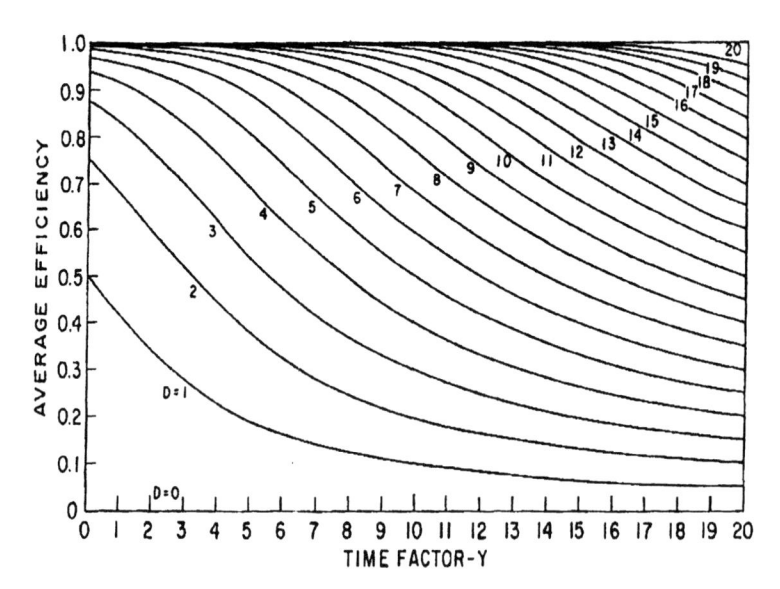

Figure 10.14. Average efficiency vs. number of time units for Hukill analysis.

Final

$$m = .15 \qquad M = .15/(1 - .15) = .176$$

Equilibrium

$$1 - rh = \exp\left(-cTM_c^n\right)$$
$$\ln(1 - rh) = -cTM_c^n$$
$$M_c^n = -\ln(1 - rh)/(cT)$$
$$M_c = [-\ln(1 - rh)/(cT)]^{1/n}$$
$$M_c = [-\ln(1 - .30)/(1.98 \times 10^{-5})(305 \text{ K})]^{1/1.90}$$
$$M_c = 8.6\% = 0.086$$
$$m_c = 0.086/(1 + 0.086) = 0.079$$

Equilibrium Relative Humidity

$$1 - rh_e = \exp(-cTM_o^n)$$
$$rh_e = 1 - \exp(-cTM_o^n)$$
$$rh_e = 1 - \exp(-(1.98 \times 10^{-5})(305 \text{ K})(28.2)^{1.9})$$
$$rh_e = 0.968 = 96.8\%$$

Equilibrium Temperature

From the psychrometric chart (see sketch in fig. 10.15), the equilibrium temperature, T_G, is 19.6 C.

Drying

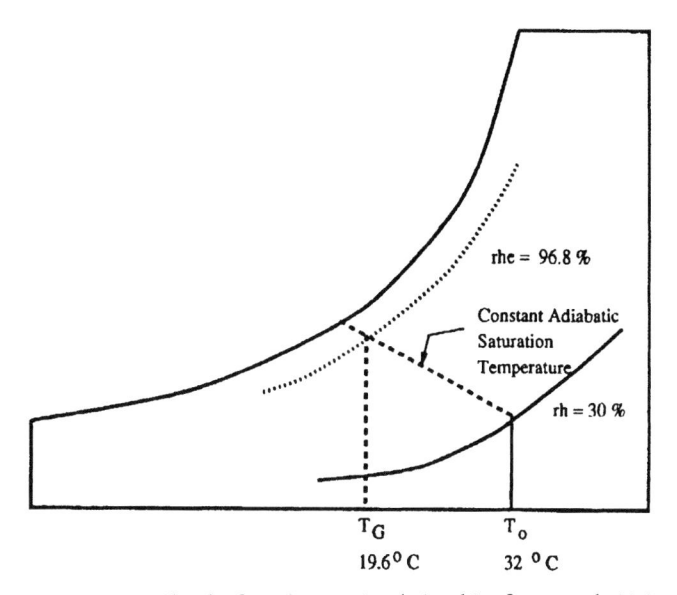

Figure 10.15. Sketch of psychrometric relationships for example 10.3.

Volume
$$V = \pi r^2 h = \pi (2.75 \text{ m})^2 (1.8 \text{ m}) = 42.76 \text{ m}^3$$

Mass Flow Rate of Air
$$\dot{m} = (0.25 \text{ m}^3/\text{s m}^3)(42.76 \text{ m}^3)/(0.876 \text{ m}^3/\text{kg}) = 12.20 \text{ kg/s}$$

Number of Depth Factors
$$D = x\rho A h_{fg}(M_o - M_e)/(c_{pa}\dot{m} H(T_o - T_G))$$

A heat of vaporization of the water of 2 720 kJ/kg is assumed. This value is greater than the heat of vaporization of pure water due to the additional forces binding the water to the grain.

$$D = (1.8 \text{ m})(605 \text{ kg/m}^3)\pi(2.75 \text{ m})^2(2720 \text{ kJ/kg})(.282 - .086)/$$
$$(1.0048 \text{ kJ/kgK})(12.20 \text{ kg/s})(3.5 \text{ h})(3600 \text{ s/h})(32 - 19.6) \text{ K}$$
$$D = 7.20$$

Thickness of Each Depth Factor
$$1.8 \text{ m}/(7.20) = 0.25 \text{ m}$$

1. Time Required for Average Moisture Content, $\overline{M}$, to Reach .176

$$@ \overline{M} = .176 \quad \overline{MR} = (.176 - .086)/(.282 - .086) = .459$$
$$\overline{MR} = .459 = (1/D \ln 2) \ln[(2^D + 2^Y - 1)/2^Y]$$
$$\ln[(2^D + 2^Y - 1)/2^Y] = \overline{MR} D \ln 2$$

$$(2^D + 2^Y - 1)/2^Y = \exp(\overline{MR}\ D \ln 2)$$
$$2^Y(1 - \exp(\overline{MR}\ D \ln 2)) = 1 - 2^D$$
$$2^Y = (1 - 2^D)/(1 - \exp(\overline{MR}\ D \ln 2))$$
$$Y \ln 2 = \ln[(1 - 2^D)/(1 - \exp(\overline{MR}D \ln 2))]$$
$$Y = (1/\ln 2) \ln[(1 - 2^D)/(1 - \exp(\overline{MR}D \ln 2))]$$
$$Y = (1/0.693) \ln[(1 - 2^{7.20})/(1 - \exp(0.459(7.20)(.693))]$$
$$Y = 4.04$$
$$\tau = YH = (4.04)(3.5\ h) = 14.1\ h$$

2. Moisture Content at Bottom of Bin @ 14.1 h
   at bottom of bin, $D = 0$
   $$MR = 2^D/(2^D + 2^Y - 1) = 2^0/(2^0 + 2^Y - 1) = 1/2^Y = 1/2^{4.04} = 0.061$$
   $$M = M_e + MR(M_o - M_e) = .086 + .061(.196) = .098$$
   $$m = .098/(1 + .098) = 0.089$$

3. Moisture Content at Top of Bin @ 14.1 h
   $$MR = 2^D/(2^D + 2^Y - 1) = 2^{7.20}/(2^{7.20} + 2^{4.04} - 1)$$
   $$= 0.905$$
   $$M = M_e + .905(M_o - M_e) = .086 + .905(.196) = .263$$
   $$m = .263/(1 + .263) = .208$$

4. Time for Top of Bin to Reach 15% w.b.
   $$MR = .459, \quad D = 7.20$$
   $$MR = 2^D/(2^D + 2^Y - 1) = .459$$
   $$MR(2^D + 2^Y - 1) = 2^D$$
   $$MR\, 2^Y = 2^D - MR\, 2^D + MR$$
   $$2^Y = (1 - MR)2^D/MR + 1$$
   $$Y \ln 2 = \ln[(1 - MR)2^D/MR + 1]$$
   $$Y = (1/\ln 2) \ln[(1 - MR)2^D/MR + 1]$$
   $$= (1/.693) \ln[(1 - .459)2^{7.20}/(.459) + 1] = 7.45$$
   $$\tau = YH = 7.45(3.5\ h) = 26.1\ h$$

5. Moisture Content @ bottom of bin @ 26.1 h
   $$D = 0, \quad Y = 7.45$$
   $$MR = 2^0/(2^0 + 2^{7.45} - 1) = 1/2^{7.45} = 1/174.26 = 0.006$$
   $$M = M_e + 0.006(M_o - M_e) = .086 + 0.006(.196) = 0.087$$
   $$m = 0.087/(1 + 0.087) = 0.080$$

Drying

6. Efficiency @ 26.1 h

$$\begin{aligned} E &= 1 - 2^Y/(2^D + 2^Y - 1) \\ &= 1 - 2^{7.45}/(2^{7.20} + 2^{7.45} - 1) \\ &= 1 - 174.26/(147.0 + 174.26 - 1) = 1 - .544 = .456 \\ &= 45.6\% \end{aligned}$$

7. Average Drying Rate @ 26.1 h

$$\begin{aligned} \dot{MR} &= (1/D)[2^Y/(2^D + 2^Y - 1) - 1] \\ &= (1/7.20)[-E] \\ &= (1/7.20)(-.456) = -.456/7.20 = -.63333 \\ d\overline{MR}/dY &= -.06333 \\ d[(\overline{M} - M_e)/(M_o - M_e)]/dY &= -.06333 \\ d\overline{M}/dY &= (M_o - M_e)(-.06333) = .196(-.06333) \\ &= -.01241 \\ d\tau &= HdY \\ dY &= (1/H)d\tau \\ d\overline{M}/dY &= d\overline{M}/(1/H)d\tau \\ d\overline{M}/d\tau &= (1/H)(-.01241) = -.01241/3.5 \text{ h} \\ &= -.00355 \text{ h}^{-1} \end{aligned}$$

8. Mean drying efficiency until $\tau = 26.1$ h

$$\begin{aligned} \overline{E} &= 1 - (1/(\ln 2Y)) \ln[(2^D + 2^Y - 1)/2^D] \\ &= 1 - (1/(\ln 2Y)) \ln[1/MR] \\ &= 1 - (1/(7.45 \ln 2)) \ln[1/.459] \\ &= 1 - [1/((7.45)(.693))] \ln(2.18) \\ &= 1 - .779/((7.45)(.693)) = 1 - .151 = .849 = 84.9\% \end{aligned}$$

10.6.4 *Numerical Deep-Bed Calculations*

Numerical solutions of equations describing bulk drying may be made by considering the deep-bed of material to be composed of a number of layers of finite thickness which may be assumed to dry according to thin-layer drying equations during a small but finite increment of time. A procedure for describing bulk drying in this manner will be described here for the case where the exponential drying equation is assumed to describe the thin-layer drying. Let us consider a layer of material of depth dx and area, A, for which the temperature, relative humidity, and mass flow rate of the air passing through it are known (fig. 10.16).

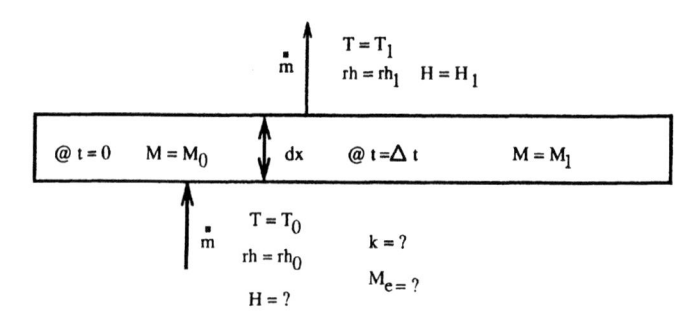

Figure 10.16. Differential depth of material being dried in bulk drier.

10.6.4.1 Humidity Ratio of Entering Air

The humidity ratio of the air entering the drying layer of material may be obtained from the psychrometric chart or it may be calculated using the equation of Chapter 9:

$$H_o = rh_o p_{sato}/[1.605(p_{atm} - rh_o p_{sato})] \qquad (10.82)$$

where

H_o = humidity ratio of the entering air-vapor mixture (kg_{H_2O}/kg_{dm})
rh_o = relative humidity of entering air-vapor mixture (decimal)
p_{sato} = saturation vapor pressure for the temperature of the entering air-vapor mixture (Pa)
p_{atm} = total pressure of the entering air-vapor mixture (Pa)

10.6.4.2 Equilibrium Moisture Content

The equilibrium moisture content of the material being dried must be determined for the conditions of the drying air stream. This may be obtained from experimental data or from an equation for equilibrium moisture content as a function of temperature and relative humidity. If Henderson's equation is used then:

$$M_e = [-\ln(1 - rh_o)/(cT_o)]^{(1/n)} \qquad (10.83)$$

where

M_e = equilibrium moisture content (dry basis percent)
T_o = temperature of entering air-vapor mixture (K)
c and n = constants for Henderson's equilibrium moisture content equation for the material being dried

10.6.4.3 Thin-Layer Drying Parameter

The drying parameter, k, for the exponential thin-layer drying equation is a function of the drying air temperature and the velocity of the air. Past experiments have shown k to vary for many materials according to the Arrhenius equation:

$$k_o = A \exp(-E'/RT_o) \qquad (10.84)$$

Drying

where

k_o = drying parameter at temperature T_o (h^{-1})
A = pre-exponential factor (h^{-1})
E' = activation energy (J/kg-mole K)
T_o = absolute temperature of entering air (K)

Using this equation, the value of k may be determined at any temperature if it is known at two temperatures for calculating A and E'. Barre et al. (1971) suggested a method for approximating k at any temperature when it is known at only one temperature. The suggested relationship is:

$$k_o = k_r (p_{sato}/p_{satr}) \tag{10.85}$$

where

k_o = drying parameter at temperature T_o (h^{-1})
k_r = drying parameter at a reference temperature T_r (h^{-1})
p_{sato} = saturation water vapor pressure at temperature T_o (Pa)
p_{satr} = saturation water vapor pressure at the reference temperature (Pa)

Young and Dickens (1975) suggested the following equation for describing the variation in k with changes in both drying air temperature and air velocity:

$$k_o = k_r (p_{sato}/p_{satr})^{ZM} (V_o/V_r)^{ZL} \tag{10.86}$$

where

k_o = drying parameter at temperature T_o and velocity V_o (h^{-1})
k_r = drying parameter at reference temperature T_r and reference velocity V_r (h^{-1})
p_{sato} = saturation vapor pressure at temperature T_o (Pa)
p_{satr} = saturation vapor pressure at reference temperature T_r (Pa)
V_o = superficial air velocity (m/s or m^3/s m^2)
V_r = reference superficial air velocity (m/s or m^3/s m^2)
ZM = 0.46 and ZL = 0.7 (constants)

The superficial air velocity can be expressed in terms of the depth of the drying bed and the volumetric air flow rate as follows:

$$V = xQ \tag{10.87}$$

where

V = superficial air velocity (m/s)
x = depth of drying bed (m)
Q = volumetric air flow rate per unit volume of grain (m^3/m^{3}s)

Then, equation (10.86) may be written as:

$$k_o = k_r (p_{sato}/p_{satr})^{ZM} (x_o Q_o / x_r Q_r)^{ZL} \tag{10.88}$$

where

x_o = depth of drying bed (m)
x_r = reference depth of drying bed (m)
Q_o = volumetric air flow rate (m³/m³ s)
Q_r = reference volumetric air flow rate (m³/m³ s)

In order to use equation 10.88, the drying parameter must be experimentally determined at some reference temperature, air flow rate, and depth. It can then be estimated at other temperatures, air flow rates, and depths if the values of ZM and ZL are known. Young and Dickens (1975) suggested the values: $ZM = 0.46$ and $ZL = 0.7$. Since the saturation vapor pressure varies in an exponential fashion with temperature, this expression would suggest that k also varies in an exponential fashion with temperature.

10.6.4.4 Moisture Content of Layer at End of Time Increment

The moisture content of the layer of material at the end of the finite time increment may be determined by solving equation 10.25 to obtain:

$$M_1 = M_e + (M_o - M_e)\exp(-k_o \Delta \tau) \tag{10.89}$$

where

M_1 = dry basis moisture content of material in layer at end of time increment (decimal)
M_e = equilibrium moisture content, dry basis (decimal)
M_o = dry basis moisture content of material at beginning of time increment (decimal)
k_o = drying parameter (h^{-1})
$\Delta \tau$ = time increment (h)

10.6.4.5 Average Rate of Change in Moisture Content

The average rate of change in moisture content of the material during the time increment is given by:

$$dM/d\tau = (M_1 - M_o)/\Delta \tau \tag{10.90}$$

10.6.4.6 Humidity Ratio of Air Leaving Layer

The moisture given up by the material being dried must equal the increase in moisture in the drying air while passing through the layer. Thus, by a balance on the mass of water:

$$\dot{m}\Delta H = -(dM/d\tau)\,dx\,\rho A \tag{10.91}$$

or

$$\Delta H = -(dM/d\tau)\rho A\,dx/\dot{m} \tag{10.92}$$

where

$\dot{m}$ = mass flow rate of air (kg/s)

Drying

ΔH = change in humidity ratio of air-vapor mixture in passing through layer (kg$_{H_2O}$/kg$_{dm}$)
$dM/d\tau$ = average rate of change in moisture content of layer during time increment (kg$_{H_2O}$/kg$_{dm}$ s)
ρ = dry matter density (kg$_{dm}$/m^3)
A = cross-sectional area of drying bin (m^2)
dx = depth increment (m)

The humidity ratio of the air leaving the layer is then given by:

$$H_1 = H_o + \Delta H \tag{10.93}$$

10.6.4.7 Temperature of Air Leaving Layer

The energy required for vaporization of the water from the drying material must be supplied by a decrease in the temperature of the air-vapor mixture passing through the layer. An enthalpy balance yields:

$$\dot{m}(c_{pa} + c_{pv} H_o)\Delta T = -\dot{m}\Delta H h_{fg} \tag{10.94}$$

or

$$\Delta T = -\Delta H h_{fg}/(c_{pa} + c_{pv} H_o) \tag{10.95}$$

where

ΔT = change in temperature of air-vapor mixture in passing through layer (K)
h_{fg} = heat of vaporization of water (J/kg$_{H_2O}$)
c_{pa} = specific heat of dry air (J/kg K)
c_{pv} = specific heat of water vapor (J/kg K)
H_o = humidity ratio of entering air-vapor mixture (kg$_{H_2O}$/kg$_{da}$)
ΔH = change in humidity ratio of air-vapor mixture in passing through layer (kg$_{H_2O}$/kg$_{da}$)

Then, the temperature of the exiting air is:

$$T_1 = T_o + \Delta T \tag{10.96}$$

10.6.4.8 Relative Humidity of Air Leaving Layer

In order to determine the relative humidity of the air-vapor mixture leaving the layer, the partial pressure of water vapor must be determined. This may be calculated from the humidity ratio as follows:

$$p_{v1} = 1.605 H_1 p_{atm}/(1 + 1.605 H_1) \tag{10.97}$$

where

p_{v1} = partial pressure of water vapor in air-vapor mixture leaving the layer (Pa)
H_1 = humidity ratio of air-vapor mixture leaving layer (kg$_{H_2O}$/kg$_{da}$)
p_{atm} = total pressure of air-vapor mixture (Pa)

The relative humidity is then calculated by:

$$rh_1 = p_{v1}/p_{sat1} \tag{10.98}$$

where p_{sat1} = saturation pressure of water at the temperature of the exiting air-vapor mixture (Pa).

10.6.4.9 Calculations for Successive Layers During Same Time Increment

After completing the calculations discussed in sections 10.6.4.1 through 10.6.4.8, all the conditions of the air-vapor mixture leaving the first finite depth increment are known. These are the conditions of the air entering the second depth increment. By repeating the previous calculations for the second layer, the conditions of the air entering the third layer may be determined. Thus, the process may be repeated until the drying of all layers during the time increment has been determined.

10.6.4.10 Iterative Calculations for Successive Time Steps

After completion of calculations discussed in sections 10.6.4.1 through 10.6.4.8 for all depth increments as discussed in section 10.6.4.9, the process may be repeated for the next time increment with the final moisture contents calculated for each layer in the previous time increment being used as the initial moisture contents for the next time increment. The process may thus be continued until the entire drying period has been simulated by the numerical process. The simulation technique described here may easily be programmed for solution on a digital computer.

10.7 Drying Procedures

Agricultural materials are dried using a variety of procedures due to differences in the material characteristics. Characteristics which affect the procedure to be used include:

1. *Temperature Tolerance.* High temperatures may reduce germination, partially cook the product, or change its chemical or physical characteristics.
2. *Humidity Response.* Materials that undergo physiological or other change during drying, e.g., tobacco, lumber, prunes, may have to be dried with air of a specific relative humidity.
3. *Compression Strength.* Materials that crush or deform under pressure, e.g., fruit and vegetables, must be dried in thin layers; ear corn can be dried in deep beds, tobacco must be suspended.
4. *Fluidity.* Loose hay, ear corn, and other poor-flowing materials cannot be dried in a continuous-flow drier. The angle of repose (section 2.21) affects drier type and design.

The procedure and type of equipment recommended for drying a specific material will depend upon the factors listed above, the quantity to be dried, the drying rate required, weather conditions, and various economic factors.

10.7.1 Batch or Bin Driers

The material to be dried is placed in a bin or container, and air is forced through the mass until dry. Arrangements such as shown in figure 10.17 are frequently used.

The systems are simple, moderately inexpensive, and serve as storage units after drying is completed. Materials to be dried by these systems must have sufficient compressive resistance to resist crushing under load and to maintain the normal void space so that

Drying   311

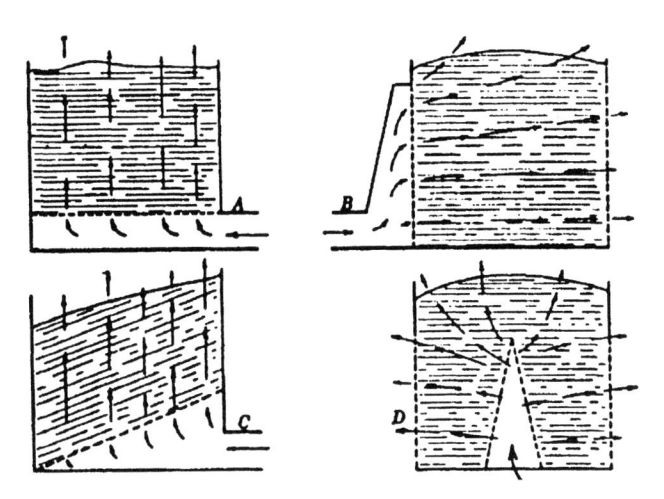

Figure 10.17. Some deep-bed drying arrangements.

proper air rates can be maintained. Resistance to air flow limits the depth for highly resistant materials since adequate air rates are possible only with excessively large power units.

The mass dries progressively in the direction of air flow. The part of the mass in the air discharge region is subject to high relative humidities and moderate temperatures and may spoil from mold before the moving drying front has reached it. Adequate drying of the mass in the air discharge region is accompanied by overdrying of the mass in the air entrance region. This undesirable feature can be minimized by (1) drying with the lowest practicable air temperature, (2) using the highest practicable air rate, and (3) transferring the material to another bin when the average moisture content for the mass is that which is desired; the mass must be mixed uniformly when moved.

The performance of batch or bin driers can be estimated by the deep-bed simulation procedures of sections 10.5 and 10.6. In general, the effect of increasing the air flow rate through a batch of material is to (a) increase the rate of drying of the batch, (b) increase the uniformity of drying throughout the batch, and (c) to reduce the thermodynamic efficiency of the drying process. The effect of increasing the drying air temperature in a batch drying system is to (a) increase the rate of drying of the batch, (b) decrease the uniformity of drying throughout the batch, and (c) to increase the thermodynamic efficiency of the process.

10.7.2 Continuous Gravity-Flow Driers

Granular materials that flow readily, permit air to flow through them, and are not damaged in handling, can be dried in a gravity-flow drier such as the cross-flow dryer of figure 10.18. The wet material is placed in the hopper and flows by gravity between the perforated retaining walls and is discharged at the bottom by a continuously operating metering valve. Heated air is forced across the column at right angles to the direction of grain motion. The column may be inclined to simplify construction or fitted with baffles to stir the mass as it progresses through the drier.

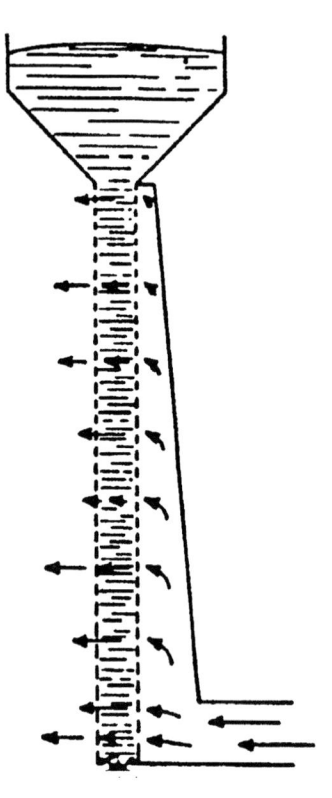

Figure 10.18. A cross-flow drier.

The capacity of a cross-flow drier is directly proportional to the column width and material movement rate through the column. The retention time in the column is the drying time for the material as defined by its drying indices, the required moisture reduction, and state factors of the drying air. The retention time required may be estimated using the deep-bed drying relationships of sections 10.5 and 10.6. Since the retention time is fixed for each individual situation, drier capacity is proportional to height.

Other types of continuous-flow driers include counterflow and concurrent-flow driers. A counterflow drier is similar to a counterflow heat exchanger in that the air and the drying material flow in opposite directions through the dryer. A concurrent-flow drier is similar to a parallel-flow heat exchanger in that the air and the drying material flow in the same direction.

10.7.3 Rotary Driers

Materials that are not free flowing and that are not damaged by continuous handling are usually dried in rotary driers. Chopped forage, fruit, and vegetable residues to be dried for livestock feed, and fertilizer components are examples of materials dried in this manner.

The rotary drier has a high initial cost and requires more floor space per unit of capacity than either the batch or cross-flow drier. Consequently, it should not be used if the batch or continuous gravity-flow type is suitable.

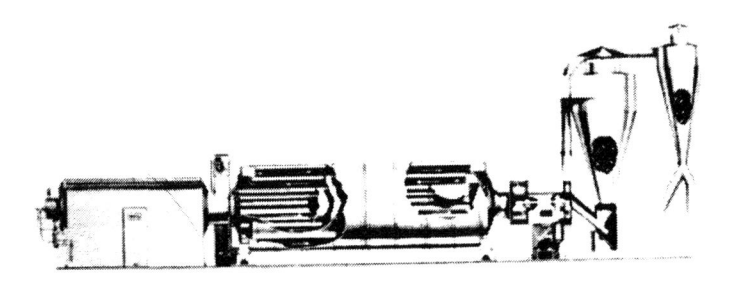

Figure 10.19. A multiple-drum drier with a direct-fired heater and cyclones for collecting and cooling the dried product. (Courtesy of The Heil Co.)

Agricultural rotary driers such as shown in figure 10.19 are direct fired and single-, double-, and triple-drum types. The multiple-drum types permit shorter overall lengths and lower heat losses by conduction and radiation.

The inside of the drum may be fitted with flights that lift the material and shower it down through the heated air. Flight design varies with the material to be dried. Chains or other dividing devices may be fitted to the inside of the drum to divide materials that tend to clump as they pass through the drier. The rate of material movement through the drum is controlled by flight design or by inclining the drum. The drum should rotate at such a speed that the material is spilled uniformly through the cross-sectional space of the drum. This procedure yields a product of uniform final moisture content.

The capacity of a rotary drier depends upon the required reduction in moisture content, the drying indices for the material, the rate of air flow, and size of the drum.

Wet materials often dried in rotary driers (e.g., green chopped alfalfa, particularly if covered with dew, and fruit and vegetable residues) go through an initial constant-rate drying period where the material approaches (or reaches) the wet-bulb temperature. At the end of this period, falling-rate drying begins and the material temperature becomes progressively hotter than the wet-bulb temperature. Finely divided wet materials dry at a fast rate, and high air temperatures may be used. Temperatures as high as 800 C may be used for chopped green alfalfa, for example.

The state of the exhaust air necessary for a rotary drier may be determined from the following moisture balance:

$$(Q/v)(H_e - H_i) = \dot{m}_{dm}(M_i - M_f) \qquad (10.99)$$

where

Q = volumetric flow rate of air through drier (m^3/s)
v = specific volume of entering air (m^3/kg)
H_e = humidity ratio of air at exhaust (kg/kg)
H_i = humidity ratio of entering air (kg/kg)

$\dot{m}_{dm}$ = flow rate of dry matter (kg/s)
M_i = entering moisture content of material, dry basis (decimal)
M_f = exiting moisture content of material, dry basis (decimal)

The volume of the drum and thus the retention time necessary to achieve a specified amount of drying will be determined by the drying characteristics of the material being dried.

10.7.4 Tray Driers

Materials that cannot be dried by any of the previously discussed methods may be dried on trays. Fruits and vegetables are examples. The material is placed in shallow trays which are stacked on cars as shown in figure 10.20. The trays are spaced to permit

Figure 10.20. Cars of trays being moved into a tunnel drier. Note the tracks used to guide the cars. (Courtesy of California Prune and Apricot Growers Association.)

Drying

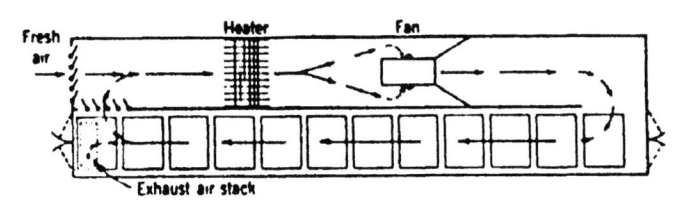

Figure 10.21. Plan of a tunnel drier. The cars move from left to right for counterflow operation, from right to left for parallel flow.

air circulation between them. The car of trays is dried in a cabinet or in a tunnel. Cabinet drying is a batch process and is used for low-rate installations. Larger capacities are provided by tunnels, figure 10.21. The cars are moved through the tunnel by a slowly moving drag chain, a ratchet ram, or manually.

Parallel air flow gives a fast initial drying rate. Counterflow gives faster drying at the dry end of the tunnel. Parallel flow is seldom used because of its poor drying ability at the dry end of the tunnel. Combination tunnels utilize the advantages of both parallel flow and counterflow but the initial cost is greater and control is more difficult. Counterflow tunnels are most extensively used. The air rate must be high enough so that the relative humidity of the discharge air is below the equilibrium relative humidity of the material at the point where the material discharges. The moisture balance of equation 10.99 for rotary driers must hold for tunnel driers as well.

10.7.5 Spray Driers

Spray driers remove the water from solutions or suspensions and dry the resulting powder to a moisture content that approaches equilibrium with the exhaust drying air. Spray driers are used extensively in the food, chemical, and pharmaceutical industries.

Design varies from a rectangular chamber fitted with spray jets, through which the drying air passes to continuous large-volume systems such as figure 10.22. Three procedures are used for breaking the material into fine drops.

1. *High-Pressure Atomization.* The liquid is forced through a nozzle under high pressure. Mixing with the drying air and the spray pattern can be controlled. Drop size and gradation are difficult to predict. Nozzle life is short when abrasive materials are sprayed.
2. *Centrifugal.* The liquid is fed at low pressure onto a horizontal disc or cup turning at speeds up to 20,000 rpm or more. The material breaks up into small drops as it leaves the edge of the rotor. The drops are of more uniform size, and materials not suitable for nozzles can be dried. Air-liquid drop mixing may be poor since the drops follow an umbrella-shaped trajectory. Disc-rotor and bearing maintenance are high.
3. *Two-Fluid Atomizing.* Air or steam under pressure breaks the liquid into fine drops by a mechanism comparable to that of paint sprayers. Operating costs are high. This system is used only for the most difficult atomizing jobs and experimental units.

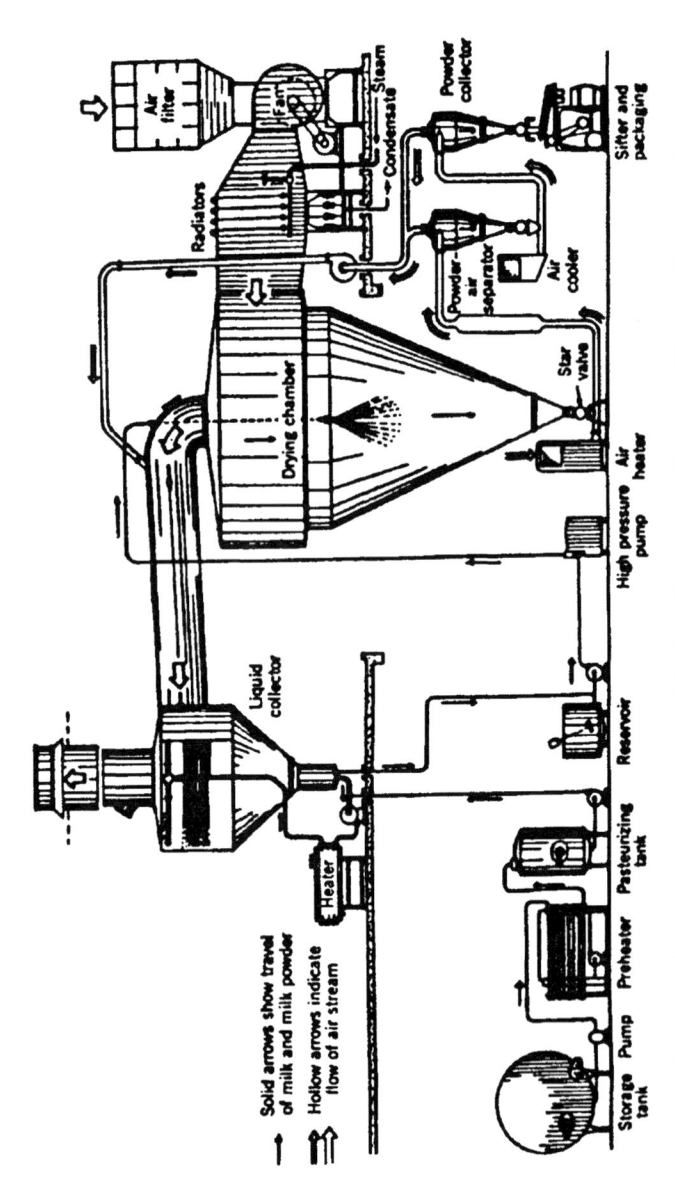

Figure 10.22. A continuous-flow spray drier for milk products. (Courtesy of Swenson Evaporator Co.)

Drying

10.7.6 Concentrators

Concentrators, also called evaporators, are used to concentrate milk, fruit and vegetable juices, jams and jellies, etc., by boiling off a portion of the water. Because of the conditions of operation, the concentrator is frequently called a vacuum pan. A "pan" shown in figure 10.23 is operated under a partial vacuum because (1) low-boiling temperatures do not damage heat-sensitive materials and (2) a large temperature difference is maintained between the steam and boiling liquid which permits a high heat-transfer rate.

The water spray condenses the water vapor removed from the concentrating liquid. The boiling temperature of the liquid being concentrated is controlled by the temperature of the water-condensate mixture leaving the condenser. The water-condensate mixture is removed by a pump or barometric leg. The vacuum is usually maintained by a steam

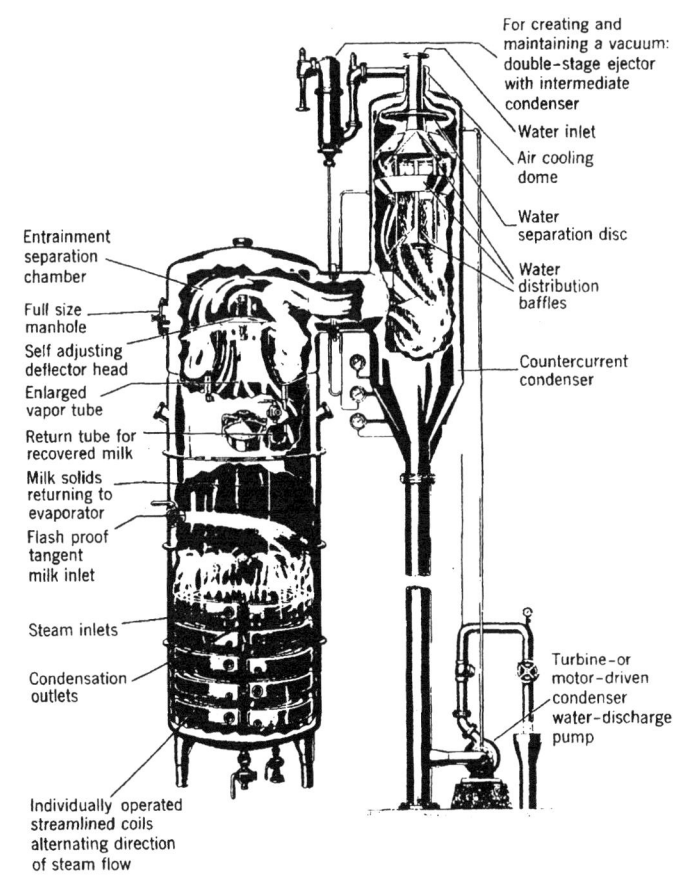

Figure 10.23. Vacuum pan with entrainment separation chamber and countercurrent condenser used for milk concentration. (Courtesy of Arthur Harris and Co.)

ejector, although a vacuum pump may be used. Since the condenser handles the vapor from the concentrating liquid, the ejector needs to handle only the non-condensable gases and air from leaking gaskets.

Consider an evaporator set up to concentrate milk on a constant flow basis. The following equations describe the energy balances in the system. In the evaporating chamber:

$$\dot{m}_s h_g + \dot{m}_m c_m t_i = \dot{m}_s h_f + \dot{m}_v h_v + (\dot{m}_m - \dot{m}_v) c_m t_v + \text{losses} \qquad (10.100)$$

where

$\dot{m}_s$ = mass flow rate of steam (kg/s)
$\dot{m}_m$ = mass flow rate of milk (kg/s)
$\dot{m}_v$ = vaporization rate of water from milk = condensation rate of water in condensing chamber (kg/s)
h_g = enthalpy of entering steam (kJ/kg)
h_f = enthalpy of steam condensate (kJ/kg)
h_v = enthalpy of evaporating water at temperature t_v (kJ/kg)
c_m = specific heat of entering milk (kJ/kgK)
t_i = entering temperature of milk (K)
t_v = temperature of vaporization of milk (K)

Equation 10.100 can be rearranged to obtain:

$$\dot{m}_s(h_g - h_f) + \dot{m}_m c_m(t_i - t_v) = \dot{m}_v h_{fg} + \text{losses} \qquad (10.101)$$

where h_{fg} = heat of vaporization of water at t_v (kJ/kg).

The following equation gives an energy balance for the steam coils in the evaporating chamber:

$$AU(t_s - t_v) = \dot{m}_s(h_g - h_f) \qquad (10.102)$$

where

A = effective steam coil area (m^2)
U = overall heat transfer coefficient of steam coils (kW/m^{2}K)
t_s = effective average temperature of steam (K)

An energy balance in the condensing chamber results in:

$$\dot{m}_v h_v + \dot{m}_w h_{f1} = (\dot{m}_w + \dot{m}_v) h_{f2} \qquad (10.103)$$

where

$\dot{m}_w$ = mass flow rate of condensing water (kg/s)
h_{f1} = enthalpy of entering condensing water (kJ/kg)
h_{f2} = enthalpy of condensed water and exiting condensing water (kJ/kg)

or

$$\dot{m}_v(h_v - h_{f2}) = \dot{m}_w(h_{f2} - h_{f1}) = \dot{m}_w c_w(t_2 - t_1) \qquad (10.104)$$

where

c_w = specific heat of liquid water (kJ/kgK)

Drying

t_1 = entrance temperature of condensing water (K)
t_2 = exit temperature of condensing water (K)

When the condenser is counterflow, as in figure 10.22, $t_v - t_2$ is usually 2 to 3 C. But if the unit is designed for parallel flow performance, the difference will probably be approximately 8 C.

When operating factor values are considered for a specific "pan", the boiling temperature, t_v, is controlled by the condensing water rate, $\dot{m}_w$.

The steam coils are usually fed with steam at 35 to 70 kPa. The boiling is extremely vigorous because of the high temperature difference. Thus, a high heat rate is possible with minimum heat exchange surface and a "cooked" flavor is improbable owing to the surface speeds of the liquid.

This unit can operate on a batch basis or continuously by means of a suitable pump that continuously removes liquid from the bottom of the "pan" and by continuous feeding in of the liquid stock.

Nomenclature

a	radius of sphere (m)
A	surface area (m^2)
c_p	specific heat of air (J/kg K)
C	concentration (kg/m^3)
D	mass diffusivity (m^2/s)
	number of depth units (dimensionless)
D_p	diameter of particle (m)
D_{liq}	diffusivity of liquid water (m^2/s)
D_v	diffusivity of water vapor (m^2/s)
E	$\exp[-(q_1 - q_2)/kT]$
f	convective heat transfer coefficient (W/m^{2}K)
f_v	water vapor transfer coefficient (kg/s m^2)
g	gravitational acceleration (m/s^2)
G	mass velocity (kg/s m^2)
h	height of column of water in capillary (m)
h_{fg}	latent heat of vaporization of water (J/kg)
H	time of half response (h)
J	flux of diffusing medium (kg/s m^2)
k	Boltzmann's constant, 1.38×10^{-23} J/molecule K
	thermal conductivity (W/mK)
	drying parameter (h^{-1})
L	half thickness of sheet (m)
Le_m	modified Lewis Number (dimensionless)
m	moisture content (wet-basis decimal)
$\dot{m}$	mass flow rate (kg/s)
M	moisture content, dry-basis (decimal)
M_o	initial moisture content (dry basis)

M_d	equilibrium moisture content for a desorption process (dry basis)
M_e	equilibrium moisture content (dry basis)
M_s	equilibrium moisture content for a sorption process (dry basis)
p	vapor pressure of gas (Pa)
p_{atm}	total atmospheric pressure (Pa)
p_{sat}	saturation vapor pressure (Pa)
p_v	water vapor pressure in air (Pa)
Pr	Prandtl Number (dimensionless)
R	universal gas constant (8.314×10^3 Pa m^3/kg-mole K)
Re	Reynolds Number (dimensionless)
q_1	heat of adsorption of bound layer of water molecules (J/molecule)
q_2	normal heat of vaporization of water molecules (J/molecule)
rh	relative humidity (decimal)
rh_{max}	maximum relative humidity with which the material has been in equilibrium prior to desorption process (decimal)
Sm	Schmidt Number (dimensionless)
t	temperature (C)
T	absolute temperature (K)
v	volume of gas adsorbed (m^3)
v_b	volume of bound water adsorbed (m^3)
v_m	volume of a unimolecular layer of the adsorbate (m^3)
V	amount of absorbed moisture at saturation (m^3)
W_m	mass of moisture (kg)
W_d	mass of dry matter (kg)
Y	number of time units (dimensionless)
α	total amount of normally condensed moisture measured in molecular layers
ε	emissivity of water surface (dimensionless)
ϕ	fraction of surface covered by at least one layer of normally condensed moisture (decimal)
γ_d	amount of absorbed moisture in a desorption process (kg)
γ_s	amount of absorbed moisture in a sorption process (kg)
μ	viscosity of air (kg/s m)
υ	porosity (dimensionless)
ρ	density (kg/m^3)
ρ_s	density of solid (kg/m^3)
σ	Stefan-Boltzmann constant (5.73×10^{-8} W/m^{2}K^4)
τ	time (s)
θ	fraction of surface covered by a layer of bound water molecules (decimal)

References

ASAE Standards. 43rd Ed. 1996. Moisture Measurement—Unground Grain and Seeds. S352.2. St. Joseph, Mich.: ASAE.

American Society of Agricultural Engineers. 1991. *Instrumentation and Measurement for Environmental Sciences*, 2nd Ed., Bailey W. Mitchell, St. Joseph, Mich.: ASAE.

Assoc. of Official Analytical Chemists. 1975. *Official Methods of Analysis*, 12th Ed. Washington, D.C.: AOAC.

Barre, H. J., G. R. Baughman and M.Y. Hamdy. 1971. Application of the logarithmic model to cross-flow deep-bed grain drying. *Transactions of the ASAE* 14(6):1061–1064.

Brooker, D. B., F. W. Bakker-Arkema and C. W. Hall. 1974. *Drying Cereal Grains.* Westport, Conn.: AVI Publishing Co.

Brunauer, S. P. H. Emmett and E. Teller. 1938. Adsorption of gases in multimolecular layers. *J. Am. Chem. Soc.* 60:309–319.

Chung, D. S. and H. B. Pfost. 1967. Adsorption and desorption of water vapor by cereal grains and their products. *Transactions of the ASAE* 10:552–575.

Day, D. L., and G. L. Nelson. 1965. Predicting performance of cross-flow systems for drying grain in storage in deep cylindrical bins. *Transactions of the ASAE* 8(2):288–292.

———. 1965. Desorption isotherms for wheat. *Transactions of the ASAE* 8(2):293–297.

Gamson, B. W., G. Thodes and O. A. Hougen. 1943. Heat, mass, and momentum transfer in the flow of gases through granular solids. *Transactions of A.I.Ch.E.* 39:1–35.

Hall, C. W. and J. H. Rodriguez-Arias. 1958. Equilibrium moisture content of shelled corn. *Agricultural Engineering* 39:466–470.

Henderson, S. M. 1952. A basic concept of equilibrium moisture. *Agricultural Engineering* 33:29–33.

Hukill, H. V. 1947. Basic principles in drying corn and grain sorghum. *Agricultural Engineering* 28(8):335–338.

Langmuir, I. 1918. The adsorption of gases on plane surfaces of glass mica and platinum. *J. Am. Chem. Soc.* 40:1361–1403.

Sherwood, T. K. 1936. The air drying of solids. *Transactions Am. Inst. Chem. Engr.* 32:150–158.

Smith, S. E. 1947. The sorption of water vapor by high polymers. *J. Am. Chem. Soc.* 69:646–651.

Thompson, T. L., R. M. Peart, and G. H. Foster. 1968. Mathematical simulation of corn drying—A new model. *Transactions of the ASAE* 11:582–586.

USDA. 1971. Oven methods for determining moisture content of grain and related agricultural commodities. Equipment Manual, Grain Div., Consumer and Marketing Service, USDA.

Van Arsdel, W. B., M. J. Copley and A. J. Morgan Jr. 1973. *Food Dehydration* 2nd Ed. Vols. 1 and 2. Westport, Conn.: AVI Publishing Co.

Young, J. H. 1966. A study of the sorption and desorption equilibrium moisture content isotherms of biological materials. Unpublished Ph.D. thesis. Oklahoma State University, Stillwater.

———. 1991. Moisture. In *Instrumentation and Measurement for Environmental Sciences,* Ed. Bailey W. Mitchell, Ch.7. St. Joseph, Mich.: ASAE.

Young, J. H. and J. W. Dickens. 1975. Evaluation of costs for drying grain in batch or cross-flow systems. *Transactions of the ASAE* 18(4):734–739.

Young, J. H. and G. L. Nelson. 1967. Theory of sorption isotherm hysteresis in biological materials. *Transactions of the ASAE* 10(2):260–263.

———. 1967. Research of hysteresis between sorption and desorption isotherms of wheat. *Transactions of the ASAE* 10(6):756–761.

Problems

1. Thirty thousand (30 000) kilograms of shelled corn at 22% (w.b.) moisture content are to be dried to 13% (w.b.). Determine:
   a. the dry basis moisture contents.
   b. the amount of dry matter in the corn.
   c. the mass of water to be removed.

2. One thousand (1 000) kilograms of shelled corn is to be dried from a moisture content of 30% (w.b.) to a moisture content of 14% (w.b.). Determine the kilograms of water which must be removed. What will be the mass of the 14% (w.b.) corn? What is the mass of the dry matter?

3. Using Henderson's equation, calculate the equilibrium moisture content of soybeans at 25 C and 80% rh.

4. Peanuts initially at 25% (w.b.) moisture are dried to 10% (w.b.). If the mass of the 10% moisture content product is 1 000 kg, how much water is removed.

5. Peanuts having an initial moisture content of 55% (d.b.) are dried in an environment for which the equilibrium moisture content is 5% (d.b.). If the peanuts reach a moisture content of 30% (d.b.) in 3.5 hours, what is the value of the drying parameter, k, assuming the exponential drying equation to hold? At what time will the peanuts reach a moisture content of 9% (d.b.)? Plot the moisture ratio $(M - M_e)/(M_o - M_e)$ versus time on semi-logarithmic paper.

6. A spherical body has a radius of 2.5 cm and a liquid diffusivity of 1.1×10^{-5} cm^2/s. If the body is initially at a moisture content of 25% (w.b.) and is placed in an environment for which the equilibrium moisture content is 10% (w.b.), what will be the average moisture content after two hours of drying?

7. Repeat problem 6 for the case of an infinite sheet having a thickness of 5 cm.

8. A steel bin 5 m in diameter and 2.5 m deep contains grain sorghum which must be dried from 18 to 14% (w.b.) moisture content. The floor is perforated and air moves vertically through the mass at 1.6 m^3/m^3 min. Unheated air at 24 C dry-bulb temperature and 13 C dew-point temperature is used for drying. If the average discharge air temperature is 19 C, how many hours will be required to dry the grain? Assume the dry matter density of grain sorghum to be 617 kg/m^3.

9. Peanuts initially at 70% (d.b.) moisture content are dried with air at 35 C dry-bulb temperature and 23 C wet-bulb temperature. The air flow rate is 1 190 kg/h m^2. The

Drying

equilibrium moisture content of the peanuts under these conditions is 6.4% (d.b.). Assume the equilibrium relative humidity for 70% (d.b.) peanuts to be 100%. The depth of the bin is 1.3 m. Other parameters are:

$A = 9.3$ m^2 = cross-sectional area of bin
$h_{fg} = 2720$ kJ/kg = heat of vaporization
$c_{pa} = 1.0$ kJ/kg K = specific heat of air
$\rho = 216$ kg/m^3 = density of dry matter
$H = 2.9$ hours = time of half response

Use Hukill's analysis to determine:
a. Time required for top layer of peanuts to reach 10% (w.b.) moisture content.
b. Average moisture content of peanuts when top layer reaches 10% (w.b.).
c. Drying rate of top layer when it reaches 10% (w.b.).
d. Efficiency when top layer reaches 10% (w.b.).
e. Average efficiency until top reaches 10% (w.b.).
f. Moisture content at bottom of bin when top reaches 10% (w.b.).

10. Consider the example problem 10.3. If the drying is to be described by dividing the bin into 10 layers of 0.18 m each and using a time increment of 20 minutes with a numerical calculation procedure, determine the following:
    a. humidity ratio of entering air.
    b. equilibrium moisture content for first layer.
    c. thin-layer drying parameter for first layer.
    d. moisture content of first layer after 20 minutes of drying.
    e. average rate of change of moisture content in first layer during first 20 minutes of drying.
    f. average humidity ratio of air leaving first layer during first time increment (20 min.).
    g. average temperature of air leaving first layer during first time increment.
    h. average relative humidity of air leaving first layer during first time increment.

11. Eggs are to be spray dried to 5% (w.b.) moisture content. Air initially at 20 C dry-bulb temperature and 15 C wet-bulb temperature is heated to 120 C.
    a. What is the lowest possible discharge air temperature?
    b. How much energy must be supplied to the air per kilogram of moisture removed if the air discharges at 55 C?

12. A rotary drier uses 20 C, 35% rh air. It is heated to 120 C and discharges at 50 C. What percent of the fuel might be saved if the latent and sensible heat in the discharge air were used to assist in heating the incoming air?

13. A tunnel dehydrator (fig. 10.20) operates with wet-bulb temperature controlled at 45 C. Outside air at 20 C and 15 C wet-bulb is mixed with some of the recirculated air. The mixture is heated to 70 C and the air is discharged at 60 C. What percent of the air must be recirculated?

14. Wet residue from a fruit canning operation is dried as noted in the sketch below. The material is reduced in moisture content from 91 to 23% (w.b.). The air surrounding

the drier is 32 C and 20% rh. The air leaves the drier at 41 C. Determine:
a. the volumetric flow rate of the 118 C air.
b. the relative humidity of the 41 C air.
c. the mass flow rate of fuel required.
d. the change in relative humidity of the 118 C air resulting from the moisture of combustion.
e. the mass of citrus residue (at 23% w.b.) dried per hour.

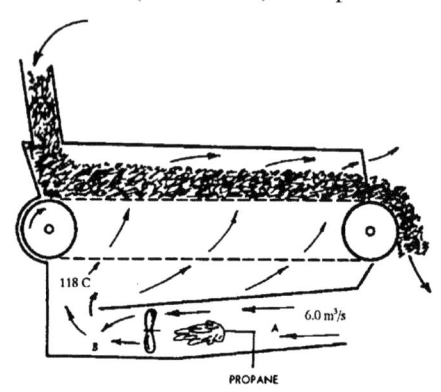

15. The drier below operates in the constant rate period and is designed to permit control of the relative humidity of the air entering the drying drum. Determine:
    a. the incoming volumetric air flow rate.
    b. the moisture removal rate.
    c. the heat input rate.
    d. the percent of the 35 C air that is recirculated.
    e. changes in heat requirements that would result if the heat source were moved to (x).
    f. changes in operating conditions if the drier relative humidity were reduced to 15%; temperatures are not to be changed. Assume that the mass flow rate through the fan does not change.

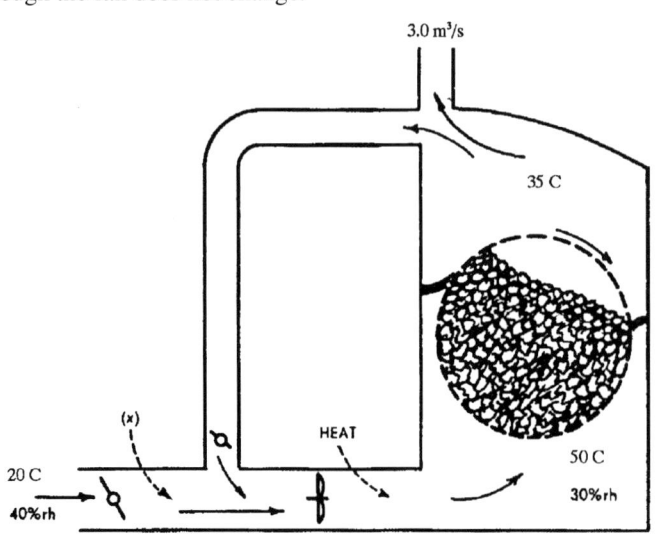

11 Refrigeration

Refrigeration may be defined as the process of removing heat from a body having a temperature below the temperature of its surroundings. Or, refrigeration may be defined as the process of transferring heat energy from a lower to a higher temperature. Natural refrigeration is that produced by the use of natural ice. Mechanical refrigeration is that accomplished by means of refrigerating engines which operate on thermodynamic principles.

11.1 Natural Refrigeration

Ice is satisfactory as a refrigeration medium for temperatures down to approximately 5 C under such conditions as (1) short annual refrigeration periods, (2) duty away from power sources, and (3) where the ice cost is nominal. Temperatures below 0 C may be produced by mixing finely divided ice and various chemicals such as those given in table 11-1.

The heat of fusion of ice is 334.944 kJ/kg at 0 C, and its specific heat is 2.0934 kJ/kgK. The cooling rate is dependent upon the temperature difference and method of air or water circulation over it. Where ice is used for cooling recirculating water, it is difficult to secure water temperatures below 4 C if the ice is floating in a tank, because the maximum density of water is reached at 4 C and circulation within the tank is poor. A shower of return water over unsubmerged blocks of ice on a rack is much more effective than a tank.

11.2 Mechanical Refrigeration

Refrigeration processes using mechanical devices and electrical or other energy are called mechanical refrigeration systems. Two broad classifications are (1) absorption and (2) vapor compression systems. The vapor compression systems are the more common and will be discussed in this chapter.

The operation of the vapor compression system is shown schematically in figure 11.1. The liquid refrigerant in the receiver or supply tank is under high pressure. Because of this pressure the liquid is forced through the liquid line to and through the expansion valve into a region of low pressure produced by the compressor. The liquid refrigerant evaporates or boils to a vapor in the evaporator. The heat required for evaporation comes from the surroundings, and cooling results. The vapor moves at low pressure through the vapor line to the compressor, is compressed to a high pressure, and passes to the

Table 11-1. Temperature of ice-freezing mixtures

Chemical	Per Cent of Chemical in Mixture, by Weight	Temperature (C)
NaCl	25	−18.7
$CaCl_2$	60	−33.1
HNO_3 (dilute)	50	−35.0
KOH	57	−39.1
HNO_3 (trace H_2SO_4)	50	−40.0

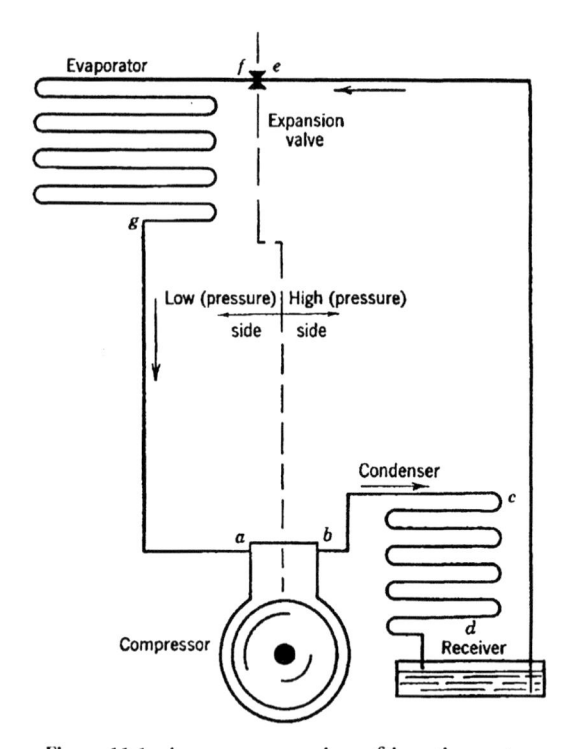

Figure 11.1. A vapor compression refrigeration system.

condenser. Here it returns to the liquid state as the heat of vaporization is transferred to the surroundings. The liquid then flows into the receiver.

The high-pressure side, called high side in the trade, is that to the right of the dotted line. The low-pressure side or low side is to the left of the dotted line.

Thermodynamically, an idealized refrigeration process can be shown by a Mollier (pressure-enthalpy) chart, figure 11.2, and the schematic system of figure 11.1. Low pressure vapor at point a on the saturated vapor curve is compressed isentropically by the compressor to point b. The high pressure vapor is cooled to point c on the saturated vapor curve and then condensed to point d by the removal of the heat of vaporization

Refrigeration

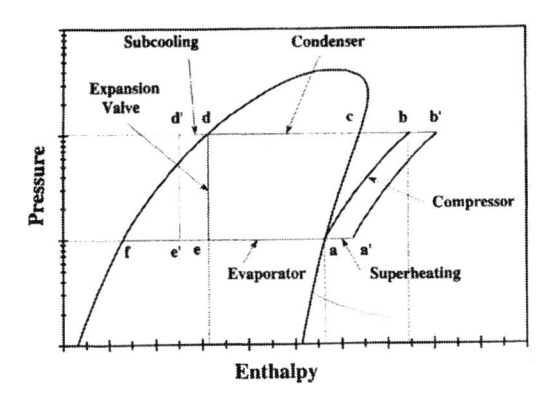

Figure 11.2. The vapor compression mechancial refrigeration process, given schematically, by a Mollier (pressure-enthalpy) chart.

from the refrigerant in the condenser. The saturated liquid at point d then undergoes an irreversible adiabatic process when it passes through the expansion valve. The state changes from d to e, a part of the liquid flashing to a gas. The portion flashing is:

$$x = (h_e - h_f)/(h_a - h_f) \tag{11.1}$$

The wet refrigerant mixture then evaporates to state a due to the absorption of energy in the evaporator.

The useful cooling or refrigeration is:

$$\text{Cooling} = \dot{m}_r(h_a - h_e) \tag{11.2}$$

where

$\dot{m}_r$ = mass flow rate of refrigerant (kg/s)
h_a = enthalpy of saturated vapor at state a (kJ/kg)
h_e = enthalpy of refrigerant entering evaporator (kJ/kg)

The energy which must be supplied by the compressor is:

$$\text{Energy supplied by compressor} = \dot{m}_r(h_b - h_a) \tag{11.3}$$

where h_b = enthalpy of compressed gas at state b (kJ/kg).

The coefficient of performance for cooling is a factor that designates the useful cooling capacity per unit of energy supplied by the compressor. Division of equation 11.2 by equation 11.3 gives:

$$\text{c.o.p.}_{\text{cooling}} = (h_a - h_e)/(h_b - h_a) \tag{11.4}$$

The Carnot or theoretical coefficient of performance for cooling is:

$$\text{c.o.p.}_{\text{theoretical}} = T_C/(T_H - T_C) \tag{11.5}$$

where

T_C = absolute temperature of the cold evaporating liquid (K)
T_H = absolute temperature of the hot condensing refrigerant (K)

The actual c.o.p. is always smaller than the Carnot c.o.p. This results from the mechanical and thermal losses of a mechanical system and the characteristics of the refrigeration cycle. The coefficient of performance will become larger as the difference between the high and low pressures in the system become smaller (or as the difference in temperature of the evaporating liquid and that of the condensing vapor becomes smaller).

A coefficient of performance for heating can also be evaluated which designates the heating capacity at the condenser per unit of energy supplied by the compressor. The heating is:

$$\text{Heating} = \dot{m}_r(h_b - h_d) \tag{11.6}$$

where h_d = enthalpy of saturated liquid at state d (kJ/kg) and the coefficient of performance for heating is found by dividing equation 11.6 by equation 11.3 to obtain:

$$\text{c.o.p.}_{\text{heating}} = (h_b - h_d)/(h_b - h_a) \tag{11.7}$$

The Carnot cycle c.o.p. for heating is:

$$\text{c.o.p.}_{\text{theoretical}} = T_H/(T_H - T_C) \tag{11.8}$$

The c.o.p. for heating may be related to the c.o.p. for cooling as follows (recognizing that $h_d = h_e$):

$$\text{c.o.p.}_{\text{heating}} = (h_b - h_d)/(h_b - h_a) = (h_b - h_a)/(h_b - h_a) + (h_a - h_e)/(h_b - h_a)$$

or

$$\text{c.o.p.}_{\text{heating}} = 1 + \text{c.o.p.}_{\text{cooling}} \tag{11.9}$$

11.3 Rating

The capacity of any refrigerating system is the rate at which it will remove heat from the refrigerated space. Prior to the era of mechanical refrigeration, ice was used as a cooling medium. Thus, it was natural that the cooling capacity of mechanical refrigerators was compared with the melting of ice. This was done by specifying the mass of ice (in tons) which would need to be melted in order to give an equivalent amount of cooling. A ton (2 000 lb_m or 907 kg) of ice absorbs 144 Btu/lb_m × 2 000 lb_m = 288 000 Btu (334.944 kJ/kg × 907 kg = 302 426.8 kJ) in melting. A mechanical system which can absorb that amount of energy (produce refrigeration) in a period of 24 hours is rated as a one "ton" unit. Thus, a one-ton refrigeration system can transfer 12 000 Btu/hr or 3.517 kW. The evaporator temperature should be specified since the capacity decreases as the temperature of the evaporating refrigerant decreases.

Standard operating conditions have been adopted for the comparison of refrigerants, systems, and components. These conditions are:

Refrigerant evaporation temperature, −15 C
Refrigerant condensing temperature, 30 C

A number of refrigerants compared on this basis are listed in table 11-2.

The *boiling point* is a general indication of the temperature at which the refrigerant would be used. The lower the boiling point (saturated temperature at 0 Pa gage), the lower

Refrigeration

Table 11-2. Properties of some refrigerants arranged in order of the boiling point at atmospheric pressure

Refrigerant	Formula	Molecular Mass	Boiling Point (C)	Saturated Pressure (kPa) −15 C	Saturated Pressure (kPa) 30 C	Refrigerating Effect (kJ/kg)
Water	H_2O	18.02	100.00	0.1653	4.246	2299.40
R-113	$CCl_2F-CCl_2F_2$	187.39	47.57	6.936	54.357	127.34
R-123	$CHCl_2CF_3$	152.93	27.87	16.31	109.52	142.30
R-11	CCl_3F	137.38	23.82	20.70	125.96	156.22
R-114	$CClF_2CClF_2$	170.94	3.80	46.54	251.61	99.19
Butane	C_4H_{10}	58.13	−0.50	56.24	283.31	292.01
R-134a	CF_3CH_2F	102.03	−26.16	163.97	770.08	150.71
R-12	CCl_2F_2	120.93	−29.79	182.57	743.79	116.58
Ammonia	NH_3	17.03	−33.30	236.37	1167.10	1102.23
R-500	12/152a Azeotrope	99.31	−33.50	214.26	879.11	140.95
R-22	$CHClF_2$	86.48	−40.76	296.35	1192.40	162.46
Propane	C_3H_8	44.10	−42.07	290.56	1078.70	279.88
R-502	12/115 Azeotrope	111.64	−45.40	348.71	1318.90	104.39
Carbon dioxide	CO_2	44.01	−78.40	2290.90	7211.10	134.24

the service temperature. For example, R-22 is preferred over R-12 for temperatures below −29 C.

The *saturation pressures* at −15 and 30 C further assist in determining the temperature operating level since a high pressure at −15 C implies that a reduction in pressure will effect a lower evaporating temperature. The pressure at 30 C is an indication of the type of design required such as line joints, shaft seals, compressors.

The *refrigeration per kilogram* of refrigerant is an inverse index of the required rate of liquid flow and the size of liquid lines needed.

11.4 General Considerations

Other system characteristics that are important in selecting a refrigerant for a job, changing refrigerants for an installation, or selecting the equipment are:

1. *Chemical Reactions.* Sulfur dioxide will not attack steel or copper if dry. If moisture is present sulfurous acid may be formed and both steel and copper and related materials will be attacked. Ammonia will not attack iron or steel even if water is present. Copper and related alloys are not attacked by dry ammonia. However, their use is not recommended since a perfectly anhydrous ammonia refrigerant is most difficult to maintain. The other refrigerants in table 11-2 are essentially chemically inert.

2. *Moisture in System.* Water is sufficiently soluble in ammonia, carbon dioxide, and sulfur dioxide that moderate amounts can move within the system without freezing occurring in the low temperature regions. Water is essentially nonsoluble in the halide refrigerants, and even minute quantities may freeze in the expansion valve or capillary tube, shutting off the flow of refrigerant. A moisture-absorbing cartridge is usually inserted in the liquid refrigerant line to remove the moisture from the refrigerant or the

Table 11-3. Some properties of saturated ammonia. Extracted from ASHRAE (1997)

Temp. (C)	Saturation Pressure (MPa)	Enthalpy Liquid (kJ/kg)	Enthalpy Vapor (kJ/kg)	Entropy Liquid (kJ/kgK)	Entropy Vapor (kJ/kgK)	Heat Vapor (kJ/kgK)	C_p/C_v
−77.67	0.00604	−147.36	1342.85	−0.4930	7.1329	1.988	1.335
−70	0.01089	−111.74	1357.04	−0.3143	6.9179	2.008	1.337
−60	0.02185	−67.67	1375.00	−0.1025	6.6669	2.047	1.341
−50	0.04081	−24.17	1392.17	0.0968	6.4444	2.102	1.346
−40	0.07168	19.60	1408.41	0.2885	6.2455	2.175	1.352
−30	0.11944	63.86	1423.60	0.4741	6.0664	2.268	1.360
−20	0.19011	108.67	1437.64	0.6542	5.9041	2.379	1.370
−10	0.29075	154.03	1450.42	0.8294	5.7559	2.510	1.383
0	0.42941	200.00	1461.81	1.0000	5.6196	2.660	1.400
10	0.61504	246.62	1471.66	1.1666	5.4931	2.831	1.422
20	0.85744	293.96	1479.78	1.3295	5.3746	3.027	1.451
30	1.1671	342.08	1485.92	1.4892	5.2623	3.252	1.489
40	1.5553	391.11	1489.82	1.6461	5.1546	3.516	1.538
50	2.0339	441.18	1491.09	1.8009	5.0497	3.832	1.602
60	2.6154	492.50	1489.32	1.9541	4.9460	4.221	1.687
70	3.3133	545.41	1483.94	2.1067	4.8416	4.716	1.801
80	4.1418	600.44	1474.20	2.2601	4.7342	5.374	1.960
90	5.1167	658.36	1459.01	2.4163	4.6209	6.302	2.192
100	6.2553	720.44	1436.53	2.5783	4.4973	7.739	2.562
110	7.5782	788.98	1403.31	2.7516	4.3549	10.331	3.247
120	9.1115	869.25	1351.08	2.9486	4.1740	16.702	4.964
130	10.8948	983.69	1246.92	3.2231	3.8760	—	—
132.2	11.333	1105.47	1105.47	3.5006	3.5006	—	—

system. Oil and refrigerant are thoroughly dried before assembling. Moisture accelerates the formation of sludge.

3. *Oil Miscibility.* Oil is not soluble with ammonia and carbon dioxide and has limited solubility in sulfur dioxide and nitrous oxide. Oil moves in these systems in drops or slugs and accumulates at low points in the system where it must be removed at periodic intervals. Oil is soluble in the halide and hydrocarbon refrigerants and moves as a solution. Circulation within the system is usually continuous, the oil moving through the vapor line as a fog. Difficulty may develop if oil-soluble refrigerants are used in flooded evaporator systems because of excessive concentration of oil by fractional evaporation of refrigerant.

Refrigerant tables and Mollier charts are comparable to steam tables and charts. Some data for ammonia, R-12, and R-134a are given in tables 11-3 to 11-5 and figure 11.3. For more complete data the Handbook of Fundamentals (ASHRAE, 1997) or a refrigeration textbook should be consulted.

Note that the enthalpy values for points a, c, d, and e can be determined directly from the refrigerant tables. Extensive tables, charts, or detailed equations are necessary in order to determine properties at state b. However, an approximation for the temperature

Refrigeration

Table 11-4. Some properties of saturated R-12. Extracted from ASHRAE (1997)

Temp. (C)	Saturation Pressure (MPa)	Enthalpy Liquid (kJ/kg)	Enthalpy Vapor (kJ/kg)	Entropy Liquid (kJ/kgK)	Entropy Vapor (kJ/kgK)	Specific Heat Vapor (kJ/kgK)	c_p/c_v
−90	0.00288	121.33	310.39	0.6526	1.6849	0.470	1.176
−80	0.00623	129.72	314.99	0.6972	1.6564	0.487	1.171
−70	0.01235	138.19	319.69	0.7400	1.6334	0.504	1.168
−60	0.02272	146.72	324.44	0.7809	1.6147	0.521	1.166
−50	0.03925	155.32	329.23	0.8203	1.5996	0.537	1.165
−40	0.06426	164.01	334.03	0.8583	1.5875	0.554	1.166
−30	0.10044	172.81	338.81	0.8951	1.5779	0.572	1.169
−20	0.15088	181.72	343.53	0.9309	1.5701	0.590	1.174
−10	0.21893	190.78	348.17	0.9658	1.5639	0.611	1.181
0	0.30827	200.00	352.68	1.0000	1.5590	0.633	1.192
10	0.42276	209.41	357.05	1.0335	1.5550	0.658	1.206
20	0.56651	219.03	361.23	1.0666	1.5516	0.687	1.224
30	0.74379	228.89	365.16	1.0992	1.5487	0.721	1.249
40	0.95909	239.03	368.81	1.1315	1.5459	0.762	1.283
50	1.2171	249.51	372.07	1.1638	1.5431	0.812	1.331
60	1.5227	260.37	374.86	1.1961	1.5398	0.880	1.399
70	1.8814	271.73	377.01	1.2288	1.5356	0.977	1.505
80	2.2991	283.75	378.26	1.2622	1.5298	1.131	1.683
90	2.7829	296.73	378.10	1.2971	1.5212	1.427	2.037
100	3.3425	311.38	375.26	1.3353	1.5065	2.252	3.039
110	3.9958	331.90	362.83	1.3874	1.4682	—	—
111.8	4.1249	347.39	347.39	1.4272	1.4272	—	—

and enthalpy at point b may be obtained by assuming an ideal gas and recognizing that the compression occurs along a constant entropy line. Then:

$$s_a = s_b \tag{11.10}$$

where

s_a = entropy of saturated vapor at condenser pressure (kJ/kgK)
s_b = entropy of compressed vapor leaving compressor (kJ/kgK)

For a constant pressure process between c and b:

$$s_a = s_b = s_c + \int_{T_c}^{T_b} \frac{c_p \, dT}{T} \tag{11.11}$$

where

s_c = entropy of saturated vapor at condenser pressure (kJ/kgK)
c_p = specific heat at constant pressure for the vapor (kJ/kgK)
T_c = absolute temperature of saturated vapor at condenser pressure (K)
T_b = absolute temperature of vapor leaving the compressor (K)

Table 11-5. Some properties of saturated R-134a. Extracted from ASHRAE (1997)

Temp. (C)	Saturation Pressure (Mpa)	Enthalpy Liquid (kJ/kg)	Enthalpy Vapor (kJ/kg)	Entropy Liquid (kJ/kgK)	Entropy Vapor (kJ/kgK)	Specific Heat Vapor (kJ/kgK)	c_p/c_v
−103.3	0.00039	71.89	335.07	0.4143	1.9638	0.585	1.163
−100	0.00056	75.71	337.00	0.4366	1.9456	0.592	1.161
−90	0.00153	87.59	342.94	0.5032	1.8975	0.614	1.155
−80	0.00369	99.65	349.03	0.5674	1.8585	0.637	1.151
−70	0.00801	111.78	355.23	0.6286	1.8269	0.660	1.148
−60	0.01594	123.96	361.51	0.6871	1.8016	0.685	1.146
−50	0.02948	136.21	367.83	0.7432	1.7812	0.712	1.146
−40	0.05122	148.57	374.16	0.7973	1.7649	0.740	1.148
−30	0.08436	161.10	380.45	0.8498	1.7519	0.771	1.152
−20	0.13268	173.82	386.66	0.9009	1.7417	0.805	1.157
−10	0.20052	186.78	392.75	0.9509	1.7337	0.842	1.166
0	0.29269	200.00	398.68	1.0000	1.7274	0.883	1.178
10	0.41449	213.53	404.40	1.0483	1.7224	0.930	1.193
20	0.57159	227.40	409.84	1.0960	1.7183	0.982	1.215
30	0.77008	241.65	414.94	1.1432	1.7149	1.044	1.244
40	0.0165	256.35	419.58	1.1903	1.7115	1.120	1.285
50	1.3177	271.59	423.63	1.2373	1.7078	1.218	1.345
60	1.6815	287.49	426.86	1.2847	1.7031	1.354	1.438
70	2.1165	304.29	428.89	1.3332	1.6963	1.567	1.597
80	2.6331	322.41	429.02	1.3837	1.6855	1.967	1.917
90	3.2445	343.01	425.48	1.4392	1.6663	3.064	2.832
100	3.9721	374.02	407.08	1.5207	1.6093	—	—
111.03	4.0560	389.79	389.79	1.5593	1.5593	—	—

If it is further assumed that the specific heat at constant pressure for the vapor remains constant over the temperature range T_c to T_b, then equation 11.11 gives:

$$s_a = s_c + c_p \ln\left(\frac{T_b}{T_c}\right) \quad (11.12)$$

which may be solved for T_b since s_a, s_c, T_c, and an estimate of c_p can be obtained from the refrigerant tables. Then,

$$T_b = T_c \exp\left(\frac{s_a - s_c}{c_p}\right) \quad (11.13)$$

The enthalpy at point b can then be estimated by:

$$h_b = h_c + c_p(T_b - T_c) \quad (11.14)$$

Example 11.1. Determine the coefficient of performance for cooling for R-134a when operating at an evaporator temperature of −15 C and a condenser temperature of 30 C if there is no superheating in the evaporator and no subcooling in the condenser.

Refrigeration

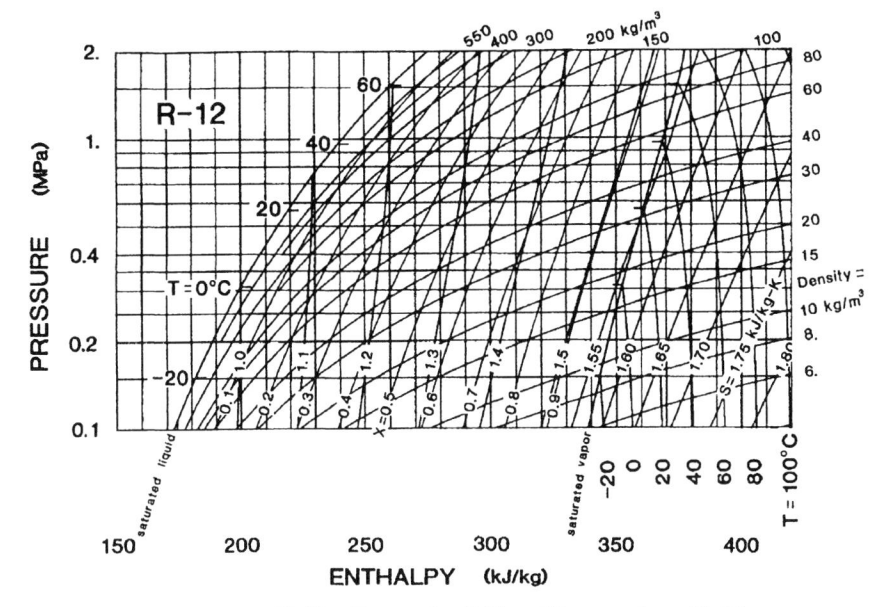

Figure 11.3. Mollier diagram for dichlorodifluoromethane (R-12).

Solution

At an evaporator temperature of -15 C, the enthalpy and entropy of saturated vapor may be obtained by interpolation from table 11-5.

$$h_a = (386.66 \text{ kJ/kg} + 392.75 \text{ kJ/kg})/2 = 389.71 \text{ kJ/kg}$$
$$s_a = (1.7417 \text{ kJ/kgK} + 1.7337 \text{ kJ/kgK})/2 = 1.7377 \text{ kJ/kgK}$$

At a condenser temperature of 30 C,

$$h_c = 414.94 \text{ kJ/kg}$$
$$c_p = 1.044 \text{ kJ/kgK}$$
$$s_c = 1.7149 \text{ kJ/kgK}$$
$$h_d = 241.65 \text{ kJ/kg}.$$

Then, by equation 11.13,

$$T_b = (303.16 \text{ K})\exp(((1.7377 \text{ kJ/kgK} - 1.7149 \text{ kJ/kgK})/1.044 \text{ kJ/kgK})$$
$$= 309.85 \text{ K} = 36.7 \text{ C}$$

Then, by equation 11.14,

$$h_b = 414.94 \text{ kJ/kg} + (1.044 \text{ kJ/kgK})(309.85 \text{ K} - 303.16 \text{ K})$$
$$= 421.92 \text{ kJ/kg}$$
$$\text{c.o.p.}_{\text{cooling}} = ((389.71 - 241.65) \text{ kJ/kg})/((421.92 - 389.71) \text{ kJ/kg}) = 4.60$$

This value compares with a value of 4.42 given by ASHRAE (1993) based upon more exact relationships for a real gas (rather than an ideal gas).

Components

11.5 Compressors

The compressor changes the gas from state *a* to state *b* (figs. 11.1 and 11.2), and is characterized by the volume rate of the gas at intake pressure and the pressure change affected. Four types of compressors are in general use.

1. *Reciprocating* or piston-type compressors (fig. 11.6) are most extensively used. Small units directly connected to electric motors are used for household refrigerating systems. Piston diameters and strokes of less than 2.5 cm are common in the latter systems. Multicylinder units are used for large industrial systems. Reciprocating compressors are used for all refrigerants and exclusively for those operating at high-pressure differentials, for refrigerants in the lower portion of table 11-2. The efficiencies (volumetric, thermal, and mechanical) are high.

2. *Rotary* compressors (fig. 4.4) are used with some success in household and other small systems where pressure differentials are small or moderate. They are mechanically

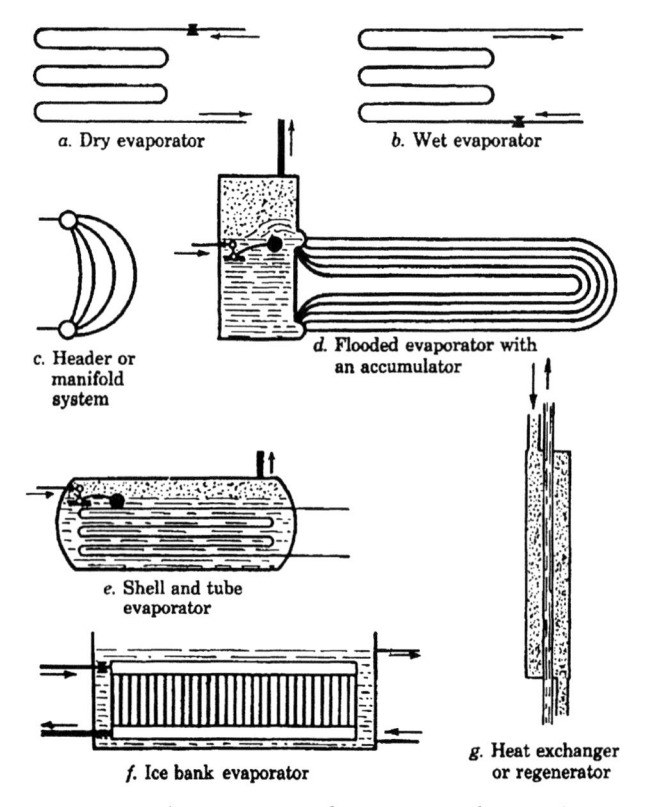

Figure 11.4. Various types of evaporators and accessories.

simpler than reciprocating units, are quiet, have high volumetric capacity with high rotative speeds, and consequently occupy small space. Starting torque is less than for reciprocating compressors since there is a smaller variation in pressure per rotative cycle. Tolerances must be exceptionally close to ensure volumetric performance. Lubrication is a problem since the vanes or other gas-confining parts are usually spring or centrifugally loaded. Wear soon increases clearances, and the volumetric capacity decreases. The mechanical efficiency may be low because of internal friction.

3. *Gear* compressors (fig. 4.1) have the same performance features as rotary compressors except that the volumetric efficiency may be less and the mechanical efficiency higher because of less starting friction. Their chief use is for boosting in compound systems.

4. *Centrifugal* compressors (Chapter 5) are used extensively for large systems using refrigerants with large specific vapor volume and small pressure differential. Air-conditioning systems using refrigerants from the upper portion of table 11-2 might employ centrifugal compressors. Single and multistaging is used, depending upon the pressure differential. Performance is comparable to a centrifugal air compressor. The volumetric capacity can be adjusted by throttling the discharge, a most convenient feature not possible with positive displacement units.

11.6 Condensers

The condenser cools the compressed gas to the saturation temperature and condenses it to a liquid, process b-c-d in figure 11.2. Some of the superheat, represented by b-c, may be removed by a special desuperheater located between the compressor and the condenser. This can be a water-cooled heat exchanger or simply a finned or extra-long bare pipe between the compressor and condenser which permits heat to escape to the room. The latent heat c-d may be removed, and the liquid is subcooled to a point d'. Significant subcooling cannot take place as long as the vapor is in contact with the liquid. Subcooling, then, results only if the liquid is in a vapor-free, heat-exchange region. Further subcooling may take place between the receiver and the evaporator.

Four types of condensers are common in agricultural work.

1. *Air-cooled condensers* which make use of finned tubes are used on systems up to 2.2 kW. Usual construction is vertical fins with horizontal tubes, the vapor being fed in at the top, the liquid flowing by gravity to the lower part of the condenser and thence to the receiver. Air is forced through the condenser by a fan.

2. *Shell and tube condensers* consist of a cylindrical drum with a series of water tubes inside. Large-capacity units are vertical, smaller units horizontal. The horizontal unit usually serves as a combination condenser and receiver. The water tubes are located in the upper portion of the cylinder so that the condensing surface will not be covered with liquid.

If the supply of water is ample and the cost low, the water is used only once and then discarded. If the supply is low or the cost high, the water may be circulated through a cooling tower where it is cooled by a portion evaporating. With a tower the actual water usage may be only 2% of that where it is wasted.

3. *Combination air- and water-cooled condensers* are available for small systems that may be required to operate when air temperatures are high, 35 C or higher. The water flow is controlled by the high side-pressure so water is used only when high temperatures of the air cause high head-pressure.

4. *Evaporative condensers* are extensively used where water supply (or disposal) or high temperatures are a problem. Water is recirculated over the pipes of the condenser in a thin film, spray, or shower. A forced draft of air over the wet pipes causes some of the water to evaporate. The heat liberated by the condensing refrigerant evaporates a small portion of the water passing over the condenser tubes. The temperature at which the evaporation takes place depends upon the air rate, temperature, humidity, the water-to-air surface area and its heat-and-vapor transfer coefficient, the water-to-refrigerant surface area and its heat-transfer coefficient, and the temperature of the make-up water. The water requirement for evaporative condensing is usually 2 to 5% that required for sensible water condensing.

11.7 Evaporators

The unit that does the cooling, that is, extracts or removes the heat from the load is called the evaporator. The cooling process is state change f-g' (fig. 11.2). The evaporator or boiler as it is sometimes called is a heat exchanger, and the principles of design and operation set out in Chapter 7 apply.

Evaporator arrangement depends upon the expected duty. Various arrangements are shown in figure 11.4 and are discussed under the letter indices.

(a) A *dry evaporator* consists of a single pipe or set of short pipes with headers (c). The liquid refrigerant is fed in at the top.

(b) The *wet evaporator* is comparable to the dry evaporator except the liquid is introduced at the bottom and flow is upward. Upward refrigerant movement gives somewhat better heat transfer than downward movement because of more vigorous mixing of the liquid and vapor. The dry evaporator facilitates oil movement. Oil flows through the unit by gravity; thus both oil soluble and nonsoluble refrigerants can be used. Oil must be continuously removed from the wet evaporators as a refrigerant-vapor oil fog for both soluble and nonsoluble refrigerants or be drained off at periodic intervals in the case of nonsoluble refrigerants. Oil removal by fog is satisfactory for the soluble refrigerants unless the refrigerating rate is low, in which case the fog action may be insufficient to carry out the oil. Fog oil removal from nonsoluble refrigerants may not be entirely satisfactory even at high refrigerant rates since the oil moves as drops or slugs, not as a solution.

(c) A *header or manifold* system is superior to the single-pipe system. Spent vapor is removed quicker, and better heat-transfer coefficients result. The pressure drop through the unit is less because of a shorter vapor-travel path. A single-pipe evaporator is used only for small installations where cost and fabricating convenience are more important than performance.

(d) A *flooded evaporator* is one designed for maximum heat-transfer effectiveness by "flooding" the inside of the heat-transfer surfaces with liquid refrigerant. The surfaces are flooded by maintaining the liquid refrigerant level in the evaporator above the transfer surfaces by a float (d, e) or other level-maintaining device. Wet evaporators are frequently called flooded evaporators. A distinction should be made between them on the basis of performance and design.

An *accumulator* is used with the flooded system to improve performance further. The coil-and-header system are designed for vigorous liquid boiling action which facilities heat transfer. Forced circulation of the liquid by a pump may be employed to provide even better performance. Because of the vigorous action, drops of refrigerant are carried toward the vapor discharge port. These drops are separated from the spent vapor and collect in the "accumulator". The liquid-free gas then returns to the compressor. This system is not completely satisfactory for oil-soluble refrigerants. The gas discharge action is not vigorous enough to carry out the oil that accumulates in the unit in solution. Non-soluble refrigerants permit the oil to settle out, and is drained off at periodic intervals.

(e) A *shell and tube evaporator* has the same operational characteristics as the accumulator system. It is used for cooling brine, water, or other liquids. The cooling material must not be permitted to freeze.

(f) The *ice-bank* evaporator is used where large quantities of "chilled" water at 0 C are needed. Evaporation takes place in a series of plates or bank of tubes which are immersed in a tank of water. During periods of low-water demand ice accumulates on the plates. The ice is then available for cooling the water during peak-demand periods. This procedure permits a smaller compressor to be used than would be required with the non-ice system.

(g) Evaporator performance can be improved by installing a *heat exchanger* or *regenerator* immediately following the evaporator. Thus, figure 11.2, state a is moved toward a' and d toward d', e in turn moving toward e'. This decreases the percentage of flash vapor at the expansion valve but increases the superheat at the compressor. Subcooling to d' can be produced by locating the heat exchanger so its trailing end is in the liquid evaporating region.

11.8 Expansion Valves

The expansion valve is used to regulate the rate of flow of liquid refrigerant into the evaporator at the evaporating rate. Four types of valves are used.

1. *Manually adjusted needle valves* may be used in large systems where loads are relatively constant and an attendant is on duty. Their advantageous features are quick adjustment, simplicity, and low first cost.

2. *Float valves* are actually automatically adjusted needle valves since they are so positioned that the incoming liquid rate equals the evaporating rate. They are used, as previously discussed, in flooded systems with accumulators.

3. *Capillary tubes* are used extensively in household refrigeration and other small systems. They are suitable only for systems composed of a single compressor and a single evaporator. The liquid passes from the high to the low side through a small tube of such a diameter and length that the rate of flow at operating pressure does not exceed the evaporating capacity at the design load. The system is simple since there are no valves, except in the compressor, and a receiver is not used. The system elements are

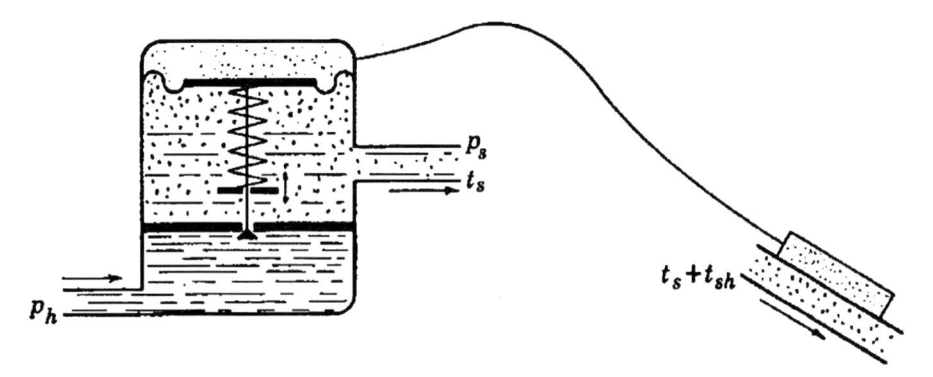

Figure 11.5. Principles of the thermal expansion valve.

located relative to each other so that when operation has ceased, the liquid will flow into the evaporator and/or condenser and pressures will equalize throughout the system. Consequently, there is no pressure differential across the compressor when it starts. This permits a low-starting torque motor to be used. The quantity of refrigerant charged into the system must be controlled carefully.

4. *Thermal expansion valve* (fig. 11.5) operates on the basis of the number of degrees of superheat in the spent vapor leaving the evaporator. Thermal expansion valves are used

Figure 11.6. Conventional compressor-condenser unit. (Courtesy the Copeland Co.)

Refrigeration

on evaporators (*a*), (*b*), and (*c*), and on systems of a wide range of sizes and those with more than one evaporator. They are especially applicable for field-assembled systems with automatic control and variable-cooling load.

The thermal sensing bulb may be filled with the same fluid used for the refrigerant. Consequently, the downward pressure on the diaphragm due to a temperature of the bulb is comparable to the upward pressure on the diaphragm due to the saturated pressure of the refrigerant within the evaporator. This force referred to the spring reaction maintains a proper opening of the valve for all temperatures and loads, the design or adjusted superheat applying at all times. Consideration of these data and the physical characteristics of the valve will show that:

1. The superheat to the sensing bulb is constant at all loads. The temperature difference between the boiling liquid and the material being cooled may be greater than the superheat but never less.
2. The liquid rate is controlled on the basis of the heat load. Thus, the amount of evaporator surface used for heat exchange is controlled by the heat load.
3. Liquid flow is stopped when the compressor stops. This feature facilitates control which is discussed later in this chapter.

System Design and Balance

11.9 The Evaporator

The heat and mass balance for an evaporator can be expressed by the following equation:

$$q = \dot{m}_m c(t_1 - t_2) = AU(\Delta t)_{\ln} = \dot{m}_r(h_a - h_e) \tag{11.15}$$

where

- q = heat transfer rate (kW)
- $\dot{m}_m$ = mass flow rate of medium to be cooled (kg/s)
- c = specific heat of medium being cooled (kJ/kgK)
- t_1 = entering temperature of medium being cooled (K)
- t_2 = exiting temperature of medium being cooled (K)
- $\dot{m}_r$ = refrigerant mass flow rate (kg/s)
- h_a = enthalpy of vapor leaving evaporator (kJ/kg)
- h_e = enthalpy of the liquid refrigerant entering the evaporator (kJ/kg)
- A = evaporator heat-exchange area (m^2)
- $(\Delta t)_{\ln}$ = log-mean temperature difference (K)
  $= ((t_1 - t_3) - (t_2 - t_3))/\ln((t_1 - t_3)/(t_2 - t_3))$
- t_3 = temperature of the evaporating refrigerant (C)
- U = overall heat-transfer coefficient between medium being cooled and the refrigerant in the evaporator (kW/m^{2}K)

The refrigerant load is defined by (1) the heat rate q, (2) original and final temperature t_1 and t_2, and perhaps (3) relative humidity. Commercial evaporators are often rated on a heat rate per degree temperature difference:

$$T.D. = t_1 - t_3 \quad (11.16)$$

Then the log-mean-temperature is replaced by T.D. in equation 11.15 and the U value is evaluated based on the T.D. This rating procedure is used because of convenience. Cooling loads which can be carried with a large T.D. can be handled with an evaporator of small effective area A. Note, however, that a low evaporator temperature will require a lower low-side-pressure and a compressor of greater volumetric capacity will be needed. The T.D. is usually limited by the possibility of freezing, frost formation, or dehumidification of the air.

Example 11.2. Determine the refrigerant mass flow rate for an R-134a refrigeration system to produce 3.5 kW of cooling if there is no superheating and no subcooling and the evaporator temperature is -15 C. If the overall heat transfer coefficient for the evaporator is 700 W/m²K, determine the evaporator surface area needed to cool air from 30 to 10 C and the mass flow rate of the air being cooled.

Solution
From example 11.1

$$h_a = 389.71 \text{ kJ/kg}$$
$$h_e = 241.65 \text{ kJ/kg}$$

Then,

$$q = \dot{m}_r(h_a - h_e)$$

or

$$\dot{m}_r = q/(h_a - h_e) = (3.5 \text{ kW})/(389.71 \text{ kJ/kg} - 241.65 \text{ kJ/kg})$$
$$= 0.0236 \text{ kg/s}$$

For an entering air temperature of 30 C, an exiting air temperature of 10 C, and an evaporator temperature of -15 C the log-mean-temperature difference is:

$$(\Delta t)_{\ln} = [(t_1 - t_3) - (t_2 - t_3)]/\ln[(t_1 - t_3)/(t_2 - t_3)]$$
$$= [(30 + 15) - (10 + 15)]/\ln[(30 + 15)/(10 + 15)] = 34.0 \text{ K}$$

Then,

$$q = UA(\Delta t)_{\ln}$$

or

$$A = q/U(\Delta t)_{\ln} = (3.5 \text{ kW})/((0.7 \text{ kW/m}^2\text{K})(34.0 \text{ K})) = 0.147 \text{ m}^2$$

Also,

$$q = \dot{m}_a c(t_1 - t_2)$$

Refrigeration

and

$$\dot{m}_a = q/[c(t_1 - t_2)] = (3.5 \text{ kW})/[(1.0048 \text{ kJ/kg K})(20 \text{ K})]$$
$$= 0.174 \text{ kg/s}$$

11.10 Defrosting

Defrosting of evaporators for air cooling above 1 C is brought about by using an evaporator of sufficient size so that the load can be handled with only part-time operation. Frost accumulating during the running cycle is melted during the off-cycle. This procedure, in addition to defrosting, helps to maintain humidity if high humidities are desired.

Evaporators for below freezing temperatures are deigned with fins a greater distance apart or with bare coils so that moderate frost or ice accumulation will not affect air flow. Defrosting is by hot water, electric heaters, or hot gas. Hot gas from the high side of the system is piped to the evaporator by a series of connecting lines and valves. Operation may be manual or automatic.

Frosting may be prevented by continuous circulation of a brine over the evaporating surface. The brine picks up moisture from the surface and must be reconcentrated or replaced at intervals.

11.11 The Compressor

$$\dot{m}_r v_g = E_v D N S \tag{11.17}$$

where

$\dot{m}_r$ = mass flow rate of refrigerant (kg/s)
v_g = specific volume of the vapor entering the compressor, at low-side-pressure (m^3/kg)
E_v = compressor volumetric efficiency, 60 to 75% for small reciprocating compressors (decimal)
D = piston displacement, m^3 per piston revolution
N = number of pistons
S = compressor speed (rev/s)

The compressor power per unit of refrigeration is given by:

$$P' = 100 * (h_b - h_a)/(E_c * (h_a - h_e)) \tag{11.18}$$

where

P' = compressor power per unit of refrigeration (kW/kW)
h_a = heat content of vapor leaving evaporator (kJ/kg)
h_e = heat content of liquid entering evaporator (kJ/kg)
h_b = heat content of vapor leaving compressor (kJ/kg)

E_c = compression efficiency, percent (65 to 85% estimated). (The compression efficiency is defined as the ratio of the work required for an isentropic compression process to the actual work required. Few data are available.)

Equation 11.11 defines the capacity of a compressor in terms of its speed, its physical characteristics, and the mass rate of the refrigerant. The vapor volume rate represented by the product $\dot{m}_r v_a$ must be the same as that for the conditions in equation 11.15. The compressor capacity must be balanced with the capacity of the evaporator. A compressor with more capacity than that specified for the desired conditions will move a greater volume of vapor at a greater specific volume. This will lower the suction pressure and evaporating temperature and increase the mass evaporation rate. The additional capacity will manifest itself by lowering the evaporating temperature t_3 in equation 11.15 with increased capacity of the evaporator and appropriate adjustments of the other factors.

11.12 Condensing Unit

A condensing unit consists of a compressor and motor and either an air-cooled finned condenser or shell and tube water-cooled condenser (fig. 11.11). Ratings in refrigeration effect at various evaporator temperatures and ambient air temperature or condensing water temperatures is given. Since most compressors are belt driven, capacity can be further adjusted by change of compressor speed.

Equations 11.16 and 11.17 are useful for estimating power requirements for conditions not covered by commercial data.

Controls

Refrigeration control may be considered from the standpoint of (1) individual component control, (2) safety of both equipment and operator, and (3) temperature and perhaps humidity control of the medium being cooled. Temperature and humidity will be considered.

11.13 Motor Circuit Thermostats

These are used on single-evaporator systems with capillary tube, thermal expansion valve, and float-controlled evaporators. Temperature control is as accurate as the thermostat. Expansion or float valves that do not completely restrict liquid flow during the off-cycle may cause compressor flooding. Compressor damage or motor overload may result when the unit starts.

11.14 Low-Side-Pressure Switch

The low-side-pressure switch opens and closes the motor circuit on the basis of the low side-pressure which is directly related to the temperature of the evaporating liquid as shown in equation 11.15. As cooling takes place, t_3 and the corresponding refrigerant saturation pressure drops. The decreased pressure opens the switch. The residual liquid refrigerant soon assumes ambient temperature, the pressure rising to the corresponding saturation pressure. A rise in ambient temperature causes a pressure rise

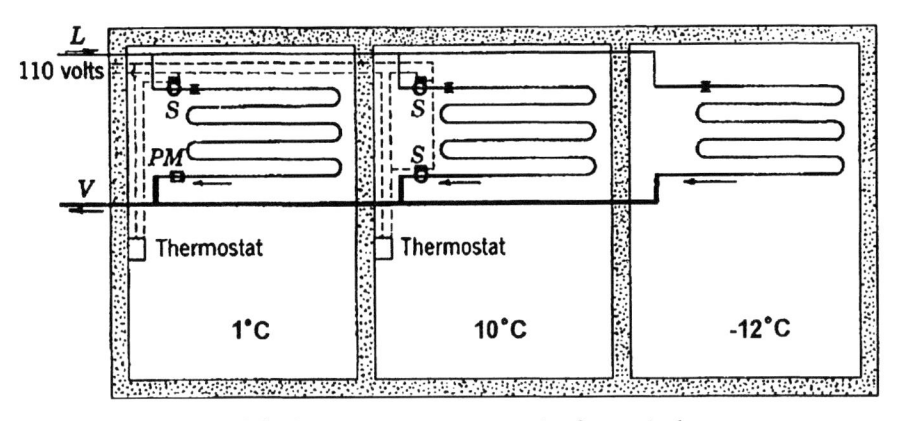

Figure 11.7. Multiple-evaporator system operating from a single compressor.

that closes the circuit. Expansion and compressor valve leakage are compensated for by automatic compressor operation. Frost accumulations decrease the U value of the evaporator, and the controlled temperature rises. The controlled temperature rises as the refrigerating rate increases because the evaporating-refrigerant and cooling-material temperature difference increases. This method of control is satisfactory only for single-evaporator systems unless additional control features such as those discussed below are added to the system.

11.15 Magnetic Valves

Magnetic valves operated by a thermostat are frequently used for control in multiple-evaporator systems (fig. 11.7). The thermostat opens the valve permitting liquid to enter the evaporator. The suction pressure switch is set for a pressure somewhat below the operating pressure of the evaporator. Magnetic valves are also used in combination with manual and float expansion valves to facilitate operation.

11.16 Evaporative-Pressure-Maintaining Valves

Also called back-pressure control valves, these can be used on multiple-evaporator systems to control the evaporating pressure in any evaporator. Without them, the evaporating pressure and temperature of all the evaporators in a multiple system are those of the lowest pressure unit with slight variations for line friction. If a -15 C room and a 1 C room are to operate on the same compressor it will probably be advisable to operate the higher temperature room with a higher pressure evaporator in order to minimize coil frosting and, if desired, to maintain a high relative humidity in the room. This is done by using a pressure-maintaining valve shown in principle in figure 11.8 and located in figure 11.7. In figure 11.7, the compressor motor is operated by a low-side-pressure switch activated by the low-side-pressure of the lowest temperature load, -12 C in this example. The solenoids of the higher-temperature loads are controlled by thermostats. Liquid refrigerant is let into the evaporator only when cooling is required. The evaporative pressure-maintaining valve PM holds the pressure in the evaporator of the 1-C load

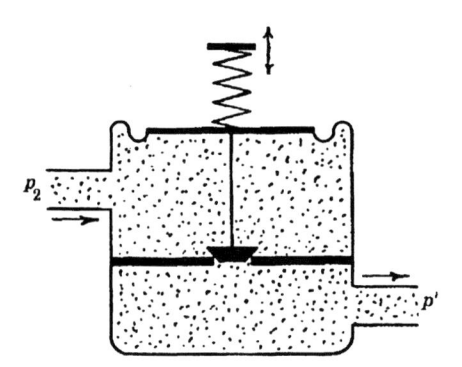

Figure 11.8. Schematic sketch of an evaporative pressure-maintaing valve.

above the pressure in the vapor return line. This permits humidity control or minimum evaporator frosting by controlling the evaporating temperature. The evaporating temperature of the 10-C load is the same as that of the -12-C load. A vapor discharge solenoid can be used as installed on the 10-C load to provide more accurate temperature control. This valve confines the residual liquid refrigerant in the evaporator, thus stopping the refrigerating action at the prescribed temperature. Without this valve the residual liquid would evaporate and the temperature might drop below the controlled temperature. This valve is used only where highly exact control is required.

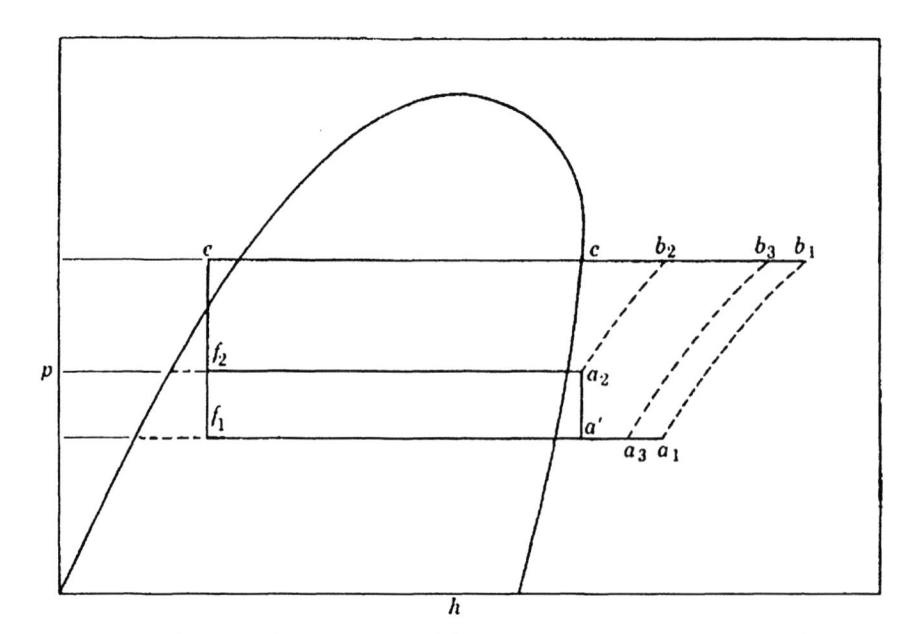

Figure 11.9. The thermodynamic process of the evaporative pressure maintaining valve and its relation to the entire system.

Refrigeration

The thermodynamics of the process are shown in figure 11.9. The flow through the valve is irreversible adiabatic, state change $a_2 - a'$. The power state path for the higher pressure unit would be $a_a - b_2$ if separate compressors were used for each load, the power energy for the system being:

$$\dot{m}_{2r}(h_{b2} - h_{a2}) + \dot{m}_{1r}(h_{b1} - h_{a1}) \tag{11.19}$$

$\dot{m}_{2r}$ and $\dot{m}_{1r}$ are the respective refrigerant mass flow rates. The total power for the system designed for a pressure-maintaining valve and a single compressor is:

$$(\dot{m}_{2r} + \dot{m}_{1r})(h_{b3} - h_{a3}) \tag{11.20}$$

The difference between equations 11.19 and 11.20 is the extra power required owing to the pressure-maintaining valve.

Multiple-evaporator systems with an operating temperature range and pressure-maintaining valves on the high-temperature evaporators are economically sound if the major portion of the load is a low-temperature load. If the major portion of the load is high temperature, independent systems may be advisable.

Heat Pump

The heat pump is a refrigeration machine installed where the heat discharged from the condenser is desired rather than the heat absorbed by the evaporator. The heat pump may be regarded as a device that lifts or "pumps" heat energy from a low-temperature source for use at a higher temperature. It is used satisfactorily for spacing heating by lifting heat from the ground, outside air, or bodies of water at a lower temperature than the space to be heated. Other uses, existing and proposed, make use of both evaporator and condenser energies.

Heat pumps may be fitted to a variety of operations, e.g.,

1. Comfort heating and cooling of buildings.
2. Water and other liquid heating, domestic, and industrial.
3. Evaporating, concentrating, and distilling.
4. Drying.
5. Simultaneous heating and cooling, e.g.
    a. Water cooling and space heating.
    b. Heating and dehumidifying, domestic and industrial.
    c. Water heating and space cooling.

The coefficient of performance of a heat pump refers high-temperature condenser energy to the driving energy so that the cycle c.o.p. is that given for c.o.p.$_{heating}$ in equation 11.7. Since the c.o.p. relates the useful heat output as refrigeration in a refrigerator or heat energy input in a heat pump, the c.o.p. for simultaneous usage could be considered as:

$$\text{c.o.p.}_{simultaneous} = [(h_a - h_c) + (h_b - h_d)]/(h_b - h_a)$$

$$= \text{c.o.p.}_{heating} + \text{c.o.p.}_{cooling} \tag{11.21}$$

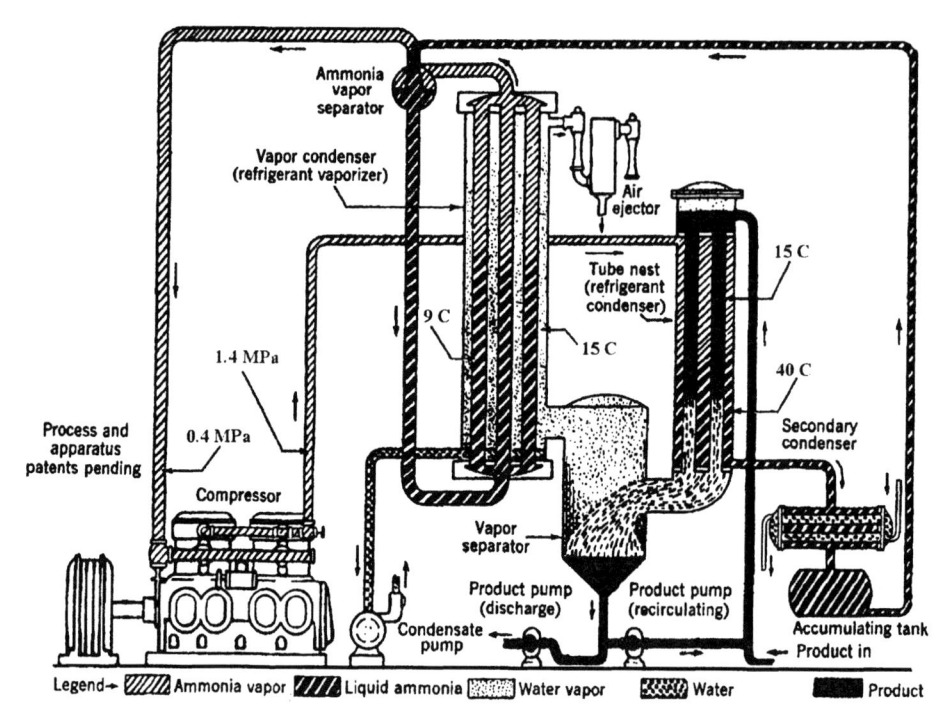

Figure 11.10. A heat pump used for concentrating orange juice. (Courtesy the Majonnier Co.)

The smaller the difference in temperature between the evaporator and condenser, the greater the c.o.p. and the greater will be the output per unit of mechanical input.

A decision to install a heat pump in lieu of a conventional gas, oil, or other heat energy source should be made on the basis of an economic and convenience study. Factors that must be considered are initial cost, length of life, upkeep, operating cost, and continuity of power source, operating attention, etc. For example, a gas-fired hot-water heater might be 70% efficient thermally; thus, 1.4 kW must be supplied for each 1.0 kW taken up by the water. A heat pump with a c.o.p. of 3.5 would need only a mechanical power of 0.29 kW.

Sources and Sinks

The "source" of heat for a heat pump can be a steady-state source such as air or water. The mass rate would be controlled so that the difference in temperature between the evaporating refrigerant and the cooling medium is nearly constant for various heat rates. Air, well water, stream water, liquid manufacturing wastes, etc., can be used.

Systems installed in areas where air and water temperatures are low or where a sufficient supply of water or other heat-source medium is unavailable may require a transient source. A transient source is a pond, well, or series of wells, the earth, or other stationary heat source with sufficient capacity and transfer properties to supply the heat at a satisfactory rate. Poor performance may occur if, when designing the source, decrease in heat transfer due to shallow thermal gradients as operation continues is not recognized.

A "sink" is a heat disposal for condenser heat. The same characteristics apply as for the source.

A heat pump used for concentrating orange juice is an example of an agricultural heat-pump installation. Figure 11.10 shows a schematic arrangement of the system and the operating conditions. The unique feature is the use of the refrigeration evaporator for condensing the vapor removed from the orange juice. The compressor is a 27.9 × 25.4 cm, 4-cylinder, 300-rpm unit with a volumetric efficiency of 91%. The orange juice is concentrated from 11 to 55 degrees Brix (specific gravity of 1.03 to 1.16).

The power required is nearly an inverse function of the heat-exchange area. If the heat-exchange areas were increased the compressor power could be reduced. The optimum size of these factors must be based upon a cost analysis.

Nomenclature

A	heat-exchanger surface area (m^2)
c	specific heat (kJ/kgK)
c.o.p.	coefficient of performance (dimensionless)
E	efficiency (percent or decimal)
h	enthalpy (kJ/kg)
$\dot{m}$	mass flow rate (kg/s)
$\dot{m}_r$	mass flow rate of refrigerant (kg/s)
P'	compressor power per unit of refrigeration (kW/kW)
q	heat transfer rate (kW)
S	compressor speed (rev/s)
t	temperature (C)
T	absolute temperature (K)
T.D.	temperature difference between evaporator cooling medium and evaporating refrigerant (C)
U	overall heat transfer coefficient (W/m^2K)

References

Am. Soc. Heating, Refrigerating, and Air Conditioning. 1997. *Handbook of Fundamentals*. Atlanta, Ga.: ASHRAE.

Am. Soc. Heating, Refrigerating, and Air Conditioning. 1990. *Refrigeration*. Atlanta, Ga.: ASHRAE.

Dossat, R. J. 1961. *Principles of Refrigeration*. New York: John Wiley & Sons.

Stoecker, W. F. 1958. *Refrigeration and Air Conditioning*. New York: McGraw-Hill Book Co.

Woolrich, W. R. 1965. *Fundamentals. Handbook of Refrigerating Engineering*, 4th Ed., Vol. I. Westport, Conn.: Avi Publishing Co.

———. 1966. *Applications. Handbook of Refrigerating Engineering*, 4th Ed., Vol. II. Westport, Conn.: Avi Publishing Co.

Problems

1. Ammonia is required for a 10.55 kW refrigeration system with an evaporating temperature of −23 C. The condenser will be water-cooled and will operate on a 6-C water temperature differential. Tap water temperature is 20 C. Determine the following. Assume saturated conditions.
   a. High- and low-side-pressures.
   b. Compressor displacement rate; assume volumetric efficiency of 85%.
   c. Compressor power per unit of refrigeration, assume thermal efficiency of 90%.
   d. The liquid ammonia rate (kg/s).
   e. The capacity in kilowatts if the evaporating temperature is raised to −9 C. The compressor capacity is unchanged.

2. Work problem 1 with R-12 as the refrigerant. Compare compressor displacement and liquid rate. Which refrigerant would you recommend for a large installation? Why?

3. A 3.5-kW R-12 system has an evaporator temperature of 0 C and a high side-pressure of 620 kPa gage. Liquid refrigerant enters the expansion valve at 15 C. Vapor enters the compressor at 10 C. Use the Mollier chart and determine:
   a. Percent flashing into vapor as the liquid passes through the expansion valve. Calculate from enthalpy values.
   b. The coefficient of performance.
   c. The compressor power per unit of refrigeration.
   d. The compressor power per unit of refrigeration if the vapor enters the compressor with no superheat.
   e. The compressor displacement, assume 80% efficient.

4. An evaporator pressure-maintaining valve holds a 210-kPa gage evaporating pressure against a 35-kPa gage low-side-pressure. The vapor enters the valve saturated. Determine the energy in kilojoule per kilogram of refrigerant made unavailable in the process.

5. Show by a Mollier chart the necessity for the secondary condenser of figure 11.10.

6. Determine the coefficient of performance of the heat pump of figure 11.10.

7. An R-12 system operating at 725- and 70-kPa gage feeds liquid refrigerant into the expansion valve at 10 C.
   a. What percent of the liquid flashes to gas as it passes through the valve?
   b. What percent of the evaporator space immediately past the valve is occupied by vapor? Solve with data from tables 11-4 and 11-5.

Refrigeration

8. Air at 25 C and 80% relative humidity passes through an evaporator at 2 C and discharges at 11 C. Determine the humidity (H) of the discharging air.

9. A finned tube NH evaporator is cooling air from 10 to 6.5 C with a 8 C T.D. The compressor head pressure is controlled at 1 275 kPa gage. Provide good estimates for:
   a. Low side pressure.
   b. T.D. if the air rate through the evaporator is reduced 50% (see section 11.9–11.10).
   c. Reduction in capacity, percent, under condition *b*.

Index

Absorptivity, 220
Adiabatic processes, 246
Adiabatic saturation temperature, 246–48
Air, properties, 196
Air lift, 103
Ammonia, saturated properties, 330
Anemometer, 78, 81, 82, 84
Angle of repose, 41

Bernoulli equation, 14
Biot number, 165–66, 170, 171–76
Bourdon tube, 54
Brown-Duvel moisture tester, 282
By-pass factor, 261–63

Coefficient of performance, 327–28, 345
Coefficients, friction, table, 42
Combustion, fuel, 263–65
Concentrators, 317
Conduction, 149–84
    cylinder, 161–66, 174
    plane wall, 153–61, 173
    sphere, 166–70, 175
    steady-state, 153–70, 177–84
    transient, 170–76
Conservation, energy, 13–15
    mass, 11–13
Convection, 149–50, 156–61, 163–66, 168–84, 187–212
    forced, 150, 191–92, 197–98
    free, 150, 189–91, 193–97
    inside pipes, 191–92, 197
    outside pipes, 193, 197
    plane surfaces, 160, 193
    spheres, 197
Conversion factors, general, table, 4, 48

Dalton's law, 238–39
Darcy's formula, 22–24
Dew-point temperature, 243
Diffusivity, mass, 289
    thermal, 153
Drag coefficient, 77
Driers, 310–19
    cross-flow, 311–12
    deep-bed, computations, 295–310
    rotary, 312–14
    spray, 315–16
    tray, 314–15
    tunnel, 314–15
Drying, 273–320
    constant-rate, 286–87
    falling rate, 288–92
    heat and mass balance, 294–95
    Hukil's analysis, 295–305
    numerical simulation, 305–10
    thermal efficiency, 300–2
    thin-layer, 286–92

Electromagnetic wave spectrum, 215–16
Emissivity, 216–20
    monochromatic, 216–19
    various surfaces, table, 218
Energy, fluid, 13–15
    elevation, 13
    kinetic, 13–15
    potential, 13–15
    pressure, 14
    total hydraulic, 13–15
    velocity, 14
Engineering, art, 6–9
    description, 1
    science, 1–6
    uncertainties, 7–8
Enthalpy, 243–46
Equilibrium moisture content, 275–80
    BET equation, 276–77
    Henderson equation, 277–79
    Langmuir equation, 275
    Smith equation, 277
    Young equations, 279–80
Error analysis, 8

Fans, 119–46
    axial-flow, 120–23, 130–31
    backward-curved blade, 124, 132–33
    centrifugal, 123–25
    classification, 119–20
    compression effect, 136
    cross-flow, tangential, 124–25
    forward-curved blade, 123, 134
    in series and in parallel, 140–43
    laws, 134–36
    performance 125–29,
    propeller, 120–21
    tube-axial, 120–22
    vane-axial, 122–23

Fluid flow, branching system design, 37–40
  compressibility error, 39
  laminar, 16–18
  streamlined, 16–18
  through granular materials, 26–29
  turbulent, 16–18
Fluids, Bingham, 19
  classification, 11
  dilatent, 21
  Newtonian, 19
  non-Newtonian, 19–22
  pseudoplastic, 19–20
  rheopectic, 21–22
  thixiotropic, 21
Fourier-Biot law, 149–50
Fourier modulus, 172–77
Friction, coefficients for small grain, table, 42
Friction factor, 22–24, 39
Friction losses, 22–31
  agricultural products, 26–30
  contractions, 24–25
  Darcy's formula, 22
  Ergun equation, 26–27
  enlargements, 25–26
  floors, 29–30
  fittings, 24
  heat exchangers, 31–32
  pipe, 22–24

Gas constant, 237–38
Gas law, 237–38
Granular materials, flow, 41–42
Grashof number, 191, 193
Greenhouse effect, 230–33

Heat exchangers, 198–212
  effectiveness ratio, 204–12
  log-mean temperature, 199–203
  pressure drop, 31–32
  types, 198–99
Heat pump, 345–46
Heats of combustion, fuels, table, 263
Heisler charts, 173–75
Hot-wire anemometer, 82–84
Hukil's analysis, 295–305
Humidity, ratio, 241–42
  relative, 242

Ice mixtures, temperatures, 326
Ideal gas law, 237–38

Kirchoff's law, 221–22

Laminar flow meter, 69–72
Lewis number, 292

Manometer, 50–55
  micro, 54–55
Mixing, 258–60
Moisture content, 273–74
  determination, 280–86
  dry-basis, 273
  equilibrium, 275–80
  wet-basis, 273
Mollier chart, 252–56

Net positive suction head, 111–13
Newtonian heating or cooling, 170–72
Nozzle meter, 65–66
Nusselt number, 191–98

Orifice, 66–69

Piezometer, 49–51
Pipe fittings, resistance, 24
Pitot tube, 57–61
Planck's law, 215–16
Prandtl number, 191–98
Pressure, absolute, 48–49
  conversion factors, table, 48
  differential, 48–49
  dynamic, 49
  gage, 48–49
  observations, 48–57
  saturation, 239–41
  static, 49
  vacuum, 49
  velocity, 49
Pressure drop, air flow through agricultural products, 26–30
  air flow through perforated floors, 29–30
  heat exchangers, 31–32
Psychrometer, wet-bulb, 248–50
Psychrometric chart, 252–68
  combustion process, 263–65
  cooling process, 257–58, 260–63
  dehumidification process, 260–63, 268
  drying process, 265–66, 293–95
  heating process, 257–58
  mixing process, 258–60
  uses, 257–68
Pumps, 89–115
  centrifugal, 95–101
  displacement, 90–95
  dynamic, 95–105

Index

efficiency, 89–90
gear, 93
jet, 101–3
lobe, 94
net positive suction head, 111–13
performance, 105
performance on system, 113–14
regenerative turbine, 96
rotary, 93–95
viscosity effect on performance, 114–15

R-12, saturated properties, 331
R-134a, saturated properties, 332
Radiant heat exchange, 150
  black surfaces, 223–25
  gray surfaces, 225–27
  solar, 230–33
  transparent layers, 227–30
Radiant shape factor, 224–25
Radiation, 150, 215–34
  emissive power, 222–23
  intensity, 222–23
  solar, 230–33
Rayleigh number, 193
Reflectivity, 220
Refrigerants, properties, table, 329
Refrigeration, 325–47
  accumulator, 337
  compressors, 334–35, 341
  condensers, 335–36, 342
  evaporators, 336–37
  expansion valves, 337–39
  mechanical, 325–29
  natural, 325
  sources and sinks, 346–47
  system balance and design, 339–41
  system rating, 328–29
Relative roughness, 22–24
Reynold's number, 16–18, 23, 192, 197–98
Rheology, 22
Roughness factor, 22–24
Rotameter, 75–78

Schmidt number, 287
Specific speed, 109–11, 129–30
Specific volume, 242–43
Steam table, saturated pressure-temperature, 240
Stephan-Boltzmann constant, 150, 216

Temperature, dew point, 243
  dry-bulb, 239
  wet-bulb, 248–50
Thermal conductivity, table, 151
Thermal diffusivity, 153
Thermal expansion valve, 338–39
Thermocouple anemometer, 84
Thomas meter, 81–82
Transmissivity, 220
  various materials, table, 231

Unit factor method, 5
Units, basic, 1–2
  derived, 2–3

Venturi meter, 61–65
Viscosity, 18–22
  fluids, table, 20

Water, properties, 194
Wet-bulb temperature, 248–50